Process
Dynamics
and Control

PRENTICE-HALL INTERNATIONAL SERIES
IN THE PHYSICAL AND CHEMICAL ENGINEERING SCIENCES

NEAL R. AMUNDSON, EDITOR, *University of Minnesota*

ADVISORY EDITORS

ANDREAS ACRIVOS, *Stanford University*
JOHN DAHLER, *University of Minnesota*
THOMAS J. HANRATTY, *University of Illinois*
JOHN M. PRAUSNITZ, *University of California*
L. E. SCRIVEN, *University of Minnesota*

AMUNDSON   *Mathematical Methods in Chemical Engineering*
ARIS   *Elementary Chemical Reactor Analysis*
ARIS   *Introduction to the Analysis of Chemical Reactors*
ARIS   *Vectors, Tensors, and the Basic Equations of Fluid Mechanics*
BALZHISER, SAMUELS, AND ELIASSEN   *Chemical Engineering Thermodynamics*
BERAN AND PARRENT   *Theory of Partial Coherence*
BOUDART   *Kinetics of Chemical Processes*
BRIAN   *Staged Cascades in Chemical Processes*
CROWE, HAMIELEC, HOFFMAN, JOHNSON, SHANNON, AND WOODS   *Chemical Plant Simulation*
DOUGLAS   *Process Dynamics and Control, Vol. 1*
DOUGLAS   *Process Dynamics and Control, Vol. 2*
FREDRICKSON   *Principles and Applications of Rheology*
HAPPEL AND BRENNER   *Low Reynolds Number Hydrodynamics*
HIMMELBLAU   *Basic Principles and Calculations in Chemical Engineering, 2nd ed.*
HOLLAND   *Multicomponent Distillation*
HOLLAND   *Unsteady State Processes with Applications in Multicomponent Distillation*
KOPPEL   *Introduction to Control Theory*
LEVICH   *Physicochemical Hydrodynamics*
MEISSNER   *Processes and Systems in Industrial Chemistry*
PERLMUTTER   *Stability of Chemical Reactors*
PETERSEN   *Chemical Reaction Analysis*
PRAUSNITZ   *Molecular Thermodynamics of Fluid-Phase Equilibria*
PRAUSNITZ AND CHUEH   *Computer Calculations for High-Pressure Vapor-Liquid Equilibria*
PRAUSNITZ, ECKERT, ORYE, O'CONNELL   *Computer Calculations for Multicomponent Vapor-Liquid Equilibria*
WHITAKER   *Introduction to Fluid Mechanics*
WILDE   *Optimum Seeking Methods*
WILLIAMS   *Polymer Science and Engineering*
WU AND OHMURA   *The Theory Scattering*

PRENTICE-HALL, INC.
PRENTICE-HALL INTERNATIONAL, INC., UNITED KINGDOM AND EIRE
PRENTICE-HALL OF CANADA, LTD., CANADA

Process
Dynamics
and Control

Volume 1
Analysis of
Dynamic Systems

J. M. Douglas
Department of Chemical Engineering
University of Massachusetts

PRENTICE-HALL, INC., Englewood Cliffs, New Jersey

CHEMISTRY

Repl. acc. # 9797

PRENTICE-HALL INTERNATIONAL, INC., *London*
PRENTICE-HALL OF AUSTRALIA, PTY. LTD., *Sydney*
PRENTICE-HALL OF CANADA, LTD., *Toronto*
PRENTICE-HALL OF INDIA PRIVATED LIMITED, *New Delhi*
PRENTICE-HALL OF JAPAN, INC., *Tokyo*

10   9   8   7   6   5   4   3   2   1

Library of Congress Catalog Card Number: 70-171041

ISBN: 0-13-723049-4

Printed in the United States of America

To
My mother and father
My lovely wife, Betsy
and to
Lynn and Bobby

Contents

*The asterisk indicates sections or chapters considered to provide a basis for an under-
graduate course.

Preface

A number of experts on process control held a roundtable discussion at the 1956 Annual AIChE Meeting in Boston.[1] Perhaps it is worth quoting one of the comments concerning the "state of the art" at that time

> Well, I think we're not far from having information which would be usable for probably 75 or possibly 90 percent of typical plant control problems. There are four types of problems which constitute about 95 percent of all process control problems. Those are temperature, pressure, flow, and liquid level.

The statement continues by acknowledging that methods available in the literature on acoustics could be used to solve three of the four basic types of control problems, providing that the control engineer had an adequate mathematical background to be able to understand the published material. Another quotation of interest, concerned with the future of the profession, is

> First, in industry we don't have enough people trained in this kind of work, and second, there does not exist avery strong activity at the academic level.

A little later in the article, the panel puts forth the recommendation

> We should encourage universities and technical shcools to include process control as an important part of chemical engineering curricula.

[1] *Chem. Eng. Prog.*, **53**, No. 5, 209 (1957).

At about this same period in time, Amundson, along with Bilous and Aris, were demonstrating that the classical techniques of nonlinear mechanics could be used to study the stability and control of chemical reactors. The epic paper by Aris and Amundson[2] caught the imagination of a number of investigators and led to a flurry of activity at several universities. A major change in emphasis took place, from the control of pressure, temperature, flow, and liquid level to the control of unit operations equipment such as reactors, heat exchangers, distillation columns, etc. Until very recently, it seems as if the growth of the field was an autocatalytic phenomenon. However, at present such an abundance of theoretical results exist which have not been applied that it is difficult to identify significant problems. In other words, it is hard to tell whether the limiting factor in applying the present theory will be the development of on-line measuring devices, an improved understanding of the process dynamics, fallacious assumptions inherent in the theory, the cost associated with implementing more sophisticated control systems, a lack of adequately trained engineers in industry, and so forth.

As a specific illustration of the dichotomy that exists in the current application of control technology, we can cite the development of direct digital controllers (DDC). With DDC systems we use either a very large or a special-purpose digital computer, along with pneumatic measuring instruments and control valves, to implement some desired control policy. The large computers make it possible to undertake on-line optimization of the process, to calculate production rates and other information of interest to management, and to provide dynamic control of the unit. Most of the DDC equipment is programmed to implement the classical ideas of three-mode control theory. Originally this situation was desirable, in order that the performance of the DDC could be directly compared with conventional control systems. However, it is widely recognized now that it is probably inefficient to make a very powerful computing tool mimic the behavior of a simple pneumatic device.

The purpose of this book is to present a survey of the various ideas used as a basis for developing control systems. Thus the problems of optimum steady state control, classical servomechanism control, modern (or optimal) control, and purposeful periodic operation are discussed in some detail. Similarly, the multivariable and nonlinear features of most chemical processes are emphasized. The book is intended to serve as an undergraduate text (at the junior or senior level), as well as an introductory graduate text. More important, however, it is also intended to be an experiment in continuing education, so that control engineers in industry can become familiar with some of the new approaches to process control. Hopefully, attempts to implement some of these ideas will reveal the limitations of the theory, and in this way new challenges will be offered to the theoreticians.

[2] R. Aris and N. R. Amundson, *Chem. Eng. Sci.*, **7**, 121, 132, 148 (1958).

Unfortunately, an attempt to cover so much ground in a text, to provide as much qualitative discussion as possible of the advantages and shortcomings of various approaches used to design control systems, and to include a sufficient development of the required mathematical tools so that undergraduates and practicing engineers can follow the analysis, means that the book is lengthy. For this reason it has been divided into two volumes. The first emphasizes the procedures that can be used to develop an understanding of the dynamics of chemical processes and techniques for evaluating the effect of process dynamics on normal steady state operation. Therefore this volume is primarily concerned with analysis. The second volume treats the design of dynamic control systems. It describes the methods that can be used to improve the dynamic characteristics of a plant by adding some kind of a controller. Thus the emphasis here is on synthesis.

To be more specific, the first chapter discusses various kinds of control problems in terms of the operation of an automobile, with the hope that even the ideas of feedforward and optimal control can be formulated on an intuitive basis. After this brief introduction, the main part of the text starts with a treatment of the optimum steady state design problem. Despite the fact that many plants are so complicated that it has not been practical actually to determine this optimum design, improved techniques for handling optimization problems, improved computing tools, and increasing competition will make this approach more popular in the future. Moreover, the achievement of the optimum steady state is the commonly accepted goal of most of the process engineering effort; therefore it should provide the basis for judging the performance of any control strategy, or control system. By considering some oversimplified examples, it is easy to introduce the basic concepts involved and then extend the approach to include optimum steady state operation (or control).

The application of steady state equations to describe how the plant responds to various kinds of input changes will be valid only if the process dynamics are negligible, so we must be able to estimate the dynamic response. Chapter 3 discusses dynamic model building, Chapter 4 describes the dynamic characteristics of lumped parameter processes, Chapter 5 extends the results to distributed parameter processes, and Chapter 6 presents a technique for evaluating the importance of the nonlinear terms in the system equations. The main emphasis in these chapters is on linear systems analysis, although methods for approximating complex systems and for evaluating plant parameters from dynamic, experimental data are considered.

In the second volume of the text, we suppose that the results of a dynamic analysis indicate that the method of optimum steady state control is not applicable and that it is desirable to look for some way of modifying the dynamic characteristics of the original plant. We consider a control system to be anything that can be attached to a unit in an attempt to make the overall system have an improved dynamic performance. Some of the most commonly used control system design techniques from servomechanism theory are described in Chapter 7. Because these techniques are limited to plants with a single input and a single output, and because most chemical plants are multivariable processes, Chapter 8 introduces the methods that can be used to extend or modify the classical theory. Next, Chapter 9 describes how optimal control theory can be used to synthesize multivariable controllers.

The basic goal of the control system design methods presented in these three chapters is the development of a regulating control system. That is, we assume that either we want to operate as close as possible to the optimum steady state design conditions (despite the fact that disturbancs enter the plant) or we want to move to some new, desired, steady state operating condition in the best possible way. Chapter 10 reviews some of the evidence which indicates that this approach might not always lead to the most profitable performance of the plant. In other words, an attempt is made to show that in some cases a deliberate periodic operation can be superior to the optimum steady state operation. The reason for including this material was to point out that even though we know a great deal about process dynamics and process control, and have been highly successful in designing and operating complex chemical plants, we are still tremendously ignorant of what it might be possible to achieve by taking a new look at the system from a more fundamental and imaginative point of view. Thus it is an attempt to open up the field of control again and stimulate interest in exploring the effects of new kinds of control laws.

The material in the Table of Contents considered to provide a basis for an undergraduate course is designated by an asterisk. Similarly, the exercises at the end of each chapter prefixed by the letter A should not be beyond the scope of an undergraduate's ability. Those labeled B are of intermediate difficulty, and those with a C are at the graduate level. Also, exercises marked by an asterisk appear in more than one chapter (several methods are applied to the same physical system), while those designated with a double dagger (‡) are "open-ended" to some extent.

Many instructors will find it advantageous to supplement the text with some additional material. In particular, since the focus in the text is on the development of control laws and since the discussion of hardware that can be used to implement these laws has been minimized (because this is such a rapidly changing field), it might be desirable to present some lectures, or required reading assignments, on the different kinds of measuring instruments, control valves, special purpose analog or digital computers, and so on. Similarly, on a more advanced level, it would be useful to describe the techniques for modifying the dynamic analysis and control laws to apply to sampled-data systems, to present an introduction to stochastic processes, to extend significantly the treatment of the control of batch processes, and so on. This material was not included in the text because of space considerations.

No attempt has been made to include a complete bibliography of the process dynamics or control literature, for the annual reviews published by *Industrial and Engineering Chemistry* are readily available. A reader who is interested in expanding his knowledge in these fields should definitely become familiar with these articles. The references listed in the text have been selected on a somewhat arbitrary basis, and it should be noted that there is no particular relationship between the significance of a paper and the fact that it has been mentioned or not. The references are listed as footnotes, which, like the tables and graphs, have been numbered consecutively within each section.

The author is grateful to Professor M. M. Denn and Professor Lowell B. Koppel for their suggestions concenring the manuscript, to several of my graduate students (in particular, Mr. N. Y. Gaitonde and C. S. R. K. Prasad) for their help with several of the calculations, and to the numerous secretaries who suffered with the manuscript.

As mentioned earlier, one of the underlying purposes of the book is to encourage investigators to look for new approaches to process control. The text describes the techniques that have been tried and suggests some alternatives. In many respects it is representative of the motto on Rhush Rhees Library at the University of Rochester. . . .

Here is the history of human ignorance, error, superstition, folly, war, and waste, recorded by human intelligence for the admonition of wiser ages still to come.

J. M. Douglas

Steady
State
Considerations

The Concepts of Process Dynamics and Control

1

Modern control theory is possibly the fastest-growing field in today's technology. A great number of new approaches for controlling chemical and petroleum units have been developed, but because of the rapid growth of the field, many of these new concepts have not been applied to industrial systems. Consequently, the translation of the modern theory into practical control devices poses a great challenge to practicing engineers. Some of the factors that make this translation difficult are an incomplete understanding of the dynamic behavior of many chemical process units, a lack of facility by many practicing engineers with the advanced mathematical tools used to describe the theories, and the difficulty of finding new research results in the almost overwhelming number of journals and periodicals.

The purpose of this book is to provide a perspective for modern control theory. Thus a survey of the various methods used to design control systems is presented, and the interrelationships between these approaches is discussed in some detail. With this kind of background, a practicing engineer should be able to classify a particular control problem of interest, recognize what additional information must be gathered before a solution can be attempted, select the theoretical approaches that might be fruitful, and evaluate the final design. Moreover, he should be in a position to update existing control systems if the modifications will lead to an improved performance of the plant, and he should be aware of future developments that might lead to major changes in the operating policies of processes.

Before attempting to develop dynamic models for chemical units or the fundamental principles associated with the design of control systems, we will consider the more common example of the operation of an automobile. Using this simple illustration, we can define most of the important terms that arise in the study of process dynamics and control and gain some insight into the scope of our investigation.

Steady State Operation of an Automobile

Initially we will restrict our attention to the steady state operation of an automobile, for this process would be analogous to the steady state operation of a chemical plant. We can visualize a situation where we are driving across a very flat desert and are maintaining the position of the accelerator pedal at a fixed setting. For this case, we would expect to move along at approximately a constant speed until we ran out of gasoline. The actual value of the speed obviously would depend on the position of the accelerator pedal and the characteristics of the car.

In order to determine the relationship between these two quantities for any particular car, we could attempt to derive and then solve a complete set of equations describing the fuel and air flow into the carburetor, the motion of various valves, the combustion process in the cylinders, the motion of the crank and drive shafts, and so on, until the motion of the wheels was established. It is quite apparent that this theoretical approach would be difficult. Alternatively, it would be possible to conduct a set of experiments where the steady state speed of the car was measured as a function of the throttle position. These data could then be plotted or, if desirable, used to develop an empirical equation relating the input and output variables—that is, we could develop a process correlation (see Figure 1.1-1).

The first approach is much more general, for all we would have to do is

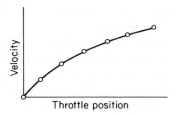

Figure 1.1-1. Empirical correlation.

to change some of the parameters used in the model (the size of the wheels, the mass of the pistons, the geometrical description of the carburetor, etc.), and we should be able to predict the speed versus throttle position for any kind of automobile. However, as mentioned previously, this procedure has the drawback of being extremely complex. Although the empirical method is conceptually much more straightforward, it is only applicable for the particular car under consideration, and the whole experimental procedure must be repeated even if a minor change is made (e.g., wheel size). Thus this approach can be quite expensive if numerous cases are to be studied or if the cost per experiment is high. As with any problem concerning the chemical industry, the choice between a theoretical approach and an empirical approach must be based on the relative costs.

System Dynamics

Although the steady state operation of an automobile has some interesting analogies to the steady state operation of a chemical plant, it is apparent that this phenomenon is not very important in the actual operation of a vehicle. The major problems encountered in everyday driving are maneuvering the car around curves in the road, avoiding potholes and other obstructions, starting and stopping at traffic lights, and so forth. Most of these maneuvers involve speeding the car up or slowing it down, and our primary concern is with the dynamic rather than the steady state operation. As any hot rodder knows, neither the theoretical model describing steady state performance nor the steady state correlation will be applicable to the dynamic system. Thus if we are at steady state conditions initially and then suddenly change the throttle position to some new value, we do not achieve the new steady state velocity instantaneously. Instead, it takes a finite time for the car to accelerate or decelerate to the new constant speed.

In order to predict the dynamic operation of the car, we would have to go back and insert accumulation terms in all the mass, energy, and momentum balances we used in the theoretical model. Also, we would have to specify all the system inputs in order to make a valid prediction of the system velocity at any time. Doing so would require a complete knowledge of the curvature of the roads, the gradients, how we would manipulate the accelerator pedal, and so on. Similarly, in the empirical approach we would have to carry out an overwhelming number of experiments in order to be able to predict the behavior of the system. For example, a dynamic correlation describing an abrupt (step) change in throttle position could not be used to estimate the effects of very rapid throttle variations, for in the former case the car will accelerate or decelerate, whereas in the second case the velocity will remain essentially constant.

Feedforward Control

From the preceding discussion it seems as if it would be almost impossible to drive a car. This conclusion would indeed be correct if we were required to make the predictions in question. For example, if we were going to tell a blind man exactly how to drive from one city to another, we would need an exact knowledge of all the obstacles he might encounter in order to pre-calculate his stopping and starting patterns, as well as the direction he should follow, at every instant during his trip. We could call the procedure we develop a feedforward control strategy because we are using knowledge of the dynamics of the automobile, plus known information describing the inputs to the system to manipulate the control variables—that is, the throttle setting and steering wheel position—in such a way as to accomplish our objective.

The hazards associated with this kind of a control philosophy are apparent. If a strong crosswind blows the blind man's car off the road somewhere along his route, our control system will have failed. Or if the shape of a pothole has changed somewhat since our original calculations, we will again fail in our objective. Thus it is obvious that we can never completely rely on a feedforward control system to guide the operation of either an automobile or a chemical plant.

Feedback Control

The real reason driving a car becomes such a simple task, after an initial learning period, is that the driver acts as a feedback control system. He visually measures the position of the car on the road and reads the instantaneous velocity on a meter. Then he decides whether or not he is in the proper lane, moving in the correct direction (following the contour of the road), and proceeding at an aceptable speed. After comparing the measured and desired results, he adjusts the throttle position and/or the steering wheel until an agreement is obtained. Since he is always making corrections based on an instantaneous comparison between his measured position and speed and the desired values of these quantities, he is able to operate the car in a satisfactory manner, irrespective of any disturbances (crosswind, curves, hills, potholes, traffic lights) acting on the vehicle. This is the great advantage of feedback control systems.

Automatic Control Systems and Instrumentation

Up to now we have considered only certain aspects of the operation of an automobile. In addition to the factors mentioned, we know that an automobile has a cooling system and an electrical system, plus many other complex mechanisms. It would be possible to have a device that would indicate the temperature of the cooling water and also a valve that adjusts the flow of coolant to the radiator, both located on the control panel of the automobile. Then the driver could adjust the valve in such a way as to keep the engine temperature within reasonable limits. However, as we all know, this function is performed automatically in most cars by installing a thermostat in the cooling lines. Similarly, a voltage regulator and an oil pressure regulator are installed to relieve the driver of the task of making certain necessary operating adjustments. This situation is completely analogous to the one in chemical plants and oil refineries where a host of instruments are used to indicate the present conditions in certain units and a great number of automatic devices are installed to keep certain operating variables within some reasonable range or to maintain them at some specified value.

Optimal Control

We have restricted our attention to the use of an automobile as a conveyance from one location to another. However, other goals may be important in particular situations. For example, we might enter the automobile in a stock car race, a rally, or a gymkhana (an obstacle course for automobiles). Depending on the nature of the competition, we must change the strategy of our control action. Thus in a speed race we are most interested in maximizing our overall speed, whereas in some rallys we attempt to drive a prescribed course at an exactly specified speed with minimum error, and in others we try to minimize the gasoline consumption by exerting as little pressure on the accelerator pedal as possible.

In order to excell in the competition, we must do the best we can; therefore we are interested in determining the optimal control action. Since we would expect to obtain different optimal policies for the different kinds of competitions, we must specify a performance criterion carefully—that is, maximum speed, minimum error, minimum fuel—in any optimal control problem. Furthermore, it is usually necessary to consider the physical limitations of the system if we hope to find realistic solutions. For example, if we want to drive the car for a distance of exactly 100 yards in a minimum time, it is not very helpful to learn that we should proceed at an infinite speed until we arrive at the finish line and then stop instantaneously. Instead, if we realize that perhaps we should ram the accelerator to the floor first and then, after some exactly computed distance, brake the car to the maximum possible extent so that we stop precisely on the finish line (i.e., follow a "bang-bang" control trajectory), we gain a better appreciation of the effort required to determine the exact optimal path and to implement the optimal control action. Optimal control problems in the chemical and petroleum industries are similar to the preceding ones with the exception that the possibility of using profit as the performance criterion we wish to maximize must also be considered.

Summary

To summarize our observations concerning automobile operation, we have found that it should be possible to develop both theoretical and empirical models describing the steady state behavior. The theoretical model is more difficult to obtain, but it is flexible and can be used to predict the results of new systems. On the other hand, the empirical process correlation is simpler to obtain for a single case, but the model development must be repeated for every case of interest. Economic considerations are used to select the appropriate approach in each new situation.

Neither of the steady state models can be used to predict the dynamics of the system (unless, of course, the inputs are changed very, very slowly), because they do not include accumulation effects. It should be possible to modify the theoretical model so that it describes the dynamic system. In order to prevent this model from becoming overwhelmingly complex, we attempt to describe only the important dynamic effects.

If the dynamics of the system are known exactly and all the disturbances (the curvature of the road, road grades, wind velocity and direction, shape of potholes, traffic lights) acting on the system can be measured without error, it should be possible to design a feedforward control system that will automatically manipulate the control variables (the acelerator, brake, steering wheel position, gear ratio) in such a way as to make the car operate properly. Errors will accumulate in a feedforward control system, so we often augment it with a feedback control system.

In feedback control we continually measure the state of the system (the position of the car, its velocity, etc.), compare the measured values with a set of desired values, and then alter the control variables to obtain a closer agreement between these quantities. Since the input to the feedback controller is the instantaneous output of the system, it is not necessary to consider the disturbances.

In most cases we build devices that automatically provide the control action. However, to be on the safe side we also provide the operator with some instruments, such as a speedometer, temperature gauge, ammeter, pressure gauge, and tachometer, so that he can monitor the system performance.

It is also possible to design optimal control systems—that is, to determine control actions that correspond to the maximum speed, minimum error, minimum fuel, minimum travel time, and so on. In many of these optimization problems, we must consider physical constraints imposed on the system (the throttle position can only be varied between a completely closed and a fully open position, the fuel tank has some maximum capacity, etc.) in order to obtain a realistic solution.

The problems described above are similar to those encountered in the chemical and petroleum industries. Many other interesting analogies can be developed—for example, start-up problems, maintenance and failure of control systems, emergency operating procedures, adaptive or learning control functions, stochastic disturbances, information sampling—but these are left to the imagination of the reader. In the rest of this book we will attempt to develop a quantitative treatment of these problems. Since the type of control action desired depends on the objectives we set for the process, we will begin our analysis with a description of the optimum steady state design problem.

QUESTIONS FOR DISCUSSION

Consider the operation of some machine or other physical system you are familiar with, and answer the questions below.

1. What is the meaning of steady state operation of the machine?

2. How would you go about developing a theoretical model describing the steady state operation?

3. How would you develop an empirical process correlation?

4. Which steady state model would be the cheapest to develop?

5. What kind of disturbances might act on the system?

6. Which system inputs could be used as control variables? What basis did you use for your selection?

7. Can you describe a feedforward control problem for the system?

8. How would feedback control be employed?

9. Can you describe any optimal control problems for the system?

10. Does the system normally operate at steady state conditions or in a dynamic mode?

Optimum
Steady State Design
and Control

2

Although process design is primarily an art, we can gain an important understanding of many design concepts by considering some oversimplified design problems. This approach will allow us to ignore temporarily the fact that a tremendous number of unknown factors are involved in the design of any real system, such as the kinetic expressions for many complex reactions, techniques for describing the fluid mechanics in complicated flow systems, the physical properties of many substances, and procedures for selecting among the many methods that can be used to accomplish the same overall processing objectives. Instead, we will be able to concentrate on the implicit assumptions used in the formulation of an optimum design problem, the relationship between these assumptions and the dynamic characteristics of the system, and the manner in which the design problem establishes the objectives for process control.

SECTION 2.1 OPTIMUM STEADY STATE DESIGN

The accepted procedure is to design processes to operate at steady state conditions. This approach is based on the implicit assumption that some steady state system will always correspond to the most profitable plant. Although the validity of this assumption will be discussed in greater detail in a later chapter, at present we will accept it as being correct. By restricting our attention to a steady state analysis we greatly simplify the design problem, for we need only consider sets of steady state relationships describing the various pieces of process equipment. However this approach requires the assumption that all the inputs to the plant are constant, even though we know that some will vary with time. The effects of these time-varying inputs can be evaluated after the optimum design has been determined, but the optimum design is based on their average values.

Once a particular equipment configuration has been selected, and with the use of the foregoing assumptions, conceptually it is a simple matter to describe a procedure for obtaining the optimum steady state design of certain processes. The normal approach is to define a profit function that includes all the capital and operating costs, a set of equations describing the operation of the specific processing units under consideration, a statement of the desired production rate that has been established by a sales estimate, the assumed average values of certain inputs[1] (average feed concentrations, average cooling water temperature, average steam pressure available), and either the known system parameters or equations that can be used to predict these values (kinetic constants, heat transfer coefficients, plate efficiencies, etc). The equations describing the process units, as well as the production statement, can then be used to eliminate some of the operating variables or equipment sizes from the profit equation, so that the profit expression will contain only independent variables and known inputs. For simple design problems, the optimum values of the independent variables can be obtained first by setting the partial derivatives of the profit equation with respect to the independent variables equal to zero and then by solving the resulting set of coupled nonlinear equations. Next the optimum values of the remaining operating variables and equipment sizes can be determined, using the system equations and the production statement.

Several illustrations of this simple procedure are given below. Then the limitations of the approach, as well as the difficulties encountered in its application, are discussed in some detail.

---

[1] Although the classification of system variables is always somewhat arbitrary, we refer to inputs as the prespecified values of the input streams, such as the temperature of a feed stream. These fixed values tie the plant to its environment. Parameters are coefficients that appear in material and energy balances, or profit expressions, such as heat capacity or a cost factor. Design variables include all equipment sizes, flow rates, and the values of effluent compositions and temperature. Not all design variables can be chosen independently, because they are related by the mass and energy balances. Thus a normal design problem involves the maximization of profit subject to a number of constraint equations.

Example 2.1-1 Design of an isothermal reactor

As an elementary example of the optimization procedure, we will consider the design of an isothermal reactor for producing G lb moles per hour of a product B by the reaction $A \longrightarrow B$, where the reaction rate is given by the expression

$$r = kA^2 \qquad\qquad 2.1\text{-}1$$

and $k = $ (ft)^3(lb mole)$^{-1}$(hr)$^{-1}$, $A = $ (lb mole)(ft^3)$^{-1}$. This desired production rate can be achieved (1) by converting most of the material (low raw material cost) entering a large reactor (high capital cost) or (2) by converting only a little of a large amount of feed material (high raw material cost) entering a small reactor (low capital cost). Consequently, we must find the reactor size, feed rate, and conversion that correspond to the most profitable system.

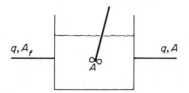

q, A_f q, A

Figure 2.1-1. Continuous-stirred-tank reactor.

From elementary reactor design considerations, we know that an iso-thermal tubular reactor gives a higher conversion per unit volume than a continuous-stirred-tank reactor.[2] However, the capital costs of the two types of system are different; therefore we might want to determine the optimum design for both cases. Moreover, we should consider the optimum design of the whole plant rather than just the reactor, for the separation costs will depend on both the reactor flow rate and conversion. In order to avoid this additional complexity, we will assume that for the system in question, the costs required to separate A and B in the effluent stream are negligible in comparison with the reactor and raw material costs and that component A cannot be recycled.

In the following example we will limit our attention to the design of a continuous-stirred-tank reactor having a cost C_v($)(ft^3)$^{-1}(hr)^{-1}$, which includes all the operating costs associated with the reactor and the capital costs on a depreciated basis.[3] The feed mixture is assumed to have an average composition A_f(lb mole)(ft^3)$^{-1}$, and the cost of this feed is taken as C_f($)(lb mole of A)$^{-1}$. The rate of profit obtained from the system, P_r ($)(hr)$^{-1}$, is merely the rate of income, or the value of the product, C_B ($) (lb mole of B)$^{-1}$, multiplied by the production rate G, minus the total cost of

[2] O. Levenspiel, *Chemical Reaction Engineering*, Ch. 6, Wiley, N. Y., 1962.

[3] Like automobiles, process equipment wears out, and therefore its value changes with time. The simplest method of estimating a cost on a depreciated basis is to divide the total cost by the expected life. Some additional discussion is given in the next section.

the operation, $C_T(\$)(hr)^{-1}$, or the cost of the reactor on a per unit time basis, $C_V V$, and the cost of the reactants, $C_f q A_f$, where V (ft³) is the reactor volume and q (ft³)(hr)$^{-1}$ is the volumetric flow rate to and from the reactor. Thus

$$P_r = C_B G - C_T \qquad\qquad 2.1\text{-}2$$

Since the production rate G is a constant, it is apparent that the design which maximizes the profit must be the same as that which minizes the total cost.

$$C_T = C_V V + C_f q A_f \qquad\qquad 2.1\text{-}3$$

The operating and design variables must satisfy certain constraints; namely, the material balance for the reactor

$$q(A_f - A) - kVA^2 = 0 \qquad\qquad 2.1\text{-}4$$

and the specified production rate

$$G = q(A_f - A) \qquad\qquad 2.1\text{-}5$$

Now, the problem is to determine the design variables q, V, and A, which will minimize the total reactor cost, assuming that the system parameters and inputs, C_V, C_f, k, A_f, and G, have been specified. There are three design variables, but these must satisfy the two constraints, Eqs. 2.1-4 and 2.1-5, so that there is only one degree of freedom in the system (i.e., only one of the design variables is an independent variable).

Solution

If we solve Eq. 2.1-4 for V and Eq. 2.1-5 for q, and substitute the results into the cost expression, Eq. 2.1-3, we obtain

$$C_T = \frac{C_V G}{k A^2} + \frac{C_f G A_f}{A_f - A} \qquad\qquad 2.1\text{-}6$$

or letting

$$x = \frac{A}{A_f} \qquad \alpha = \frac{C_V}{C_f k A_f^2} \qquad\qquad 2.1\text{-}7$$

we find that

$$\frac{C_T}{C_f G} = \frac{\alpha}{x^2} + \frac{1}{1 - x} \qquad\qquad 2.1\text{-}8$$

Then the optimum value of the fraction of unconverted material, x, is determined by

$$\frac{\partial(C_T/C_f G)}{\partial x} = 0 = \frac{-2\alpha}{x^3} + \frac{1}{(1 - x)^2}$$

or the solution of the cubic equation

$$x^3 - 2\alpha(1 - x)^2 = 0 \qquad\qquad 2.1\text{-}9$$

The positive value of x between zero and unity that satisfies this equation can be determined graphically, numerically, using either the standard formula for the roots of a cubic equation or Figure 2.1-2. The composition of unreacted material in the effluent, the optimum feed rate, and the optimum reactor volume can then be calculated by using Eqs. 2.1-7, 2.1-5 and 2.1-4, respectively.

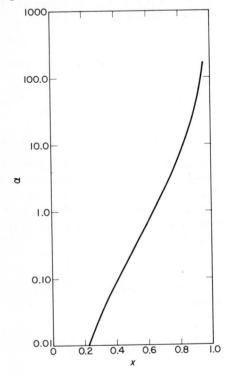

Figure 2.1-2. Solution of optimum design equation, Eq. 2.1-9.

If we consider a specific example where $G = 75$ lb mole/hr, $C_f =$ \$0.60/lb mole, $C_V = \$0.01/(\text{hr})(\text{ft}^3)$, $k = 1.2$ ft^3/(lb mole)(hr), and $A_{fs} = 1.0$ lb mole/ft^3, then

$$\alpha = \frac{C_V}{C_f k A_f^2} = \frac{0.01}{(0.60)(1.2)(1)} = 0.0139$$

and from Figure 2.1-2

$$x = 0.25$$

Since $x = A/A_f$, then $A = 0.25$ lb mole/ft^3. From Eq. 2.1-5

$$q = \frac{G}{A_f - A} = \frac{75}{1 - 0.25} = 100 \text{ ft}^3/\text{hr}$$

and from Eq. 2.1-4

$$V = \frac{q(A_f - A)}{kA^2} = \frac{100(1 - 0.25)}{1.2(0.25)^2} = \frac{75}{0.075} = 1000 \text{ ft}^3$$

Thus it is a simple matter to establish the optimum design of an isothermal, continuous-stirred-tank reactor. This solution can be compared with the results for a tubular reactor, a series of continuous-stirred-tank reactors, or other designs, and the reactor system having the lowest cost determined. It is interesting to note that the optimum reactor conversion is independent of the production rate and that the total cost or profit is merely a linear function of the production rate.

Example 2.1-2 Design of a nonisothermal reactor

In the previous example we assumed that the reaction rate constant k was a known value. Since this parameter depends on the reactor temperature, we have implicitly assumed that the reaction proceeds at the feed temperature. This assumption would be approximately correct either if the heat of reaction was very small or if the reactant material was dissolved in a large excess of solvent so that the reactor temperature was essentially independent of heat generated (or consumed) by the reaction. However, an examination of the total cost expression, written in terms of the single independent design variable and the known system parameters and inputs, Eq. 2.1-6, indicates that the total cost decreases as the reaction rate constant k is increased. This result is to be expected, for it is known from elementary reactor design theory that the highest conversion per unit volume is obtained for single irreversible reactions when the reactor is operated isothermally at the highest allowable temperature.

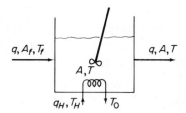

Figure 2.1-3. Continuous-stirred-tank reactor with a heating coil.

In order to assess the possible savings resulting from operation at elevated temperatures, we must consider the design of a heat exchanger as well as the reactor. For simplicity, we will consider only the case where a heating coil is installed directly in the vessel. As we increase the heat duty supplied by the coil, we would expect to cut down the required reactor size and cost of the reactor. However, the cost of the exchanger will increase, and this factor will more than compensate for the reactor cost at some point. Thus we expect to find an optimum design. Similarly, any particular heat duty can be achieved either by having a high flow rate of heating material (high operating cost) and a corresponding low value of the temperature driving force ΔT and exchanger area (low capital cost) or by having a low flow rate of hot fluid (low operating cost) and a corresponding large ΔT and heat-transfer area (high capital cost). Consequently, an optimum heat-exchanger design exists for any specific heat duty.

We will consider a case where the heat-transfer area of the coil is A_H (ft²), the cost of this coil on a depreciated basis is C_A ($)(ft²)$^{-1}(hr)^{-1}$, heating fluid is available at a temperature T_H, the cost of q_H lb/hr of heating fluid is C_H ($/lb), and the hot fluid is available at sufficient pressure to achieve any desired flow rate in the coil. We suppose that we still desire to produce G lb mole/hr of product by the reaction $A \longrightarrow B$, but in this case we assume that the reaction rate is first-order. The temperature of the reactant stream is taken as T_f, and the heat of reaction for the exothermic process is $-\Delta H$ (Btu)(lb mole of A)$^{-1}$. For simplicity, we again assume that the separation costs are negligible and that the reactant material cannot be recycled. Using this information, we want to determine the quantitative relationships describing the optimum reactor design.

Since the production rate is a specified quantity, the design that corresponds to a minimum total cost will also correspond to the maximum profit rate (see Eq. 2.1-2). The total cost of the reactor plus exchanger system is

$$C_T = C_V V + C_f q A_f + C_A A_H + C_H q_H \qquad 2.1\text{-}10$$

where the first two terms are the capital and operating costs for the reactor and the last two terms are the corresponding costs of the heat exchanger. The expression for the production rate is still

$$G = q(A_f - A) \qquad 2.1\text{-}11$$

The steady state material and energy balances for the reactor, for a first-order reaction, are

$$q(A_f - A) - kVA = 0 \qquad 2.1\text{-}12$$

where

$$k = k_0 e^{-E/RT} \qquad 2.1\text{-}13$$

and

$$qC_p\rho(T_f - T) + (-\Delta H)kVA + Q_H = 0 \qquad 2.1\text{-}14$$

where Q_H is the heat duty supplied by the exchanger, (Btu)(hr)$^{-1}$, k_0 is the frequency factor (hr)$^{-1}$, E is the activation energy (Btu)(lb moles)$^{-1}$, R is the gas constant, T is the reactor temperature, C_p is the heat capacity (Btu) (lb)$^{-1}$(°R)$^{-1}$, and ρ is the mass density of reactants and products (lb)((ft³)$^{-1}$. An energy balance for the heating fluid gives

$$Q_H = q_H C_{pH}(T_H - T_0) \qquad 2.1\text{-}15$$

and the heat transfer between the coil and the reactor is

$$Q_H = UA_H\left(\frac{T_H - T_0}{\ln\left[(T_H - T)/(T_0 - T)\right]}\right) \qquad 2.1\text{-}16$$

where T_0 is the outlet temperature of the heating fluid, C_{pH} is its heat capacity, and U is the overall heat-transfer coefficient $(\mathrm{Btu})(\mathrm{ft}^2)^{-1}(\mathrm{hr})^{-1}(^\circ\mathrm{R})^{-1}$. We can simplify the problem somewhat if we are willing to approximate the last expression by the equation

$$Q_H = UA_H(T_{H\,\mathrm{av}} - T) \qquad 2.1\text{-}17$$

and by making the assumption that

$$T_{H\,\mathrm{av}} = \tfrac{1}{2}(T_H + T_0) \qquad 2.1\text{-}18$$

so that

$$Q_H = UA_H[\tfrac{1}{2}(T_H + T_0) - T] \qquad 2.1\text{-}19$$

Equations 2.1-11 through 2.1-15, and either Eq. 2.1-16 or 2.1-19, provide six relationships between the nine design variables, q, A, T, k, Q_H, q_H, T_0, A_H, V, and the specified system parameters and inputs, k_0, E, C_p, ρ, $(-\Delta H)$, C_{pH}, U, and G, A_f, T_f, and T_H, respectively. Hence any three of the design variables may be chosen arbitrarily and the remaining six calculated by using the constraint equations. Of course, we want to select the set of values for these three degrees of freedom that lead to a minimum total cost for the system, Eq. 2.1-10.

Solution

One procedure[4] for eliminating six of the design variables from the problem is to solve Eq. 2.1-11 for q

$$q = \frac{G}{A_f - A} \qquad 2.1\text{-}20$$

and Eq. 2.1-12 for V

$$V = \frac{q(A_f - A)}{kA} = \frac{G}{kA} \qquad 2.1\text{-}21$$

and then substitute these expressions into the profit expression, Eq. 2.1-10, and the reactor energy balance, Eq. 2.1-14, giving

$$C_T = \frac{C_V G}{kA} + \frac{C_f G A_f}{A_f - A} + C_A A_H + C_H q_H \qquad 2.1\text{-}22$$

and

$$Q_H + (-\Delta H)G + \frac{GC_p \rho (T_f - T)}{A_f - A} = 0 \qquad 2.1\text{-}23$$

[4] N. Y. Gaitonde and J. M. Douglas, *AIChE Journal*, **15**, 902 (1969).

Similarly, Eqs. 2.1-15 and 2.1-19 can be solved for q_H and A_H,

$$q_H = \frac{Q_H}{C_{pH}(T_H - T_0)} \qquad 2.1\text{-}24$$

$$A_H = \frac{2Q_H}{U(T_H + T_0 - 2T)} \qquad 2.1\text{-}25$$

and the results used to eliminate these variables from the profit equation

$$C_T = \frac{C_V G}{kA} + \frac{C_f G A_f}{A_f - A} + \frac{C_A 2Q_H}{U(T_H + T_0 - 2T)} + \frac{C_H Q_H}{C_{pH}(T_H - T_0)} \qquad 2.1\text{-}26$$

Now, Q_H can be eliminated from this expression by solving Eq. 2.1-23 for heat duty

$$Q_H = -\left[(-\Delta H)G + \frac{GC_p\rho(T_f - T)}{A_f - A}\right] \qquad 2.1\text{-}27$$

and substituting this result into the preceding equation,

$$C_T = \frac{C_V G}{kA} + \frac{C_f G A_f}{A_f - A} - \left[(-\Delta H)G + \frac{GC_p\rho(T_f - T)}{A_f - A}\right]\left[\frac{2C_A}{U(T_H + T_0 - 2T)}\right.$$
$$\left. + \frac{C_H}{C_{pH}(T_H - T_0)}\right] \qquad 2.1\text{-}28$$

In this last equation A, T, and T_0 are the only unknowns, once the Arrhenius equation, Eq. 2.1-13, is used to eliminate k, and these variables will be determined by a simple optimization procedure. Since

$$\frac{\partial C_T}{\partial T_0} = 0 = -\left[(-\Delta H)G + \frac{GC_p\rho(T_f - T)}{A_f - A}\right]\left[\frac{-2C_A}{U(T_H + T_0 - 2T)^2}\right.$$
$$\left. + \frac{C_H}{C_{pH}(T_H - T_0)^2}\right]$$

and since the first term in brackets is simply the heat duty Q_H (see Eq. 2.1-27), which will not be zero, the second term in brackets must vanish, leading to

$$C_H U(T_H + T_0 - 2T)^2 = 2C_A C_{pH}(T_H - T_0)^2 \qquad 2.1\text{-}29$$

which is a quadratic relationship between T_0 and T. Similarly, we must have

$$\frac{\partial C_T}{\partial A} = 0$$

or

$$\frac{C_V(A_f - A)^2}{kA^2} = C_f A_f + C_p\rho(T - T_f)\left[\frac{2C_A}{U(T_H + T_0 - 2T)} + \frac{C_H}{C_{pH}(T_H - T_0)}\right] \qquad 2.1\text{-}30$$

and that $\partial C_T/\partial T = 0$, or

$$\frac{C_V E}{kART^2} = -\left[(-\Delta H) + \frac{C_p\rho(T_f - T)}{A_f - A}\right]\left[\frac{4C_A}{U(T_H + T_0 - 2T)^2}\right]$$
$$+ \frac{C_p\rho}{A_f - A}\left[\frac{2C_A}{U(T_H + T_0 - 2T)} + \frac{C_H}{C_{pH}(T_H - T_0)}\right] \qquad 2.1\text{-}31$$

Thus Eqs. 2.1-29, 2.1-30, and 2.1-31 give us three equations that we can use to calculate the three unknowns T_0, A, and T (using the Arrhenius equation to write k explicitly in terms of T). Since this is a set of coupled, nonlinear, transcendental equations, we will have to use a numerical or graphical method in order to obtain a solution.

One possible approach we can use to solve the equations is first to write an expression for T_0 in terms of T by taking the square root of both sides of Eq. 2.1-29. Next, a graph showing the optimum values of T_0 for any reactor temperature could be prepared. A number of values of T, and the corresponding optimum values of T_0, could be substituted into Eq. 2.1-30, and this quadratic equation used to calculate the optimum values of A between zero and A_f. These values can be plotted against the specific values of temperature previously chosen. Now, arbitrary values of T, and the corresponding optimum values of A and T_0, can be substituted into the left- and right-hand sides of Eq. 2.1-31, and the functions on the left- and right-hand sides plotted against T. The intersection of these two curves will give the optimum temperature, which is the value of temperature that satisfies the optimization equations. Now the optimum values of T_0 and A can be found by using the set of graphs prepared earlier. Once the optimum values of these three independent variables have been determined, the corresponding optimum values of the other design variables, k, q, V, Q_H, q_H, and A_H, can be calculated using Eqs. 2.1-13, 2.1-20, 2.1-21, and 2.1-23 through 2.1-25.

If we consider a particular example where the system parameters are given in Table 2.1-1, then Eq. 2.1-29 can be written as

$$T_0 = \frac{1}{1 + \sqrt{\dfrac{2C_A C_{pH}}{C_H U}}} \left[2T - \left(1 - \sqrt{\frac{2C_A C_{pH}}{C_H U}} \right) T_H \right] \qquad \text{2.1-29}$$

This quadratic equation was solved for T_0, for a number of values of T, and the results are plotted in Figure 2.1-4.

TABLE 2.1-1

SYSTEM PARAMETERS*

Cost data:	$C_A = 0.0979$ ($)(sq cm)$^{-1}(hr)^{-1}$, $C_f = 0.0236$ ($)(g mole)$^{-1}$
	$C_H = 0.00002$ ($)(gram)$^{-1}$, $C_V = 0.0148$ ($)(hr)$^{-1}$(liter)$^{-1}$
System constants:	$C_p \rho = 1.0$ (cal)(cc)$^{-1}$($^\circ$K)$^{-1}$, $C_{pH} = 1.0$ (cal)(gram)$^{-1}$($^\circ$K)$^{-1}$
	$E = 59{,}800$ (cal)(g mole)$^{-1}$, $(-\Delta H) = 12{,}900$ (cal)(g mole)$^{-1}$
	$k_0 = 3.01 \times 10^{30}$ (sec)$^{-1}$
Design data and system inputs:	$G = 1285$ (g mole)(hr)$^{-1}$, $T_f = 300$ ($^\circ$K), $T_H = 373$ ($^\circ$K)
	$U = 1.0$ (cal)(sq cm)$^{-1}$(sec)$^{-1}$($^\circ$K), $A_f = 0.005$ (g mole)(cc)$^{-1}$

* Reproduced from N. Y. Gaitonde and J. M. Douglas, *AIChE Journal*, **15**, 902 (1969), by permission.

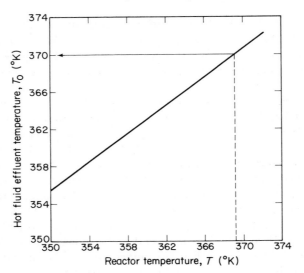

Figure 2.1-4. Optimum values of hot-fluid effluent temperature T_0 as function of T.

Similarly, Eq. 2.1-30 can be written as

$$\left\{1 - \frac{C_f}{C_V}kA_f - \frac{kC_p\rho(T - T_f)}{C_V}\left[\frac{2C_A}{U(T_H + T_0 - 2T)} + \frac{C_H}{C_{pH}(T_H - T_0)}\right]\right\}A^2$$
$$- 2A_fA + A_f^2 = 0 \qquad\qquad 2.1\text{-}30$$

The values of T selected above, and the corresponding optimum values of T_0, were substituted into this equation and then the optimum values of A were determined in terms of these quantities. A graph of the results is given in Figure 2.1-5. Next, the left- and right-hand sides of Eq. 2.1-31 were plotted against T, using the values from Figures 2.1-4 and 2.1-5 to eliminate A and T_0 from this equation. The two curves are given in Figure 2.1-6. Since the functions on the two sides of the equation must be equal to each other at the point where the two curves intersect, the optimum value of T is found to be $T = 369 \,°\text{K}$. Then the optimum values of T_0 and A can be obtained by using Figures 2.1-4 and 2.1-5 and are found to be $T_0 = 370\,°\text{K}$ and $A = 0.00143$ (g mole)(sc)$^{-1}$. The corresponding optimum values of the remaining design variables are calculated via the constraint equations, Eqs. 2.1-13, 2.1-20, 2.1-21, and 2.1-23 through 2.1-25. The results are given in Table 2.1-2.

Even though this procedure does give us a solution to the set of optimization equations, it does not indicate the sensitivity of the solution; that is, we would like to have some feeling for how much difference it makes if we change the values of the optimum design variables a small amount. Now that the optimum values are known, we can determine the change in the total cost in the neighborhood of the minimum for the three independent variables,

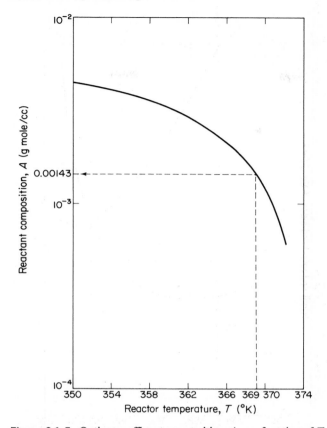

Figure 2.1-5. Optimum effluent composition A as a function of T.

T, T_0, and A, simply by plotting C_T in Eq. 2.1-28 versus each one of these variables, with the others fixed at their optimum values. Figures 2.1-7 and 2.1-8 show the results for T and T_0. In order to make a similar plot for the other design variables, it is necessary to write the total system cost explicitly in terms of these quantities. For example, we can use Eq. 2.1-21 to eliminate A from Eq. 2.1-28 and in this way write the total cost explicitly in terms of V. A graph showing how C_T varies with changes in V is given in Figure 2.1-9. It is clear from these graphs that the total cost is very sensitive with respect to changes in T and T_0 but is fairly insensitive with respect to V. In other words, the multidimensional surface obtained when C_T is plotted against the design variables is quite steep in the neighborhood of the optimum values of T and T_0, but it is relatively flat in the neighborhood of the optimum value of V.

Of course, the method just described for obtaining a solution to the optimization equations is not unique. It would be possible to solve Eq. 2.1-29 for T_0 in terms of T and to use this result to eliminate T_0 from Eqs. 2.1-30 and

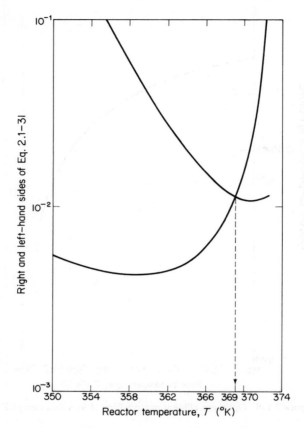

Figure 2.1-6. Graphical solution of Eq. 2.1-31.

TABLE 2.1-2

OPTIMUM DESIGN VALUES*

Design variables:	$T = 369$ (°K), $T_0 = 370$ (°K), $A = 0.00143$ (g mole)(cc)$^{-1}$
	$k = 0.072$ (hr)$^{-1}$, $q = 100$ (cc)$^{-1}$(sec)$^{-1}$, $V = 12{,}500$ (liters)
	$Q_H = 2310$ (cal)(sec)$^{-1}$, $q_H = 770$ (gram)(sec)$^{-1}$, $A_H = 920$ (sq cm)
Optimum costs:	$C_V V = 185.0$ (\$)(hr)$^{-1}$, $C_f q A_f = 42.5$ (\$)(hr)$^{-1}$
	$C_A A_H = 90.8$ (\$)(hr)$^{-1}$, $C_H q_H = 55.4$ (\$)(hr)$^{-1}$
	$C_T = 373.7$ (\$)(hr)$^{-1}$

* Reproduced from N. Y. Gaitonde and J. M. Douglas, *AIChE Journal*, **15**, 902 (1969), by permission.

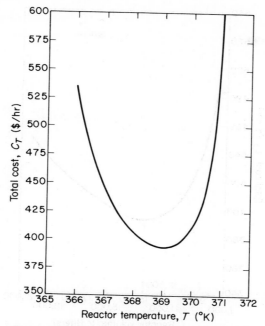

Figure 2.1-7. Total cost vs. reactor temperature T.

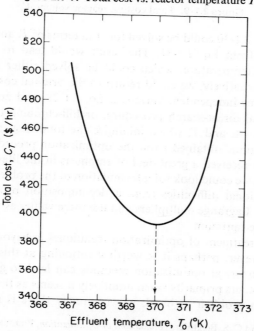

Figure 2.1-8. Total cost vs. exchanger effluent temperature T_0.

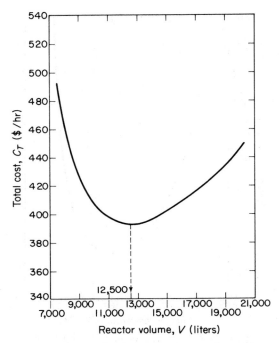

Figure 2.1-9. Total cost vs. reactor volume V.

2.1-31. Then Eq. 2.1-30 could be solved for A in terms of T, and the composition eliminated from Eq. 2.1-31. The result would lead to a relationship containing only temperature, which could be solved either numerically or graphically. Alternatively, we could return to the original cost expression in terms of the three independent variables, Eq. 2.1-28, and attempt to use a three-dimensional direct-search procedure, or hill-climbing routine, to find the values of T_0, A, and T, which minimize the total cost, and completely ignore the equations obtained from the optimization procedure. This kind of approach has received a great deal of emphasis in the literature.[5] As still other alternates, we could look for a formulation of the problem that involves fewer computational difficulties (e.g., a decomposition procedure), or we could introduce Lagrange multipliers and use these values as measures of the sensitivity of the optimum.

A detailed treatment of optimization techniques is beyond the scope of this book. However, perhaps it is worth mentioning at this point that the blind use of numerical optimization methods can lead to great difficulties. This situation occurs primarily when intuitively it seems as though there were an optimization problem at hand, when actually there is not—that is, in

    [5] D. J. Wide and C. S. Beightler, *Foundations of Optimization*, Prentice-Hall, Englewood Cliffs, N.J., 1967.

some cases the set of optimization equations may have no solution. Since it is an extremely difficult task to establish when a set of coupled, nonlinear, transcendental equations does in fact have a solution, it is sometimes informative to look at the equations from a completely opposite point of view. For example, in the reactor design problem discussed above, we know that all the equipment costs must be positive quantities; therefore we can check the equations to make certain that they will always predict this kind of behavior. An inspection of Eq. 2.1-29 indicates that when C_A is positive, we always expect to obtain a positive value for C_H. Similarly, Eq. 2.1-31 shows that positive values of C_A and C_H will always yield positive values of C_V, for the first term in brackets is merely the heat duty (see Eq. 2.1-27). However, even if these three cost factors are positive, Eq. 2.1-30 reveals that there is no guarantee that C_f will always be positive. After some manipulation, it can be shown that a solution will not exist unless

$$1 > \frac{(-\Delta H)(A_f - A)}{C_p \rho (T_H - T_f)} + \left(\frac{E}{RT^2}\right)\left(\frac{A}{A_f - A}\right)\frac{(T - T_f)(T_H - T)}{(T_H - T_f)} \qquad \text{2.1-32}$$

A similar approach can be used to demonstrate that a continuous-stirred-tank reactor containing a cooling, rather than a heating, coil can never correspond to an optimum economic design (in the sense defined above). This conclusion might explain why Westbrook and Aris[6] found that the optimum conditions for a somewhat similar design problem were obtained with adiabatic operation—that is, no cooling. Of course, it is impossible even to attempt to assess the existence of a solution if a direct attack is made on the cost equation, instead of developing the final set of optimization equations.

Example 2.1-3 Design of a Reactor and Separation System

As a final example of economic considerations, we will consider the optimum steady state design of a very simple chemical plant where we have an isothermal, continuous-stirred-tank reactor followed by an extraction unit. Again we desire to make G lb moles/hr of a product B by the first-order reaction $A \longrightarrow B$. For this example we will let x_f be the mole fraction of component A in the feed, x_A be the mole fraction of A in the reactor, k be the reaction rate constant $(\text{lb mole})(\text{hr})^{-1}(\text{ft}^3)^{-1}$, C_f be the cost of reactant material $(\$)(\text{lb mole of } A)^{-1}$, and C_V be the cost of the reactor on a depreciated basis $(\$)(\text{hr})^{-1}(\text{ft}^3)^{-1}$. We assume that the flow rate, $F (\text{lb mole})(\text{hr})^{-1}$, through the reactor is constant and that the reactor effluent is fed directly into the first stage of a countercurrent extraction unit.

The extracting solvent, which is fed into stage N of the extraction unit at a constant rate of $S (\text{lb mole})(\text{hr})^{-1}$ is immiscible with the carrier liquor of the feed and removes only component B from the mixture. We assume that

[6] G. T. Westbrook and R. Aris, *Ind. Eng. Chem.*, **53**, 181 (1961).

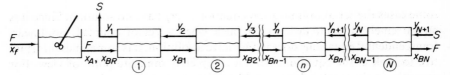

Figure 2.1-10. Reactor and extraction unit.

the extraction process does not affect the solvent or raffinate flow rates, that the inlet solvent stream does not contain any B, that the equilibrium relationship is linear, and that unreacted A is not recycled. Letting the total hourly cost of each extraction stage on a depreciated basis be C_N ($)(stage)$^{-1}(hr)^{-1}$ and the cost of solvent, including recovery from the final extract and makeup of spillage losses, be C_S ($)(lb mole of solvent)$^{-1}$, we want to determine the optimum steady state design of the system.

Since the production rate has been specified, the system having the minimum cost will also produce the maximum profit. The total cost of the system is

$$C_T = C_V V + C_f F x_f + C_N N + C_S S \qquad \text{2.1-33}$$

The constraints on the system are the material balance for the reactor

$$F(x_f - x_A) - kVx_A = 0 \qquad \text{2.1-34}$$

the stoichiometric relationship between reactant A and product B

$$x_{BR} = x_f - x_A \qquad \text{2.1-35}$$

the material balance for the extraction unit, that is, the Kremser equation

$$x_{BN} = x_{BR}\left(\frac{M^{N+1} - M^N}{M^{N+1} - 1}\right) \qquad \text{2.1-36}$$

where

$$M = \frac{F}{mS} \qquad \text{2.1-37}$$

and m is the slope of the equilibrium relationship

$$y_n = mx_{Bn} \qquad \text{2.1-38}$$

and the equation for specified production rate

$$G = F(x_{BR} - x_{BN}) \qquad \text{2.1-39}$$

The unknown design variables in this problem are V, F, N, S, x_A, x_{BR}, M, and x_{BN}; whereas the variables specified for design purposes are x_f, k, m, G, and y_{N+1}.

Since there are only five constraint equations, Eqs. 2.1-34 through 2.1-37 and Eq. 2.1-39, available to calculate the eight design variables, there are three degrees of freedom in the system—three variables that must be determined by an optimization procedure. Of course, the compositions of both phases at intermediate points in the extraction train, y_n and x_{Bn}, could also be considered as design variables. However, since the Kremser equation, Eq. 2.1-36, can be written for each stage and since the equilibrium relationship, Eq. 2.1-38, can be used to calculate the corresponding compositions in the solvent phase, there will be as many new equations as unknown design variables and the number of degrees of freedom will not be changed.

Solution

The necessary set of optimization equations can be obtained by substituting the Kremser equation, Eq. 2.1-36, into the production equation, Eq. 2.1-39, and then solving for x_{BR}

$$x_{BR} = \frac{G}{F}\left(\frac{M^{N+1} - 1}{M^N - 1}\right) \qquad 2.1\text{-}40$$

The stoichiometric relationship, Eq. 2.1-35, can now be written as

$$x_A = x_f - \frac{G}{F}\left(\frac{M^{N+1} - 1}{M^N - 1}\right) \qquad 2.1\text{-}41$$

Solving the reactor material balance, Eq. 2.1-34, for V

$$V = \frac{F(x_f - x_A)}{kx_A}$$

and then substituting Eq. 2.1-41 to eliminate x_A, gives the result

$$V = \frac{FG(M^{N+1} - 1)}{k[Fx_f(M^N - 1) - G(M^{N+1} - 1)]} \qquad 2.1\text{-}42$$

Now, V can be eliminated from the total cost equation, Eq. 2.1-33,

$$C_T = \frac{C_V FG(M^{N+1} - 1)}{k[Fx_f(M^N - 1) - G(M^{N+1} - 1)]} + C_f Fx_f + C_N N + C_S S \qquad 2.1\text{-}43$$

which, when Eq. 2.1-37 is substituted, is an equation depending only on the three variables F, S, and N. The optimum values of these quantities can then be found, using the equations below.

$$\frac{\partial C_T}{\partial F} = \frac{\partial C_T}{\partial S} = \frac{\partial C_T}{\partial N} = 0 \qquad 2.1\text{-}44$$

After some manipulation, the results of this procedure are

$$\frac{C_S}{C_N} = \frac{M^{N+1} - (N+1)M + N}{S(M-1)\ln M} \qquad 2.1\text{-}45$$

$$\frac{C_V}{C_N} = \frac{k[Fx_f(M^N - 1) - G(M^{N+1} - 1)]^2}{F^2 Gx_f M^N (M-1)\ln M} \qquad 2.1\text{-}46$$

$$\frac{C_f}{C_S} = \frac{G(M^{N+1} - 1)^2 - Fx_f M^N [M^{N+1} - (N+1)M + N]}{F^2 x_f^2 M^N [M^{N+1} - (N+1)M + N]/S} \qquad 2.1\text{-}47$$

It is apparent that some numerical or graphical technique will be required to find the values of F, S, and N that satisfy this set of coupled, nonlinear

equations. Once these values have been determined, the corresponding optimum values of x_{BR}, x_A, and V can be calculated by using Eqs. 2.1-40 through
2.1-42, respectively.

Of course, the foregoing procedure is not really valid. The last optimization equation in the set, Eqs. 2.1-44, requires that N, the number of plates
in the column, be a continuous variable, which we know is never true in
practice. In order to avoid this contradiction, we must develop a new optimization procedure that will treat N as a discontinuous variable. A case-study
approach would be the most straightforward: selecting various integral values
of N and repeating the optimization analysis for the two remaining independent design variables S and F for each value of N until the optimum design
is established. However, normally it is possible to simplify the amount of
work involved by solving the original set of optimization equatons to obtain a
first estimate of the optimum design and then selecting integral values of N
in the neighborhood of this solution.

SECTION 2.2 DISCUSSION OF OPTIMUM STEADY STATE
DESIGN PROCEDURE

Although the preceding problems are greatly simplified illustrations of the
optimum design procedure, they do provide a vehicle for examining some of
the concepts and implicit assumptions involved in the formulation of an
optimum steady state design problem. The main ingredients required in the
problem statement are a profit equation or cost function, a set of steady state
equations describing the input-output relationships for each unit in the process, a specified production rate established by a sales forecast, and the a
priori specification of certain system parameters and input variables, such as
kinetic constants, heat or mass transfer coefficients, feed composition and
temperature, and average coolant temperature. Certain difficulties can be
associated with each of these four main areas, and therefore we must make
certain that we recognize the limitations involved.

Process Economics

The profit function, used to describe the process economics, must include
all the capital and operating costs. However, the operating costs are on a
per unit time basis, whereas the capital costs represent fixed expenditures.
A number of methods[1] can be used to compute the time value of money,
such as venture profit, cash flow, rate of return, and projected worth. Different techniques are used by different industrial corporations and, in some

---

[1] D. F. Rudd and C. C. Watson, *Strategy of Process Engineering*. Ch. 4, Wiley, N.Y.,
1968.

cases, by different departments in the same company. Since it is not too difficult to construct problems that have opposing solutions, depending on the technique employed, obviously it is not possible to establish a unique way of putting capital and operating costs on the same time basis.

Another major difficulty encountered in the economic analysis is the discontinuities often found in the sales prices of products and raw materials. Similarly, most equipment costs are not continuous functions of equipment size. Thus the costs of vessels or pipes undergo discontinuities at certain diameters because of increased wall-thickness requirements, and many kinds of equipment, like heat-exchanger tubes, are available only in certain specified diameters and lengths. Hence the implicit assumption of continuity of the variables, required for the calculus optimization procedure to be valid, might not always be satisfied.

System Equations

A basic assumption implicit in the optimum steady state design procedure is that an optimum steady state process will always be the most profitable. The fact that this assumption is not always valid will be demonstrated in Chapter 10. However, in order even to attempt a steady state analysis, a set of steady state equations describing the operation of each processing unit must be available. In the foregoing examples simple algebraic expressions were used for this purpose. Normally more complex units are described by sets of nonlinear ordinary or partial differential equations, which often cannot be solved to give explicit relationships between the input and output variables. Also, in many cases, the appropriate equations for industrial processes are not available in analytical form. For example, the input-output relationships for catalytic cracking units, butane alkylation reactors, and so on, generally are only available as a set of empirical correlations. Of course, these correlations could be expressed as continuous mathematical functions, which then could be used as the system equations in an optimization procedure. However, as is well known, the correlations developed from pilot plant data seldom give an exact prediction of actual plant operation. Usually it is necessary to adjust the level of the correlation to match the plant characteristics, although even the predictions of the trends of the major variables must be reversed in some cases.

A more severe limitation in attempting to specify the system equations required to accomplish a certain processing objective is that normally a unique set does not exist. Instead, there are numerous alternatives between processing equipment, all leading to the same final goal and all having a different set of system equations and costs. Thus the optimization procedure must be repeated for each alternative and the various cases then compared. Sometimes it is possible to eliminate many of the alternatives, using order-of-magnitude calculations, but it is seldom possible to find only a single processing scheme that must be optimized.

Other drawbacks to the method are that physical constraints (flooding velocities, pressure and temperature limitations, etc.) are often associated with certain kinds of equipment and these constraints introduce additional discontinuities into the problem. Moreover, the system equations are sometimes transcendental and therefore cannot always be used to eliminate some of the design variables from the cost function without introducing Lagrange multipliers and increasing the dimensionality of the problem. A more serious obstacle occurs when the design variables are actually functions— for example, when it is possible to have an optimal temperature profile, an optimal feed distribution—so that the calculus of variations, rather than elementary calculus, must be used to solve the optimum steady state design problem. An introduction to this type of analysis is presented later on, for this approach is also useful for designing control systems.

Production Rate

Although at first glance it would seem that the most precise piece of information used in the formulation of an optimum steady state design problem is the specification of the production rate, this figure, as well as the selling price of a new product, is usually established by a market analysis. The hazards of forecasting are well known, and experience has shown that the demand for a new product is often grossly underestimated or catastrophically over-estimated in some cases. An interesting discussion of methods that can be used to estimate the optimum initial production and planned expansion of a new plant is available in Rudd and Watson's book.[2] Perhaps it is sufficient to say here that the specified production rate, like the other elements of the problem, represents the best engineering estimate available.

Another complicating factor should be mentioned, however, for there are plants in existence where the desired product, and therefore production rate, is deliberately changed during certain time intervals. For example, attempts are sometimes made to maximize the furnace oil production from a catalytic cracking unit during the fall and winter months, whereas gasoline production is maximized during the spring and summer months. Similarly, sometimes the same equipment is used for the hydrodealkylation of toluene to produce benzene and the hydrodealkylation of catalytic gas oil to produce napthalene, although the operating conditions are altered. Obviously this kind of problem, where there is a variable production rate, is more complicated than those described previously.

[2] See footnote 1, p. 28.

System Parameters and Inputs

The specification of the system parameters and some of the process inputs also poses certain problems. Mass transfer coefficients, physical properties, plate efficiencies, reaction rate constants, and so on are seldom avilable to a high degree of accuracy. Consequently, there will be some uncertainty associated with the final design. In addition, it is well known that some of the system inputs will generally vary with time. For example, it is to be expected that the raw materials used for the process will not be exactly uniform (the coke-producing tendency of the various kinds of gas oil processed in a catalytic cracking unit change when the feed tank is changed). Similarly, cooling-water temperatures vary throughout day, the available steam pressure in a plant often changes as steam demand changes, prices of raw materials and products vary with market conditions, and so forth. It is commonly assumed, in the optimum steady state design procedure, that these variables are all known and that they remain constant at their average values. However, a procedure for compensating for these effects has been developed and will be outlined later in this chapter.

Other Difficulties

The optimum steady state design procedure also assumes that it is, in fact, possible to solve the final set of optimization equations. The magnitude of this task should not be underestimated, however. Even for the simple problems discussed earlier, the optimization procedure led to a coupled set of nonlinear algebraic equations. Techniques for handling this type of problem are neither simple nor straightforward. In fact, in many cases it is much easier to make a direct attack on the final cost equation and, by a direct-search procedure or a hill-climbing method, attempt to find the unknown design variables leading to a minimum cost. Such methods have been discussed in detail by Wilde and Beightler.[3] Similarly, an attempt to show that the solution of the optimization equations actually does correspond to a minimum by considering the second derivative terms will probably not be successful because of the complexity of most problems. Instead, it is usually simpler to evaluate the profit in the neighborhood of the calculated optimum point in order to ensure that it is at least a local minimum.

Obviously, as we consider more realistic design problems involving many more pieces of much more complicated process equipment, the attempt to optimize the whole process at one time probably will fail because of the overwhelming complexity—that is, chemical plants normally have a large number

[3] See footnote 5, p. 24.

of recycle streams that interconnect the various units. However, in the last few years techniques for handling problems of this nature have been developed. For simple plants having a sequential structure, Bellman's dynamic programming approach will yield fruitful results. In more complicated cases, a decomposition procedure must be carried out before applying dynamic programming. Excellent reviews of much of the available material have been presented by Rudd and Watson[4] and Himmelblau and Bischoff.[5]

Future Outlook

Despite the array of apparently overwhelming obstacles encountered in the formulation of an optimum steady state design problem (which obviously are also present in any other kind of an approach), this procedure will undoubtedly receive much greater attention in the near future. Increasing development costs and increasing competition, together with decreasing profit margins, will make it necessary to decrease the uncertainties associated with present design techniques; that is, safety factors will have to be decreased by an order of magnitude in order to cut capital costs. The only hope of accomplishing this goal is to quantify the design procedure. There will be numerous unknown factors in the initial attempts at quantification, but these difficulties should be resolved as more experience is gained. Better methods for handling the mathematics problems associated with design, both with regard to computational algorithms and to computing equipment, are rapidly being developed. Thus major advances in the optimum steady state design approach are to be expected in the near future.

At the same time, however, there will always be uncertainties associated with a design, and for this reason the designer should attempt to incorporate as much flexibility as possible. Of course, increased flexibility means higher costs, but it is far better to have a slightly suboptimum unit that works than an optimum system (?) that does not operate. Again, the approximate balance between maximum flexibility and minimum cost is a matter of "engineering judgment" on the part of the designer.

SECTION 2.3 OPTIMUM STEADY STATE CONTROL

Once a process has been designed and put into commercial operation, the optimum operating procedure must be determined, even though the plant was designed in an optimum manner. It was pointed out earlier that some of the

[4] See footnote 1, p. 28.

[5] D. M. Himmelblau and K. B. Bischoff, *Process Analysis and Simulation*, Ch. 8, Wiley, N.Y., 1968.

inputs to the plant and/or the materials and utilities costs will vary with time, and even if the optimum steady state design is sufficiently flexible to handle these disturbances, they will, nevertheless, cause a change in the operating costs—and hence the profit—obtained from the plant. The method for finding the optimum operating policy of the plant as a function of the input variables is essentially the same as that for the optimum steady state design procedure except that in this case all the equipment sizes will be fixed at their optimum design values. Some elementary examples of this approach are discussed below.

Example 2.3-1 Optimum steady state control of an isothermal reactor

If we consider the operation of the optimum isothermal reactor described in Example 2.1-1 but admit the possibility that the feed composition A_f or the operating temperature, and hence the reaction rate constant k, might fluctuate with time, then it is apparent from Eqs. 2.1-3 through 2.1-5 that the total cost, the production rate, and the conversion would all change with time. The reactor volume V has been fixed by the steady state design, and we must still maintain the specified production rate G. Assuming that the changes in A_f and k take place slowly, so that accumulation effects are negligible and we can continue to use the steady state material balance to describe the reactor, we can attempt to determine the value of the feed rate that will maintain G at the desired value. Solving the material balance equation, Eq. 2.1-4, for the outlet composition A, and substituting this result into the production statement, Eq. 2.1-5, gives the result

$$G = q\left[A_f - \frac{q}{2kV}\left(-1 + \sqrt{1 + \frac{4kVA_f}{q}}\right)\right] \qquad 2.3\text{-}1$$

Thus for any value of A_f or k, we can adjust the feed rate q in such a way as to maintain G constant. The profit obtained from the process will depend on this value of q, along with A_f and k, and therefore will vary with time. In fact, for certain time intervals this profit may be higher than that corresponding to the optimum steady state design.

At first it may seem advantageous to operate the reactor at a higher feed composition and temperature. However, doing so would require additional separation equipment to improve the purity of the feed stream or heat-transfer equipment to raise the reactor temperature, which would increase the capital cost of the plant. It has been assumed in this simple problem that these additional costs would lead to a lower overall profit rate. As an illustration of a commercial unit that has a time-varying feed composition, we could consider a catalytic cracking unit where the coke-producing materials in the gas-oil feed stream are usually different in every feed tank, and in some cases, because of stratification, there are variations within each feed tank.

Example 2.3-2 Optimum steady state control of a nonisothermal reactor

In the previous example, so few degrees of freedom were available in the system that the control variable had to be used to maintain the production rate constant at the specified level. However, if we consider the system described in Example 2.1-2, a true optimum operating policy is required. We will examine a case where the feed composition A_f, the feed temperature T_f, and the temperature of the heating fluid T_H, all vary with time, but we will assume that these fluctuations are slow enough so that the steady state equations remain valid. Again, we consider that all the equipment sizes, V and A_H, have been fixed at their optimum steady state design values (based on the time average values of A_f, T_f, and T_H), and we attempt to find the settings of the feed rate q and the flow rate of heating fluid q_H, which minimize the total operating cost.

For this case, the expression for C_T reduces to

$$C_{OP} = C_f q A_f + C_H q_H \qquad 2.3\text{-}2$$

since V and A_H are fixed. Now we can eliminate q, using Eq. 2.1-20, and then A, using Eq. 2.1-21, to give

$$C_{OP} = \frac{C_f G A_f kV}{A_f kV - G} + C_H q_H \qquad 2.3\text{-}3$$

Equations 2.1-15 and 2.1-19 can be combined and solved for T_0

$$T_0 = \frac{(q_H C_{PH} - \frac{1}{2} U A_H) T_H + U A_H T}{q_H C_{pH} + \frac{1}{2} U A_H} \qquad 2.3\text{-}4$$

and this expression used to eliminate T_0 from Eq. 2.1-15,

$$Q_H = \frac{q_H C_{pH} U A_H (T_H - T)}{q_H C_{pH} + \frac{1}{2} U A_H} \qquad 2.3\text{-}5$$

The substitution of this result into the reactor energy balance, Eq. 2.1-14, and the simultaneous elimination of q and A from this expression, using Eqs. 2.1-20 and 2.1-21, give a result that can be solved explicitly for q_H in terms of T

$$q_H = \frac{-U A_H [GkVC_p\rho(T_f - T) + (-\Delta H)G(A_f kV - G)]}{2C_{pH}[GkVC_p\rho(T_f - T) + (-\Delta H)G(A_f kV - G) + U A_H(T_H - T)(A_f kV - G)]}$$

$$2.3\text{-}6$$

Next q_H can be eliminated from the cost equation, thereby leaving an expression with T as the only unknown (after k has been eliminated by using the Arrhenius equation, Eq. 2.1-13)

$$C_{OP} = \frac{C_f G A_f kV}{A_f kV - G}$$

$$- \frac{C_H U A_H [GkVC_p\rho(T_f - T) + (-\Delta H)G(A_f kV - G)}{2C_{pH}[GkVC_p\rho(T_f - T) + (-\Delta H)G(A_f kV - G) + U A_H(T_H - T)(A_f kV - G)]}$$

$$2.3\text{-}7$$

The optimum steady state temperature, for arbitrary values of A_f, T_f, or T_H, is determined from the equation

$$\frac{\partial C_{OP}}{\partial T} = 0 \qquad\qquad 2.3\text{-}8$$

Once this value has been established, the optimum values of q_H, Q_H, T_0, A, and q can be determined, using Eqs. 2.3-6, 2.3-5, 2.3-4, 2.1-21, and 2.1-20, respectively.

Although the optimum steady state control problem is simpler than the optimum steady state design problem, because it involves a smaller number of degrees of freedom, the algebra is sufficiently complicated that it is usually necessary to use a numerical or graphical approach even for small problems. However, these calculations can be carried out on off-line computing equipment. For our problem we could prepare a set of tables, or, better yet, a pair of nomographs, telling us how to set the values of q and q_H for various measured values of A_f, T_f, and T_H. It might also be desirable to tabulate the corresponding values of A and T, in case we might want to change the operating conditions in the separation equipment. The extension of the method to more complex problems is straightforward in concept, but the numerical work involved can increase tremendously.

SECTION 2.4 DISCUSSION OF THE OPTIMUM STEADY STATE CONTROL PROCEDURE

In the optimum steady state control problem, we consider the operation of a plant having fixed equipment sizes and attempt to manipulate certain inputs (the control variables) in such a way that we always minimize the operating costs and maintain the desired production rate constant even though some of the other process inputs, the disturbances, fluctuate with time. It is apparent that this procedure is simpler than the optimum steady state design analysis, for all the equipment sizes have been fixed so that there are fewer degrees of freedom in the system. Similarly, the difficulty of putting capital and operating costs on the same basis does not arise; discontinuities in raw material, product, and equipment costs probably will not be encountered; it is not necessary to consider processing alternatives; forecasting problems are minimized; and so forth. However, we must still find ways of handling sets of complicated equations or process correlations in the optimization procedure; be able to treat cases where there are discontinuities caused by pressure, temperature, or flooding constraints on the equipment; be able to assess the accuracy of our model,[1] the model parameters, and the measurements of the disturbances and control variables; devise techniques for

---

[1] The term model refers to the set of equations used to describe the behavior of the plant.

measuring all the disturbances; develop methods for solving the final set of optimization equations; and recognize that our analysis is only valid for steady state operation.

Although it again seems as if the application of the optimum steady state control procedure to any realistic process would be overwhelmingly difficult, this approach is the one taken in most of the existing computer control systems. Thus large on-line digital computers are used to examine up to 75 process variables, to calculate the optimum control settings every 10 to 30 minutes, and to adjust between 10 and 20 control variables every 30 minutes to 2 hours. These control computers have been installed on catalytic cracking units, crude-oil distillation columns, catalytic polymerization reactors, catalytic reforming units, ammonia synthesis reactors, ethylene reactors, open-hearth furnaces, paper-making processes, cement kilns, and a great number of other processes. In some very large and complex plants, a hierarchy of computers is used, with an executive computer supervising the operation of a number of control computers that guide the operation of one or more units. Additional information about these systems is available in the references listed below[2,3] and in the annual reviews appearing in *Industrial and Engineering Chemistry*. A complete discussion of these problems is beyond the scope of this text. However, in order to ensure that the concepts involved are clear, so that we will be able to understand their relationship to dynamic operation, we will discuss a few additional aspects of the optimum steady state control problem in further detail.

Measurement of Disturbances

In the previous illustrations of optimum steady state control systems, we assumed that it was possible to measure all the disturbances entering the system. Even for the simplest problem, Example 2.3-1, it is apparent that if we cannot measure one of the disturbance variables, say A_f, we cannot use Eq. 2.3-1 to determine the value of the feed rate that will keep the production rate at its desired level. Thus we must find some way of modifying the procedure. The technique commonly used to overcome this difficulty is first to measure one of the output variables, say A, and then use one or more of the constraint equations to calculate the unknown input. Of course, this approach requires that the steady state equations used to describe the system remain valid during the transient period. The new value of the control variable that leads to the best system performance can then be calculated and the control settings adjusted to correspond to the new optimum conditions. A similar procedure can be developed for the system described in Example 2.3-2. Alternatively, the manipulations required to obtain the final set of optimiza-

[2] E. S. Savas, *Computer Control of Industrial Processes*, McGraw-Hill, N.Y., 1965.
[3] G. L. Farrar, *Oil & Gas J.*, October 22, 1962.

tion equations could be changed so that the final cost expression and control variables are written in terms of the time-varying system variables that can be measured. For true steady state operation, all these procedures are equivalent; but as we shall see later, different results are obtained for dynamic systems.

Alternate Methods for Determining the Optimum Control Action

Since most industrial units are fairly complex pieces of equipment, a set of steady state process correlations are normally used to describe their operating characteristics. These are developed either by solving the set of theoretical equations (usually sets of nonlinear algebraic and/or ordinary and partial differential equations) and then developing a simpler input–output correlation for the results in the region of interest or by making a series of test runs on an operating unit and then developing an empirical process correlation. Although simpler than the original set of system equations, these correlations are still complex enough so that is it difficult to employ the straightforward optimization procedure described earlier. Instead, the common practice is to make a more direct attack on the cost function.

One procedure in extensive use is referred to as sectional linear programming.[4] In this approach all the nonlinear terms appearing in the cost function and the system equations or constraints are linearized over a small interval by expanding the functions in a Taylor series around some starting point and neglecting all second- and higher-order terms in this expansion. The simplex algorithm, or some modified linear programming routine, is used to find the optimum conditions at the boundary of this region. Next, the performance index and system equations are linearized once more around these optimum conditions, and the linear programming procedure is used again to obtain a new estimate of the optimum. This approach can be repeated until an adequate approximation of the optimum conditions has been determined.

A number of other methods that can handle the nonlinear equations directly have been developed. For example, Lagrange multipliers can be used to couple the system equations to the cost function so that an augmented performance index is obtained. An elementary mathematical analysis indicates that the minimum of this augmented function is identical to the minimum of the original constrained cost equation, which means that the problem can always be put into the form of a simple, but multidimensional, optimization problem. Then one of the direct-search methods—the use of a grid, a Monte Carlo technique, etc.—or a hill-climbing routine can be used to find the optimum.

---

[4] See p. 135 of the reference given in footnote 2 on p. 36.

Experimental Optimization Methods

In all the preceding discussions we assumed that either a theoretical or an empirical model of the input–output relationships for the various pieces of processing equipment was available. However, in certain situations this assumption is not valid. Of course, it would be possible to develop either a theoretical or empirical model for these units and to use one of the optimization methods described earlier, but this development cost must be weighed against the cost of finding the optimum operating conditions experimentally. Several techniqes for accomplishing this task are available. The simplest approach for a steady state system having only a single manipulative input— that is, control variable—is to make some small change in this variable and to compute the new operating costs. If the operating costs decrease, an additional change is made in the same direction as before. This procedure is repeated until the operating costs increase, and then the control variable is varied in the opposite direction. Changes in the control setting are made continually so that the optimum operating conditions will always be approached, even if some of the process inputs change with time.

The procedure described by Draper and Li[5] is a modification of the foregoing method. They vary the control variable at a constant rate and continuously monitor the system output or cost function. If the proper direction for a change is chosen initially, the operating costs will eventually pass through a minimum and start to increase. After some prespecified increase is observed, the direction of change of the control variable is reversed, and this procedure is continually repeated. Hence the system will again approach the optimum operating costs, even if disturbances cause the operating level to change.

A somewhat more complicated method was described by Chanmugan and Box[6] for chemical processes, although the original idea was developed earlier. The fundamental principle is to vary the control variables sinusoidally with a very low frequency so that the steady state equations describing the system remain applicable, and then to use the resulting periodic output information to locate the direction of the minimum of the cost function. If we are at point A on Figure 2.4-1 and first increase and then decrease the value of the control variable, the cost function will first decrease and then increase. However, a similar change in the control variable when the system is at point B will cause the cost function to increase and then decrease. Thus we find that if the input and output signals are in phase, we must decrease the control setting to approach the optimum; whereas if the signals are out of phase, we must increase the value of the control variable. Schindler and

[5] C. S. Draper and Y. T. Li, "Principles of Optimalizing Control Systems and an Application to an Internal Combustion Engine," *ASME Publications*, N.Y., 1951.

[6] G. E. P. Box and J. Chanmugan, *Ind. Eng. Chem. Fundamentals*, **1**, 2 (1962).

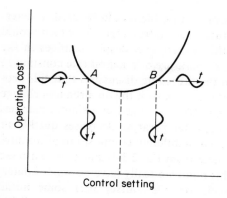

Control setting

Figure 2.4-1. Questing control.

Aris[7] studied several aspects of the method and referred to it as "Questing control."

Although the various methods are described in terms of systems having only a single control variable, it is possible to extend them to multivariable processes. An extensive literature discussing these techniques is available,[8] and therefore we will not treat them in further detail. Our main interest is to obtain an understanding of the fundamental concepts involved and to recognize the effect that the system dynamics will have on the results.

Dynamic Considerations

From our discussion on the operation of an automobile (Chater 1), we know that if we change the position of the accelerator pedal very, very, slowly, the accumulation effects in the system will never be significant and the speed of the car will increase according to the curve given by the process correlation, Figure 1.1-1. However, if we suddenly make a large change in the setting of this control variable, it will take some finite time for the system to arrive finally at the steady state value given by the correlation, and the intermediate points on this curve will have no relationship with the actual speed of the vehicle during the transient period. Also, we know that if we fluctuate the position of the accelerator pedal around some fixed position very rapidly, the car speed will remain essentially unchanged. Providing that we are willing to extrapolate this kind of knowledge of the characteristics of dynamic systems to the operation of chemical plants and to assume that the dynamic effects of disturbances are similar to those of changes in the control variables, we can at least make some qualitative statements about the effects of system dynamics on the optimum steady state control problem.

[7] R. N. Schindler and R. Aris, *Chem. Eng. Sci.*, **22**, 319, 337, 345 (1967).
[8] E. M. Grabbe, S. Ramo, and D. E. Wooldridge, *Handbook of Automation, Computation, and Control*, Vol. 3, Ch. 13, Wiley, N.Y., 1961.

In Example 2.3-1 we developed a procedure that could be used to determine the value of the reactor feed rate q, a control variable, which would maintain the production rate G at the desired level despite changes in the feed composition A_f and the reactor temperature or reaction rate constant k, the disturbances. It is apparent from the preceding discussion that this procedure will only be valid on an instantaneous basis if the disturbances change very, very slowly and if the control variable is changed in a similar manner. Thus if it suddenly rains and the reactor temperature decreases quickly or if we change to another feed tank that has a different composition of reactant material, the value of q that we calculate from Eq. 2.3-1 cannot be expected to maintain G constant. In fact, at this point we really have no way of determining the significance of this calculation. Of course, after some initial transient period, the system will eventually reach a new steady state condition and our original results will be valid once again, continuing to be correct until a new disturbance enters the system.

With this elementary insight into the problem, we see that the optimum steady state control policy will give optimum operation during the time intervals when the accumulation effects in the system are negligible and the steady state equations are valid but will give some suboptimum output during the remaining time periods. Clearly, then, we must be able to determine whether or not the process is normally at steady state or is in some dynamic condition a major portion of the time. Doing so requires a knowledge of how fast the disturbances enter the system (the frequency of the disturbances) and how long it takes the system to arrive at a new steady state after a disturbance has entered the plant. Obviously if new disturbances keep entering the process before it has had a chance to reach the steady state condition corresponding to the preceding disturbance, the system will never achieve a steady state, and thus the concept of optimum steady state control is meaningless. Also, if the disturbances enter the system rapidly, so that the system does not have a chance to move away from its steady state condition, we will be better off maintaining the control variables at their optimum steady state design values rather than the optimum steady state control settings, for these variations will either cause the system to move away from the optimum condition or will have to be changed so rapidly that they do not alter the system output.

A similar difficulty is encountered in the experimental optimization methods. When any change is made in a control variable setting, it is necessary to wait until the system reaches steady state before computing the new value of the performance index in order to see if an improvement has been obtained. Of course, if a disturbance enters during this period, the results are no longer meaningful and another experiment has to be performed. Similarly, depending on the dynamic characteristics of the system, the questing control technique can give completely incorrect results, and it will always force the system away from the optimum operating condition. Therefore these experimental techniques are applicable only when the time between disturbances is much longer than the time it takes the system to reach a new steady state.

It appears as if we will have to know the complete future behavior of the disturbances, as well as the dynamic characteristics of the plant, in order to develop an alternate procedure for the optimum steady state control policy. Since the problem of predicting the exact nature of the disturbances and their effect on the system is essentially the same as instructing a blind man how to drive a car from one city to another, we know that we will not be capable of developing a new method in a rigorous manner. There have been some attempts in recent years to use statistical descriptions of the disturbances and then to look for control settings that maximize the expected value of the profit. Much work remains to be done in this area. The problem of determining the dynamic characteristics of the plant provides the subject matter for Part II.

SECTION 2.5 SUMMARY

In order to determine the optimum steady state design of simple processes, we must first develop an expression for the profit obtained from some particular equipment configuration. This profit equation includes both the operating costs and the capital costs of the equipment on a depreciated basis. Since the operating costs normally depend on the various inlet flow rates, feed compositions, feed temperatures, and so on, and the values of the product streams depend on the effluent flow rates, compositions, temperatures, and the like, we see that the rate of profit obtained from the system must be a function of the various cost factors, the equipment sizes, the system inputs, and the system outputs. In mathematical terms we can write this functional dependence in general terms as

$$P = f(C_\alpha, V_\beta, u_j, v_m, x_i) \qquad \text{2.5-1}$$

where P = rate of profit

C_α = cost of reactants, products, and equipment $\alpha = 1, 2, \ldots, L$

V_β = equipment sizes $\beta = 1, 2, \ldots, N$

u_j = manipulative input variables (control variables) $j = 1, 2, \ldots, J$

v_m = other input variables (disturbances) $m = 1, 2, \ldots, M$

x_i = system outputs (state variables) $i = 1, 2, \ldots, I$

and L represents the total number of reactant and product streams in addition to the total number of pieces of equipment, N is the total number of equipment size variables, J is the total number of control variables, M is the total number of disturbance variables, and I is the total number of system outputs. If any terms in this general functional relationship are additive constants, they may be neglected because they affect the level of the profit only, not the shape of the curve in the neighborhood of an optimum.

Normally we design the plant to produce a certain amount of desired product. Expecting the production rate to depend on at least some of the sytem inputs and outputs, we can write

$$G = g(u_j, v_m, x_i) \qquad\qquad 2.5\text{-}2$$

Also, we know that there is a set of input–output relationships describing the various pieces of equipment, which, in some cases, can be written in the general form

$$h_s(V_\beta, u_j, v_m, x_i, k_\gamma) = 0, \qquad s = 1, 2, \ldots, S \qquad 2.5\text{-}3$$

where

$$k_\gamma = \text{system parameters} \qquad \gamma = 1, 2, \ldots, K$$

S is the total number of constraint equations, and K is the total number of system parameters, such as kinetic constants and heat–transfer coefficients. These equations represent either theoretical models of the plant or empirical-process correlations.

The cost factors and system parameters are known quantities for most design problems, and it is common practice to assume that the disturbances remain constant at some known average value. Hence in the design problem we attempt to determine the N equipment sizes, the I system outputs, and the J control variables. Since we have $S + 1$ equations relating these design variables, the number of degrees of freedom F in the system is

$$F = N + I + J - (S + 1) \qquad\qquad 2.5\text{-}4$$

In other words, we can use Eqs. 2.5-2 and 2.5-3 to eliminate $S + 1$ of the unknown variables from the problem and rewrite the profit expression, Eq. 2.5-1, in terms of just F independent design variables. Denoting these F independent variables by w_r, $r = 1, 2, \ldots, F$, we can determine their optimum values by solving the set of coupled, nonlinear equations.

$$\frac{\partial P}{\partial w_r} = 0, \qquad r = 1, 2, \ldots, F \qquad\qquad 2.5\text{-}5$$

This set of F equations, along with the $S + 1$ constraint equations, provide sufficient information to calculate the $N + I + J$ design variables.

Of course, there are many implicit assumptions and certain limitations in the procedure outlined above. These have been discussed in some detail in Section 2.2. Despite the drawbacks, however, often we can obtain at least an engineering estimate of the optimum steady state design for a particular equipment configuration. Then by repeating the procedure for alternate processing schemes, the most profitable system design can be established.

The optimum steady state design of a process does not necessarily correspond to the optimum steady state operation. In the design problem we assumed that all the system inputs remain constant even though we know that some of them—the disturbance variables—will vary with time. Since the operating costs for the plant depend on the instantaneous values of these inputs, both directly and through the constraint equations, the operating costs will also fluctuate. Thus in some cases it pays to adjust some of the other input variables—the control variables—in such a way as to ensure minimization of the operating costs, thereby obtaining the maximum profit from the system. The problem of determining the optimum steady state settings of the control variables is similar to the foregoing design problem. Exactly the same equations, Eqs. 2.5-3, are used to describe the operating characteristics of the various pieces of equipment, and maintaining the production rate at its original value is still desired. However, all the equipment sizes have been fixed by the optimum steady state design procedure and thus there are N fewer degrees of freedom in the system. We can follow exactly the same approach as outlined earlier, but the only independent variables in the problem—that is, in the set w_r—will be the control variables, and there will be

$$F = I + J - (S + 1) \qquad\qquad 2.5\text{-}6$$

simultaneous equations that must be solved in order to determine their optimum values.

One of the major limitations of this last procedure is the assumption that steady state equations can be used to describe the operation of systems that are changing with time. Although this assumption will be approximately correct in certain instances, we cannot make an a priori judgment as to its validity until we obtain at least some estimate of the dynamic characteristics of the plant. In the remainder of this book we will consider methods for developing dynamic models of process units and procedures for designing control systems that compensate for the effects of disturbances during the transient operation of the system.

QUESTIONS FOR DISCUSSION

It is common practice to include a 15 to 20 percent safety factor in the design of each piece of process equipment. (Actually, 100 percent or larger safety factors are not uncommon for novel operations.) How are these safety factors related to attempts to introduce additional flexibility into the design? What is the relationship between a safety factor for a unit and the sensitivity of the optimum design of that unit—that is, the curvature in the neighborhood of the minimum? Are safety factors ever related to the dynamic operation or control of a plant?

EXERCISES*

1. (A) Consider the design of a condenser for a distillation column, which must have the capacity of condensing 5000 lb of vapor per hour. The vapor condenses at 170 °F and the heat of condensation is 200 Btu/lb. Cooling water is available at 70 °F and costs $0.02 per 1000 gallon. Also, assume that the heat capacity of the water is constant at 1.0 Btu/(lb.)(°F), that the overall heat-transfer coefficient is constant at 50 Btu/(hr)(ft^2)(°F), and that the depreciated cost of the exchanger is $2.333 × 10^{-4} (hr)$^{-1}(ft)^{-2}$. Find the optimum design of the heat exchanger. (This is a modification of a problem of Peters.[1])

2. (A) In the preceding problem, suppose that the condensate temperature changes to 185 °F, because of a variation in the amount of inerts present in the condensing vapor. How would you alter the operating conditions of the condenser in order to achieve the "best" performance?

3. (A) In a calculus course we prove that a function has a minimum by showing that the second derivative is positive when the first derivative is equal to zero. Is this procedure practical for the optimization problems described in Sections 2.1 and 2.3? How would you demonstrate that your optimum design did correspond to a minimum cost? How could you determine whether or not the optimum design was unique?

4. (A*) Levenspiel[2] considered the design of a reactor to produce 100 g moles/hr of a product by a first-order, irreversible, isothermal reaction. He assumed that the reactant was available at a cost $0.50/g mole for a concentration $A_f = 0.1$ g mole/liter and that the reaction rate constant was $k = 0.2$ hr^{-1}. For a continuous-stirred-tank reactor costing $0.01/(hr)(liter), he found that the optimum design conditions were $A = 0.05$ g mole/liter, $q = 2000$ liters/hr, and $V = 10,000$ liters.

What are the optimum design conditions if we use a plug flow reactor that costs $0.015/(hr)(liter)? Is the ratio of the reactor costs for the two systems a reasonable value?

5. (B*) Compare the results of the preceding problem with the optimum design of a plant where we use three continuous-stirred-tank reactors (CSTR) in series, each reactor having an arbitrary volume. Discuss your results.

6. (B*) Another case of interest in the foregoing problems would be if we used three CSTRs having equal volumes in series, but we recycled some of the unreacted material. Develop the design equations for this configuration and find the number of degrees of freedom. If we let Q be the recycle flow rate, A_3 be the

---

* The following system has been used to indicate levels of exercises: the letter A denotes material suitable for undergraduates, the letter B indicates problems of intermediate difficulty, and the letter C represents material for the graduate student. Exercises marked with an asterisk are also considered in other exercises or in later chapters, whereas those marked with a double dagger are open-ended to some extent (some basic data or assumptions may be missing, or there is no single answer to the problem).

[1] M. S. Peters and K. D. Timmerhaus, *Plant Design and Economics for Chemical Engineers*, Ch. 9, McGraw-Hill, N.Y., 1968.

[2] See footnote 2, p. 12.

composition in the third reactor, and A_0 be the composition of the mixed feed stream entering the first reactor—that is, $(q + Q)A_0 = qA_f + QA_3$—it can be shown that one of the necessary conditions to obtain a minimum cost when recycle costs are negligible is

$$2\left(\frac{A_0}{A_3}\right)^{1/3} + \left(\frac{A_0}{A_3}\right)^{-2/3} - 3 = 0$$

Find a solution of this equation and then discuss in detail the implications of this result.

7. (A*) Derive the design equations for a pair of parallel reactions, $A \longrightarrow B$ and $A \longrightarrow C$, in a nonisothermal, continuous-stirred-tank reactor. Carefully list your assumptions and define your terms. Determine the number of degrees of freedom in the design. Also, show that, in general, the design corresponding to the minimum cost will not be the same as that which maximizes the profit. How does this optimum design problem compare to selecting the design conditions that maximize the yield of component B?

8. (A) Develop the optimization equations for an isothermal, plug flow reactor followed by an extraction unit.

9. (A) Develop the optimum design equations for an isothermal, continuous-stirred-tank reactor followed by an extraction unit for the case where the unreacted material is recycled. List all the assumptions used in your analysis.

10. (A*) Consider the operation of the isothermal reactor described in Examples 2.1-1 and 2.3-1 and suppose that the feed composition suddenly increases by 10 percent for one hour, then drops 10 percent below the original level for one hour, and finally returns to its original value. Plot the reactor conversion, the production rate, and the total system cost for the case where no change is made in the feed rate and for the case of optimum steady state control. Discuss your results. How would you determine the optimum feed composition?

11. (A) What are the appropriate equations for determining the optimum steady state control of the reactor-extractor system described in Example 2.1-2? How many degrees of freedom are there in this problem?

12. (B) Although the capacities, V, of various storage and surge vessels are usually specified in a preliminary plant design, the exact dimensions of each vessel depend on economic considerations. It is common practice to assume that the cost of the vessel is proportional to the weight of steel required, and therefore the design that minimizes the weight will minimize the cost. If we consider a cylindrical vessel having $2:1$ ellipsoidal heads, which are most commonly used in industry, the relationship between the tank capacity and its dimensions is

$$V = \frac{\pi D^2}{4}\left[L + 2\left(\frac{D}{6}\right)\right]$$

However, the vessel weight depends on the surface area A

$$A = \pi DL + 2(1.16D^2)$$

the minimum wall thickness required for structural rigidity t, and the density of the material. One method for determining this wall thickness is the use of an approximate equation for hoop stress

$$t = \frac{D_i + 100}{1000} + t_0$$

where t_0 is a corrosion allowance and D_i is the vessel diameter in inches. If it is assumed that $t_0 = \frac{3}{16}$ inches, that the density of steel plate, including a 5 percent overweight tolerance, is 515 lb/ft^3, and that it is possible to account for the fact that the ends of the vessel cost approximately 50 percent more than the walls by writing the area relationship in the form

$$A = \pi DL + 3.48 D^2$$

find the optimum design of a 156-cu ft vessel.

An alternate method for determining the wall thickness is the selection of the appropriate value from the table[3] below.

TABLE 2.5-1

WALL THICKNESS

Inside diameter, inches	42 and under	42 to 60	over 60
Minimum thickness, inches	$\frac{1}{4}$	$\frac{5}{16}$	$\frac{3}{8}$

Assuming that the corrosion allowance is included in these values, find the optimum design of the vessel. Discuss your results.

13. (B) Most chemical processes are so complex that it is not possible to use the straightforward design approach described in Section 2.1. Instead, an attempt is made to break the complex problem down into a number of simpler problems and then combine the results obtained from the simple optimization studies in such a way that at least an engineering estimate of the optimum, overall design is obtained. As a very elementary example of a decomposition technique, we might reconsider the optimum design of the nonisothermal reactor described in Example 2.1-2. We see that the total cost of the system is merely the sum of the costs of the reactor and heating coil. Thus we could consider the total reactor cost, the production equation, and reactor material balance, and then determine the optimum design of an isothermal reactor as a function of reactor temperature. Similarly, we could consider the total heat-exchanger cost, the energy blalance for the heating fluid, and the equation for heat transfer across the coil, and then determine the optimum heat-exchanger design as a function of the reactor temperature and the heat flux Q_H. These two simple optimization problems can be coupled together for any particular temperature, using the reactor energy balance. Thus when the costs of the two units are added together and plotted against temperature, the total system cost should pass through a minimum. See if this procedure leads to the same results as were previously obtained in Example 2.1-2, and discuss any discrepancy in detail.

14. (B) Consider a nonisothermal reactor followed by an extraction unit for a case where the distribution coefficient m decreases with temperature. If the reaction is exothermic, it might be advantageous to cool the reactor effluent before it is introduced into the extraction train. Write the appropriate equations describing the system, and determine the number of degrees of freedom.

15. (B) Develop the optimization equations for a continuous-stirred-tank reactor containing a cooling coil. Show that this set of equations has no solution.

[3] J. Happel, *Chemical Process Economics*, pp. 179, 250, Wiley, N.Y., 1958.

16. (B) Using the numerical values given in Example 2.1-2, determine the optimum design of an adiabatic, continuous-stirred-tank reactor.

17. (B) Given the values $C_A = 0.36$ ($)(sq cm)$^{-1}(hr)^{-1}$, $C_f = 0.10$ ($)(g mole)$^{-1}$, $C_H = 0.02$ ($)(1000 g)$^{-1}$, $C_V = 0.36$ ($)(hr)$^{-1}$(liter)$^{-1}$, $C_p\rho = 1.0$ (cal)(cc)$^{-1}$(°K)$^{-1}$, $C_{pH} = 1.0$ (cal)(g)$^{-1}$(°K)$^{-1}$, $E = 15,000$ (cal)(g mole)$^{-1}$, $G = 180$ (g mole)(hr)$^{-1}$, $(-\Delta H) = 15,000$ (cal)(g mole)$^{-1}$, $k_0 = 0.22 \times 10^4$ (sec)$^{-1}$, $T_f = 300$ °K, $T_H = 373$ °K, $U = 1.0$ (cal)(sq cm)$^{-1}$(sec)$^{-1}$(°K)$^{-1}$, $A_f = 0.01$ (g mole)(cc)$^{-1}$, find the optimum design of a nonisothermal, continuous-stirred-tank reactor. Compare your results with the optimum adiabatic design. Why does the adiabatic reactor give a less-expensive design for this case?

18. (B) If we consider the design of the system described in Example 2.1-3 for a case where $G = 98.67$ lb moles/hr, $x_f = 0.2$, $k = 0.2$ (lb mole)/(hr)(ft^3), $C_f = 0.138$ ($)/(lb mole), $C_V = 0.0052$ ($)/(hr)(ft^3), $C_N = 2.0$ ($)/(stage)(hr), $C_S = 0.5$ ($)/(lb mole), $Y_{N+1} = 0$, and $m = 2.2$, it can be shown that the optimum design is obtained when $F = 765$ lb mole/hr, $S = 316$ lb mole/hr, and $N = 10$ plates. Plot the total system cost vs. both F and N in order to find the sensitivity of the solution to small changes in the optimum values. Discuss your results. Also, describe a procedure that can be used to solve the optimization equaions.

19. (B) Determine the optimum steady state control settings if the feed temperature to the nonisothermal reactor described in Example 2.1-2 suddenly changes to 35 °C. Is there any possibility that a variation in the system parameters makes it necessary to change the assumptions used in the analysis?

20. (C‡) Peters and Timmerhaus describe a problem for the optimum design of an absorption tower containing wooden grids, which is to be used for absorbing SO_2 in a sodium sulfite solution. They assume that an SO_2–air mixture of known composition enters the tower at 70,000 cfm (cubic feet per minute), at a temperature of 250 °F, and at a pressure of 1.1 atm, and that it is necessary to remove a specified fraction of the SO_2 in the tower. The molecular weight of the inlet gas stream is given as 29.1, the relationship between the number of transfer units required and the superficial gas velocity G_S (lb)/(hr)(ft^2) based on the cross-sectional area of the empty tower is given as NTU $= 0.32G_S^{0.18}$, and the height of a transfer unit is taken to be a constant value of 15 ft. The cost for the installed tower is $1 per cubic foot, annual fixed charges are 20 percent of the initial cost, and the variable operating costs for the absorbent, blower power and pumping are given by the equation

$$\text{Variable operating costs \$/hr} = 1.8G_S^2 \times 10^{-8} + \frac{81}{G_S} + \frac{4.8}{G_S^{0.8}}$$

Then you are supposed to find the height and diameter of the tower which minimize the annual cost for a case where the unit is operated 8000 hr/year.

After solving this problem, and having constructed a tower with the appropriate dimensions, how would you establish a procedure to manipulate the operating conditions to compensate for changes in the inlet gas composition, temperature, and pressure?

21. (C‡) Peters and Timmerhaus (p. 313) also present a somewhat lengthy description of the optimum design of a distillation column. How would you modify their analysis to design an optimum steady state control system for the tower?

22. (C‡) Denn and Aris[4] describe the optimum steady state design of a simple

---

[4] M. M. Denn and R. Aris, *Ind. Eng. Chem. Fundamentals,* **4,** 248 (1965).

plant where there is a pair of consecutive reactions, $A \longrightarrow B \longrightarrow C$, carried out in a sequence of three stirred-tank reactors. The effluent from the reactor battery passes through a two-stage extraction column, where the reactant A is recovered and recycled to the reactor. A schematic flow sheet is shown in Figure 2.5-1. The mate-

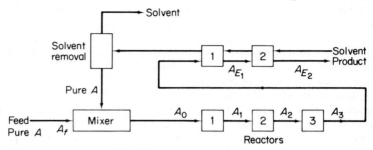

Figure 2.5-1. Flow sheet of plant.

rial balance equations for each reactor are written in the form:

$$0 = A_{n-1} - A_n - \theta k_{10} e^{-E_1/RT_n}(A_n)^2, \qquad n = 1, 2, 3$$
$$0 = B_{n-1} - B_n + \theta k_{10} e^{-E_1/RT_n}(A_n)^2 - \theta k_{20} e^{-E_2/RT_n} B_n, \qquad n = 1, 2, 3$$

and the material balances for the extractor are

$$A_{E_1} - A_3 + u[\phi(A_{E_1}) - \phi(A_{E_2})] = 0$$
$$A_{E_2} - A_{E_1} + u\phi(A_{E_2}) = 0$$

where u represents the ratio of the solvent to reactor flow rates and $\phi(A_{E_i})$ represents the equilibrium distribution between the solvent and reactor streams. It is assumed that the solvent extracts only component A and that the feed stream contains no products. Hence

$$B_{E_2} = B_3 \qquad C_{E_2} = C_3$$

and

$$A_0 = A_f + u\phi(A_{E_1}), \quad B_0 = 0, \quad C_0 = 0$$

The design problem was to choose the three reactor temperatures, T_1, T_2, and T_3, along with the ratio of the flow rates u, so that the amount of B produced was maximized subject to an allowance for the costs of raw materials and extraction— that is, to minimize the function

$$C_T = -B_3 - C_1 A_{E_2} + C_2 u$$

How would you modify their results to obtain an optimum steady state control system?

23. (C) Making the same assumptions as given in Example 2.1-1, determine the optimum design of a batch reactor that produces, on the average, G lb moles per hour of product.

Process Dynamics

PART

Dynamic
Model   **3**
Building

The theoretical models[1] of various process units are derived by using the fundamental principles of conservation of mass, energy, and momentum. In its most general form, the conservation principle states that

$$\text{Input} - \text{output} = \text{accumulation}$$

For steady state systems, the accumulation terms are always equal to zero; therefore the total input of any conserved quantity to a particular unit must be equal to the total output. If we now want to extend our theoretical models to include the dynamic operating characteristics, we merely add the accumulation terms to the material and energy balances. In practice, of course, this procedure often introduces a tremendous amount of additional complexity into the system equations. Consequently, often we are forced to make simplifying assumptions and to attempt to develop only approximate descriptions of the dynamic behavior. Just as it is necessary to test the steady state model experimentally, it is also necessary to conduct experiments to see if the dynamic model actually does predict the plant response.

[1] See footnote 1, p. 35.

This chapter will be concerned mainly with the actual development of the models used to describe process dynamics. Thus the emphasis will be on the kinds of assumptions and approximations that have been proven useful and the kinds of equations that are often obtained. Some understanding of the latter material is necessary in order to gain an appreciation of the mathematical difficulties we might encounter. In this way we can obtain some insight into the required balance between simplifying assumptions, and hence lack of rigor, and the mathematical complexity of the model, and hence the necessity of devising techniques for obtaining approximate solutions of the equations. This conflict of interests often makes it necessary to include some empirical features in the theoretical model. Since the difficulties encountered are best illustrated through the discussion of particular examples, we will focus our attention on the dynamic equations describing a few representative systems. However, an approximation procedure that generally can be used to obtain a rough estimate of the dynamic characterisitcs of many processes is developed later in the chapter.

SECTION 3.1 LUMPED PARAMETER SYSTEMS

Our previous discussion of optimum steady state design problems emphasized systems that were described by algebraic equations. A reexamination of these models reveals that we assumed that the contents of most of the units were perfectly mixed. Thus in the models of the isothermal and nonisothermal reactors, we assumed that the composition and temperature were uniform throughout the reactor volume, and in the case of the extraction unit, we assumed that the compositions of the extract and raffinate phases each were uniform, although different, in every stage. This kind of an assumption, however, was not made in the equations used to describe the heating coil in the nonisothermal reactor. Instead, the heat-transfer equation, which includes the log-mean temperature driving force, Eq. 2.1-16, is based on the assumption that the temperature of heating fluid varies along the length of the coil.

Systems that can be considered perfectly mixed are often called *lumped parameter systems*. The steady state equations describing these processes are always algebraic equations, because it is assumed that the vessel contents are uniform and the compositions and/or temperature in the vessel are equal to the effluent conditions. Hence the total change occurring in the process takes

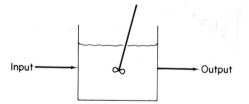

Input ⟶   ⟶ Output

Figure 3.1-1. Lumped parameter system.

place across the inlet boundary. It follows that the dynamic equations describing a lumped parameter system will be sets of ordinary differential equations. The conditions in each stage change with time, but these changes occur uniformly throughout the perfectly mixed region. Therefore time is the only independent variable[2] in the dynamic model. There can be a number of dependent variables, corresponding to the compositions and temperatures of one or more phases in one or more vessels. A separate material or energy balance is derived for each of these variables. Several examples of dynamic models for lumped parameter systems are given below.

Example 3.1-1   Dynamic model of an isothermal reactor

If we consider the dynamic operation of the isothermal reactor described in Example 2.1-1, it is a simple matter to develop a dynamic model. Following the standard convention,[3] we will always write the accumulation in the system as a positive quantity. Then the amount of reactant present in the vessel at time t is $(VA)|_t$; at time $t + \Delta t$, it increases to a value $(VA)|_{t+\Delta t}$. Therefore the accumulation of reactant is merely $(VA)|_{t+\Delta t} - (VA)|_t$. This term must be equal to the amount of material entering the reactor minus the amount leaving, both in the effluent stream and by reaction during the time interval Δt. In mathematical terms we have

$$(VA)|_{(t+\Delta t)} - (VA)|_{(t)} = q_1 A_f \, \Delta t - q_2 A \, \Delta t - kVA^2 \, \Delta t \qquad 3.1\text{-}1$$

where q_1 represents the input flow rate and q_2 the effluent flow rate. Dividing both sides of the equation by Δt, taking the limit as Δt approaches zero, and recognizing that the definition of a derivative is merely

$$\frac{dVA}{dt} = \lim_{\Delta t \to 0} \frac{(VA)|_{t+\Delta t} - (VA)|_t}{\Delta t} \qquad 3.1\text{-}2$$

our material balance equation becomes

$$\frac{dVA}{dt} = q_1 A_f - q_2 A - kVA^2 \qquad 3.1\text{-}3$$

This equation should then predict the time behavior of the reactant composition in the vessel or the effluent stream (these have been assumed to have the same value) when changes are made in the flow rates q_1 or q_2 or the feed composition A_f. An additional relationship between the inlet and outlet flow rates and the reactor volume can be obtained by writing a total material balance on the system. Making the common assumptions for liquid-phase systems, this relationship can be written as

$$\frac{dV}{dt} = q_1 - q_2 \qquad 3.1\text{-}4$$

---

[2] Here we use the term independent variable as it is defined in the theory of differential equations.

[3] R. B. Bird, W. E. Stewart, and E. N. Lightfoot, *Transport Phenomena*, Wiley, N.Y., 1960.

Obviously, for the case where $q_1 = q_2$, the system volume will remain constant, and *Eq.* 3.1-3 can be simplified by taking the volume outside the derivative sign

$$V\frac{dA}{dt} = q(A_f - A) - kVA^2 \qquad\qquad 3.1\text{-}5$$

This equation can be written

$$\frac{dA}{dt} + \frac{q(t)}{V}A + kA^2 = \frac{q(t)}{V}A_f(t) \qquad\qquad 3.1\text{-}6$$

Thus it is a nonlinear, ordinary differential equation that is nonhomogeneous —that is, qA_f is never equal to zero—and it may have variable coefficients if q varies with time. Since it is merely a particular case of the general equation

$$\frac{dA}{dt} + P(t)A + Q(t)A^2 = R(t) \qquad\qquad 3.1\text{-}7$$

it is a Riccati equation. Analytical solutions are available for certain kinds of inputs, but we will defer a discussion of these solutions until Chapter 4.

Example 3.1-2 Dynamic model of a plate, gas absorption unit

The dynamic models used to describe plate, gas absorbers, extraction units, and the rectifying and stripping sections of a distillation column are essentially the same. In all these units, two immiscible phases containing some transferable component are brought into an intimate contact, usually by dispersing one of the phases in the form of small droplets or bubbles. A concentration gradient for the transferable component exists in each stage of the unit so that mass transfer will take place and the concentrations will approach the phase equilibrium values. By contacting the two phases in a countercurrent fashion, the average driving force in the column is maximized at steady state conditions and thus the overall separation is enhanced.

Since the dispersed phase is broken up into a large number of small droplets, or bubbles, as it enters each stage and is then recombined in the stage, normally we assume that it is perfectly mixed. The stirring action that occurs when the bubbles pass through the continuous phase also tends to make it perfectly mixed. However, depending on the width of the plates, the flow rates of the two phases, and the physical properties of the fluids, it is possible to have concentration gradients in the continuous phase along the plate or poor contact (by passing) between the phases; therefore these assumptions are not always valid. Frequently we try to compensate for the discrepancy by introducing the concepts of local plate efficiency and overall plate efficiency. Although we will consider some efficiency effects later, an interested reader should refer to some of the standard texts that treat this subject in more detail. For present purposes we will consider that both phases are perfectly mixed on each plate and that the plates are ideal; that is, phase

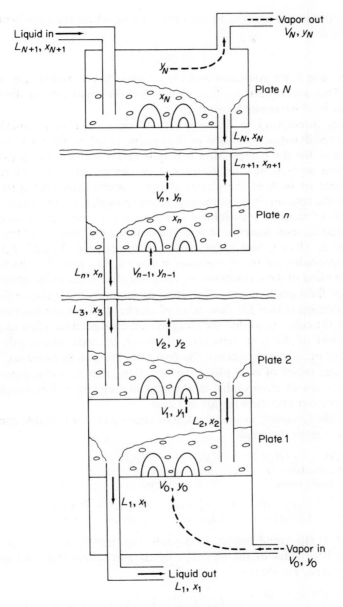

Figure 3.1-2. Plate absorber.

equilibrium is achieved. Even though equilibrium relationships may be nonlinear functions,

$$y_n = f(x_n) \qquad 3.1\text{-}8$$

we will restrict our analysis to the simplest case where the functional dependence is linear

$$y_n = mx_n + b \qquad 3.1\text{-}9$$

where m and b are constants and the composition variables are discussed below. This assumption is often valid when the separation is difficult and therefore is of particular interest.

Before attempting to derive a set of material balance equations that describe the dynamic operation of the system, it will be necessary to select a set of units for the composition and flow rate variables. For a case where a sufficient amount of mass transfer is occurring, such that the total flow rate of one or both of the phases changes significantly during its passage through the column, it is common practice to express compositions in terms of moles, or mass, of transferable component per moles, or mass, of carrier fluid—that is, total material in one phase minus the amount of transferable component. Then x_n and y_n (or, as they often appear, X_n and Y_n) refer to the compositions for the two phases in stage n. The units for the flow rates corresponding to these composition variables are chosen to be moles or mass of carrier fluid per unit time. The great advantage of this choice for steady state operation is that the flow rates of carrier fluid remain constant everywhere in the column, so that the composition changes from plate to plate are independent of the total flow rate. However, in cases where only a small amount of mass transfer occurs, the flow rates are usually expressed in terms of the total moles of each phase per unit time, and x_n and y_n represent the mole fractions of transferable component in each phase. For simplicity we will restrict our attention to this last case.

With the foregoing assumptions, a balance on the transferable component for stage n gives the result

$$(H_n x_n)|_{t+\Delta t} - (H_n x_n)|_t + (h_n y_n)|_{t+\Delta t} - (h_n y_n)|_t$$
$$\underset{\text{Accumulation in liquid phase}}{} \qquad \underset{\text{Accumulation in vapor phase}}{}$$

$$= L_{n+1} x_{n+1} \Delta t + V_{n-1} y_{n-1} \Delta t - L_n x_n \Delta t - V_n y_n \Delta t \qquad 3.1\text{-}10$$
$$\underset{\text{Liquid in}}{} \qquad \underset{\text{Vapor in}}{} \qquad \underset{\text{Liquid out}}{} \quad \underset{\text{Vapor out}}{}$$

where H_n is the liquid holdup in the nth stage and h_n is the vapor holdup. Dividing both sides of the equation by Δt and taking the limit as Δt approaches zero, we obtain

$$\frac{dH_n x_n}{dt} + \frac{dh_n y_n}{dt} = L_{n+1} x_{n+1} + V_{n-1} y_{n-1} - L_n x_n - V_n y_n \qquad 3.1\text{-}11$$

The holdups in the gas and liquid phases can be determined by making a total material balance, rather than a component balance, for the vapor and liquid phases. However, in order to keep the analysis as simple as possible, we will assume that both values remain constant, and that the holdups are the same on every plate, $h_n = h$ and $H_n = H$. In addition, we will assume that the equilibrium relationship remains valid even during dynamic operation of the column. With these assumptions, the system equation can be written

$$(H + mh)\frac{dx_n}{dt} = L_{n+1}x_{n+1} + V_{n-1}(mx_{n-1} + b) - L_nx_n - V_n(mx_n + b)$$

$$3.1\text{-}12$$

When one of the phases is a vapor, the term involving the vapor holdup is usually negligible in comparison with the liquid holdup. (This simplification is not valid for extraction columns.) Also, for most absorption columns, the pressure and temperature throughout the column are essentially constant; whereas the temperature in a distillation column varies from plate to plate in accordance with the equilibrium relationship. Since we are considering only dilute mixtures where the total vapor rate is not changed by mass transfer, and we have assumed that the vapor holdups are negligible, the vapor flow rate will be constant throughout the column, $V_{n-1} = V_n = V_0 = V$. The liquid flow rates will also be constant, $L_{n+1} = L_n = L_{N+1} = L$, unless we are interested in estimating the response of the column to changes in the inlet liquid flow L_{N+1}. In this latter case, a more detailed analysis of the column hydrodynamics should be undertaken, but for simplicity we will neglect any dynamic effects. Thus the equation for any plate n becomes

$$H\frac{dx_n}{dt} = Lx_{n+1} - (L + mV)x_n + mVx_{n-1} \qquad 3.1\text{-}13$$

which is applicable for every plate in the column—that is, $n = 1, 2, \ldots, N$. In other words, our dynamic model for the absorber is a set of N first-order, linear differential equations, which have variable coefficients if the flow rates change with time.

Other assumptions will lead to different results. For example, a nonlinear equilibrium relationship, Eq. 3.1-8, will introduce nonlinearities into the model, and variable liquid holdups will make it necessary to increase the number of dynamic equations. For the linear case with constant coefficients, it is possible to solve the equations analytically. We will discuss these solutions in detail in the next chapter.

Example 3.1-3 Dynamic model of a nonisothermal,
continuous-stirred-tank reactor

The steady state model for a continuous-stirred-tank-reactor was presented in Example 2.1-2. Since the equation used to describe the heat transfer between the reactor contents and the heating coil was based on an ordinary differential equation, where the distance along the coil was the independent variable, the dynamic equation for this transfer operation will become a partial differential equation; that is, the temperature at any point in the coil will depend on both time and distance along the coil. Consequently, without some simplifying assumption, we cannot expect the dynamic model of a nonisothermal reactor to correspond to a lumped parameter system. We will derive a very general set of equations for the process and then discuss some of the simplifying assumptions that are often introduced.

If we restrict our attention to a case where the inlet and outlet flow rates of reacting material are always maintained equal to each other, so that the reactor volume remains constant, a material balance for the perfectly mixed vessel gives

$$V(A|_{t+\Delta t} - A|_t) = qA_f\,\Delta t - qA\,\Delta t - kVA\,\Delta t \qquad 3.1\text{-}14$$

where the reaction rate constant is related to the reactor temperature by the Arrhenius equation

$$k = k_0 \exp\frac{-E}{RT} \qquad 3.1\text{-}15$$

If we again divide by Δt and take the limit as Δt approaches zero, we obtain

$$V\frac{dA}{dt} = q(A_f - A) - kVA \qquad 3.1\text{-}16$$

Similarly, an energy balance for the reactor fluid, assuming that the heat capacity of the reactant stream equals that of the effluent stream and that the physical properties are constant, gives[4]

---

[4] Actually, the derivation of the energy balance, given by Eqs. 3.1-17 and 3.1-18, is not correct, although frequently encountered in the literature. A rigorous derivation has been published by Aris[5] and a simpler derivation is discussed by Russell and Denn.[6] A summary of this last treatment is presented below.

[5] R. Aris, *Elementary Reactor Analysis*, Ch. 3, Prentice-Hall, Englewood Cliffs, N.J., 1969.

[6] T. W. F. Russell and M. M. Denn, *Introduction to Chemical Engineering Analysis*, manuscript copy of a new book.

An energy balance (actually a heat balance) must be written in terms of the enthalpy accumulated and the convective flow of enthalpy through the reactor

$$H|_{t+\Delta t} - H|_t = \rho q[\bar{H}_f|_{T_f} - \bar{H}|_T]\,\Delta t + Q\,\Delta t$$

where H is enthalpy, $\bar{H}$ is enthalpy per mass, and Q represents the heat added through the system boundaries. Taking the limit as Δt approaches zero, we obtain

$$\frac{dH}{dt} = \rho q(\bar{H}_f|_{T_f} - \bar{H}|_T) + Q$$

If we first consider the terms on the right-hand side, we see that

$$\rho\bar{H}_f|_{T_f} = \rho\bar{H}_f|_T + \rho C_p(T_f - T)$$

Also,

$$\rho\bar{H} = \sum C_i\bar{H}_i$$

$$VC_p\rho[T|_{t+\Delta t} - T|_t] = qC_p\rho T_f \,\Delta t - qC_p\rho T \,\Delta t + (-\Delta H)kVA \,\Delta t$$
$$+ h_rA_r(T_w - T)\Delta t \qquad \text{3.1-17}$$

After applying the limiting operation, we obtain

$$VC_p\rho\frac{dT}{dt} = qC_p\rho(T_f - T) + (-\Delta H)kVA + h_rA_r(T_w - T)$$

$$\text{3.1-18}$$

The last term in this equation refers to the heat transfer between the perfectly mixed reactor fluid and the metal walls of the heating coil. This expression is based on the implicit assumption that a film-heat-transfer coefficient can be used to describe the dynamic characteristics of heat transfer as well as steady state operation. Although this assumption is commonly accepted, it deserves further consideration and will be discussed in detail in the next section.

If we consider a length Δz of the coil and write an energy balance for the heating fluid, assuming that the density and heat capacity are both inde-

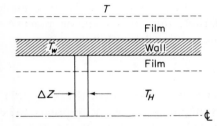

Figure 3.1-3. Section of a heating coil.

where C_i represents the concentration of the ith component and $\bar{H}_i$ is the partial molar enthalpy.

Similarly, we can write the term on the left-hand side of the energy balance as

$$\frac{dH}{dt} = \frac{\partial H}{\partial T}\frac{dT}{dt} + \sum \bar{H}_i\frac{dn_i}{dt} = VC_p\rho\frac{dT}{dt} + \sum \bar{H}_i\frac{dVC_i}{dt}$$

where n_i is the number of moles of component i. The last term in this expression is merely the accumulation in the material balance, and thus we can write

$$\frac{dVC_i}{dt} = q(C_{fi} - C_i) + \alpha_i rV$$

where α_i represents the stoichiometric coefficient of the ith component and r is the reaction rate.

After substituting the equations we developed into the energy balance, we find that

$$VC_p\rho\frac{dT}{dt} = qC_p\rho(T_f - T) + q \sum C_{fi}\bar{H}_i - q \sum C_i\bar{H}_i$$
$$- q \sum C_{fi}\bar{H}_i + q \sum C_i\bar{H}_i - rV \sum \alpha_i\bar{H}_i + Q$$

Since

$$\sum \alpha_i\bar{H}_i = (-\Delta H), \quad r = kA, \quad Q = h_rA_r(T_w - T)$$

we obtain the result given by Eq. 3.1-18.

pendent of temperature, we obtain the result[7]

$$C_{pH}\rho_H A_t \, \Delta z[(T_H)|_{t+\Delta t} - (T_H)|_t] = C_{pH}\rho_H A_t v_H \, \Delta t[(T_H)|_z - (T_H)|_{z+\Delta z}]$$

Accumulation in element of   Flow in   Flow out
volume $A_t \, \Delta z$

$$-h_c A_c (T_H - T_w) \, \Delta z \, \Delta t \qquad\qquad 3.1\text{-}19$$

Loss to coil wall

Dividing by $\Delta t \, \Delta z$ and taking the limit as both quantities approach zero, we find that

$$C_{pH}\rho_H A_t \frac{\partial T_H}{\partial t} + C_{pH}\rho_H A_t v_H \frac{\partial T_H}{\partial z} = h_c A_c (T_H - T_w) \qquad 3.1\text{-}20$$

where A_t is the cross-sectional area of the coil, A_c is the inside heat transfer area per foot of length, h_c is the inside film coefficient, and v_H is the velocity of the hot fluid.

In the preceding equations we assumed that the wall temperature is uniform. Of course, a more rigorous analysis would require a treatment of the unsteady state heat conduction in the wall, and therefore it would be necessary to add a second-order partial differential equation to the dynamic model. Since the tube wall generally is relatively thin, it is common practice to approximate the wall dynamics by the simple relationship

$$m_w C_{pw} \frac{\partial T_w}{\partial t} = h_c A_c (T_H - T_w) - h_r A_r (T_w - T) \qquad 3.1\text{-}21$$

where $m_w C_{pw}$ is the wall capacity per foot of tube length $(\text{Btu})(°\text{F})^{-1}(\text{ft})^{-1}$. The approximation is based on the assumptions that all the resistance to heat transfer through the wall can be lumped into two stagnant film regions on each side of the wall and that the wall capacitance can be lumped into a single location. Hence the assumption that the wall temperature is uniform is equivalent to a perfect mixing assumption, and the resulting model is an ordinary differential equation.

Once the systems inputs, q, A_f, T_f, and v_H, and the initial values of A, T, T_w and $T_H(z)$, have been specified, then the set of equations above, Eqs. 3.1-15, 3.1-16, 3.1-18, 3.1-20, and 3.1-21, can be used to determine the system outputs A, T, T_w, and T_H. This set of equations includes both ordinary and partial differential equations, which are nonlinear and have variable coefficients. Thus we cannot expect to obtain analytical solutions. Depending on the particular system under investigation, however, additional simplifications can be introduced into the model. For example, if the capacitance of the coil walls is very small, we might assume that the enthalpy accumulation term in Eq. 3.1-21 is negligible. Similarly, if the flow rate of heating fluid in the coil is very large, we might assume that accumulation effects in the coil are also negligible.

---

[7] This derivation contains the same error discussed in footnote 4, except in this case the enthalpy depends only temperature, not on composition.

With these assumptions, Eq. 3.1-21 can be solved for T_w, and this result used to eliminate the wall temperature from Eqs. 3.1-18 and 3.1-20. Then Eq. 3.1-20 can be solved for T_H, and the common expression for the heat flux in terms of the log-mean temperature driving force can be developed. Introducing the definition for the overall heat-transfer coefficient

$$\frac{1}{UA_H} = \frac{1}{h_c A_c} + \frac{1}{h_r A_r}$$ 3.1-22

the approximation for the log-mean driving force in terms of the arithmetic average discussed in Example 2.1-2, adding an overall energy balance on the heating fluid, and manipulating the results somewhat, our dynamic model reduces to the simpler form

$$V\frac{dA}{dt} = q(A_f - A) - kVA$$ 3.1-23

$$VC_p\rho\frac{dT}{dt} = qC_p\rho(T_f - T) + (-\Delta H)kVA + \frac{UA_H Kq_H}{1 + Kq_H}(T_H - T)$$

3.1-24

where

$$K = \frac{2C_{pH}\rho_H}{UA_H}, \quad q_H = v_H A_H, \quad k = k_0 \exp\frac{-E}{RT}$$ 3.1-25

These equations are often used to approximate the dynamic behavior of a continuous-stirred-tank reactor. They are a coupled pair of first-order, nonlinear differential equations. Again, we cannot expect to find an analytical solution, but we have reduced our model to a lumped parameter form. We will study the dynamic characteristics of this model in great detail throughout the text, the equations being relatively simple yet illustrating many of the nonlinear effects sometimes observed in more complex processes.

Example 3.1-4 Dynamic model of a catalytic cracking unit

In the previous example we developed a fairly complete description of a unit and then introduced enough assumptions to reduce it to a relatively simple form. This is the approach we prefer to follow whenever possible, for we are always aware of the assumptions we have made and thus have some indication of how to change the model if it fails to predict the behavior of an operating unit. However, in many cases we simply do not understand all the physical laws that would be required to describe a unit, and therefore we are forced to modify our objective. In these cases we often attempt to develop fairly simple models that will describe what we expect to be the most important dynamic features of the system. After comparing the model predictions with experimental data, we modify both the form of the model and the values of the parameters until an agreement is eventually obtained. As an example of this approach, we will consider the operation of a catalytic cracking unit.

Under normal operating conditions, a gas-oil feed stream is vaporized

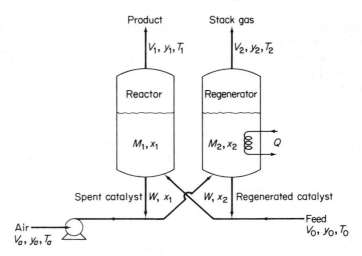

Figure 3.1-4. Catalytic cracking unit.

and fed into a fluidized bed reactor containing cracking catalyst. The high molecular weight gas-oil cracks on the surface of the catalyst and forms a wide range of lower molecular-weight compounds, including light gases, kerosene, gasoline, and furnace oil. However, at the same time, a very high-molecular-weight polymeric hydrocarbon, called coke, forms on the catalyst surface, and the catalyst activity decreases as the coke accumulates on the catalyst. In order to maintain the catalyst activity at a high level, the fluidized catalyst is transferred to a regeneration unit where the coke is burned off. An air stream is used as a fluidizing agent in this unit and as an oxygen supply for the combustion process.

It is apparent that a large number of reactions occur in the reactor; for example, there are approximately 100 pure chemical compounds in the gasoline stream alone, and since the mechanism and the kinetics of most of these reactions are not known, we will not be able to develop a detailed dynamic model of the system. Even though a rigorous treatment is not feasible, it still might be possible to develop a crude model that will describe the most important dynamic effects—that is, it will predict the direction of change of the most important dependent variables and at least give an order-of-magnitude estimate of the response time of these variables. Luyben and Lamb[8] have proposed a model of this type, although it has never been tested experimentally. They assume that only a single reaction takes place in the reactor and that the stoichiometric equation is

$$A \longrightarrow B + 0.1C \downarrow \qquad\qquad 3.1\text{-}26$$

[8] W. L. Luyben and D. E. Lamb, *Chem. Eng. Prog. Symposium Ser.*, **59**, No. 46, 165 (1963).

where A is the gas-oil feed, B is the desired gasoline product, and C is the coke material. Similarly, they assume that the only reaction taking place in the regenerator is

$$C + mO \longrightarrow P \qquad\qquad 3.1\text{-}27$$

where O represents oxygen and P the products, and m is a stoichiometric coefficient. They assume, too, that each unit is perfectly mixed, that the heat capacities are independent of composition and temperature, that the catalyst holdup remains constant in each vessel, and that the reaction rates are first-order. Based on these assumptions, they derive the material and energy balance equations given below:

Reactor

Gas phase

$$\frac{dN_1}{dt} = V_0 - V_1 \qquad\qquad 3.1\text{-}28$$

Component B

$$\frac{dN_1 y_1}{dt} = V_0 y_0 - V_1 y_1 + N_1(1 - y_1)A_1 e^{-E_1/RT_1} \qquad\qquad 3.1\text{-}29$$

Component C

$$M_1 \frac{dx_1}{dt} = Wx_2 - Wx_1 + 0.1 N_1(1 - y_1)A_1 e^{-E_1/RT_1} \qquad\qquad 3.1\text{-}30$$

Energy

$$C_{ps}M_1 \frac{dT_1}{dt} = C_{p1}V_0 T_0 + C_{ps}WT_2 - C_{p1}V_1 T_1 - C_{ps}WT_1$$
$$-(\Delta H_1)N_1(1 - y_1)A_1 e^{-E_1/RT_1} \qquad\qquad 3.1\text{-}31$$

Gas law

$$N_1 = \frac{P_1 H_1}{RT_1} \qquad\qquad 3.1\text{-}32$$

Regenerator

Gas phase

$$\frac{dN_2}{dt} = V_a - V_2 \qquad\qquad 3.1\text{-}33$$

Component O

$$\frac{dN_2 y_2}{dt} = V_a y_a - V_2 y_2 - mN_2 y_2 A_2 e^{-E_2/RT_2} \qquad\qquad 3.1\text{-}34$$

Component C

$$M_2 \frac{dx_2}{dt} = Wx_1 - Wx_2 - N_2 y_2 A_2 e^{-E_2/RT_2} \qquad\qquad 3.1\text{-}35$$

Energy

$$C_{ps}M_2\frac{dT_2}{dt} = C_{p2}V_aT_a + C_{ps}WT_1 - C_{p2}V_2T_2 - C_{ps}WT_2$$

$$-(\Delta H_2)N_2y_2A_2e^{-E_2/RT_2} + Q \qquad \text{3.1-36}$$

Gas law

$$N_2 = \frac{P_2H_2}{RT_2} \qquad \text{3.1-37}$$

where A_1 and A_2 = frequency factors in Arrhenius equation $(\text{sec})^{-1}$
C_{ps}, C_{p1}, and C_{p2} = heat capacities $(\text{Btu})(\text{mole})^{-1}(°R)^{-1}$
E_1 and E_2 = activation energies $(\text{Btu})(\text{mole})^{-1}$
H_1 and H_2 = gas holdup (ft^3)
M_1 and M_2 = catalyst holdup (lb)
N_1 and N_2 = gas holdup (moles)
P = pressure $(\text{lb})(\text{ft}^2)^{-1}$
V_0, V_1, V_2, and V_a = gas flow rates $(\text{moles})(\text{sec})^{-1}$
W = catalyst circulation rate $(\text{lb})(\text{sec})^{-1}$
x_1 and x_2 = coke on catalyst (moles C) (lb catalyst)$^{-1}$
y_0 and y_1 = gas phase mole fraction of component B
y_a and y_2 = gas phase mole fraction of oxygen
Q = heat removed from the regenerator by a cooling coil
m = stoichiometric coefficient
ΔH_1 and ΔH_2 = heat of reaction $(\text{Btu})(\text{mole})^{-1}$
t = time $(\text{sec})^{-1}$

It is possible to simplify these expressions somewhat by using Eqs. 3.1-28 and 3.1-32 to eliminate N_1 and V_1 from the first set and by using Eqs. 3.1-33 and 3.1-37 to eliminate N_2 and V_2 from the second set. This procedure leads to the following results:

Reactor

$$\frac{P_1H_1}{RT_1}\frac{dy_1}{dt} = V_0(y_0 - y_1) + \frac{P_1H_1}{RT_1}(1 - y_1)A_1e^{-E_1/RT_1} \qquad \text{3.1-38}$$

$$M_1\frac{dx_1}{dt} = W(x_2 - x_1) + 0.1\frac{P_1H_1}{RT_1}(1 - y_1)A_1e^{-E_1/RT_1} \qquad \text{3.1-39}$$

$$\left(\frac{P_1H_1C_{p1}}{RT_1} + C_{ps}M_1\right)\frac{dT_1}{dt} = C_{ps}W(T_2 - T_1) + C_{p1}V_0(T_0 - T_1)$$

$$-\frac{(\Delta H_1)P_1H_1}{RT_1}(1 - y_1)A_1e^{-E_1/RT_1} \qquad \text{3.1-40}$$

Regenerator

$$\frac{P_2H_2}{RT_2}\frac{dy_2}{dt} = V_a(y_a - y_2) - \frac{P_2H_2}{RT_2}my_2A_2e^{-E_2/RT_2} \qquad \text{3.1-41}$$

$$M_2\frac{dx_2}{dt} = W(x_1 - x_2) - \frac{P_2H_2}{RT_2}y_2A_2e^{-E_2/RT_2} \qquad 3.1\text{-}42$$

$$\left(\frac{P_2H_2C_{p2}}{RT_2} + C_{ps}M_2\right)\frac{dT_2}{dt} = C_{ps}W(T_1 - T_2) + C_{p2}V_a(T_a - T_2)$$
$$- \frac{(\Delta H_2)P_2H_2}{RT_2}y_2A_2e^{-E_2/RT_2} + Q \qquad 3.1\text{-}43$$

This is a set of six, coupled, ordinary differential equations that are nonlinear and that may have variable coefficients. Thus, unless we introduce additional simplifying assumptions, it will be necessary to solve the equations numerically in order even to estimate the dynamic characteristics of the system. Of course, this estimate might be very poor, or even misleading, because it depends on the assumptions made in the derivation. However, it does provide a starting point for developing an approximate model. Procedures for testing the model and modifying the equations will be discussed in more detail in Chapter 4.

Analogies

The dynamic response of many different kinds of very elementary systems can be approximately described by a single, first-order, linear differential equation; for example, continuous mixing in a stirred vessel, a first-order reaction in a continuous-stirred-tank reactor, a stirred heater with a steam jacket, a thermometer or thermocouple, liquid level in a tank, and a simple *RC* circuit. Since the same basic model can be used to represent the behavior of this wide range of systems, it should not be surprising that it is possible to develop analogies between the various systems. Thus the physical quantities appearing in corresponding terms in the differential equation can be considered analogous. For example, when Kirchhoff's law is applied to the simple *RC* circuit shown in Figure 3.1-5, the equation describing the voltage across the capacitor *C* can be shown to be

$$RC\frac{de}{dt} + e = v \qquad 3.1\text{-}44$$

where R is the resistance, C the capacitance, and $v(t)$ the time-variable impressed voltage. Similarly, a material balance for a salt solution entering an agitated vessel gives the result

$$V\frac{dx}{dt} + qx = qx_f \qquad 3.1\text{-}45$$

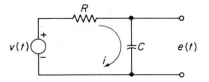

Figure 3.1-5. Simple *RC* circuit.

or after dividing by q,

$$\frac{V}{q}\frac{dx}{dt} + x = x_f \qquad\qquad 3.1\text{-}46$$

Therefore we can consider the salt composition x_f to be analogous to the impressed voltage and the coefficient V/q to be analogous to the product of the resistance and capacitance. This coefficient has the units of time and thus should define some characteristic time associated with the system. Alternately, we could study the dynamic response of the mixer using an RC circuit merely by making the voltages proportional to the compositions for a case where the coefficients and system boundary conditions were identical. Of course, the exploitation of this idea led to the development of passive analog computers, where RC circuits are used to model some other physical system.

Since much of the theory of dynamic systems was developed by electrical engineers, the RC circuit is commonly used as the basic reference system. Thus discussions of process dynamics are often phrased in terms of some stored quantity (electrical charge or mass of reactant material), a driving force (voltage or reactant concentration), and a flow (current or mass flow rate of reactant). Then the resistance is defined as the change in driving force divided by the change in flow, and the capacitance is defined as the change in storage divided by the change in the driving force.

$$\text{Resistance} = \frac{\Delta \text{ driving force}}{\Delta \text{ flow}} \qquad \text{Capacitance} = \frac{\Delta \text{ storage}}{\Delta \text{ driving force}}$$
$$3.1\text{-}47$$

For the mixer problem, these definitions give the results

$$\text{Resistance} = \frac{x}{qx} = \frac{1}{q} \qquad \text{Capacitance} = \frac{Vx}{x} = V \qquad 3.1\text{-}48$$

For a simple thermometer, or other transfer process, described by the equation

$$mC_p\frac{dT}{dt} = hA(T_f - T) \qquad\qquad 3.1\text{-}49$$

the stored quantity is the internal energy, the driving force is temperature, and the flow is the heat flow, Proceeding as before, we find that

$$\text{Resistance} = \frac{1}{hA} \qquad \text{Capacitance} = mC_p \qquad 3.1\text{-}50$$

The preceding discussion makes it clear that a thorough understanding of the dynamic behavior of electrical circuits should provide a great deal of insight into the behavior of chemical processes. Furthermore, when we speak of lumping all the resistance into one location and all the capacitance into another, as in the discussion of Eq. 3.1-21, in essense we are approximating some complex electrical network by the simple RC circuit shown in Figure

3.1-5. In fact, the terms "lumped parameter" and "distributed parameter" systems originated in the electrical engineering literature. However, instead of always attempting to translate the physics of a particular process into electrical engineering terminology, we will take more of a mathematical point of view and define a lumped parameter system as one that is described by a set of ordinary differential equations, where the only independent variable is time. For cases where we are interested in the dynamic response of some property of a fluid mixture, a lumped parameter model is obtained only if the system is perfectly mixed.

SECTION 3.2 DISTRIBUTED PARAMETER SYSTEMS

Although the assumption of perfect mixing is an important one in chemical engineering, it is only applicable for a limited number of processes. In a great many systems, compositions, temperatures, and pressure vary continuously in one or more spatial directions within the system boundaries; that is, their variations are distributed throughout space. For these systems we will need at least two independent variables, time and distance, to describe the dynamic behavior, and therefore we expect the dynamic models to consist of partial differential equations. An illustration of this type of model was presented in Example 3.1-3, although we did not actually treat the dynamics of the distributed portion of the system in that example. Some additional illustrations of distributed parameter systems are discussed below.

Example 3.2-1 Dynamic model of a steam-heated exchanger

The equation describing the dynamic operation of a steam-heated exchanger is simple to derive and is essentially the same as the model of the heating coil in an agitated tank. If steam condenses on the outside of one or more tubes of a heat exchanger, the condensate temperature in the shell will depend directly on the steam pressure (providing there are no inerts in the system), but will be independent of the position in the shell. Assuming that the only resistance to heat transfer exists in a stagnant film on the inside of

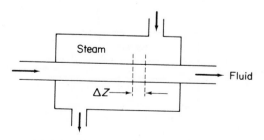

Figure 3.2-1. Steam-heated exchanger.

the tubes, that the heat capacity and density of the flowing fluid are constant, that the only significant accumulation effects are in the flowing fluid, that the velocity profile is flat, and that there are no radial temperature gradients, then an energy balance over a differential length of the exchanger, Δz, gives[1]

$$C_p \rho A \, \Delta z \, [(T)|_{t+\Delta t} - (T)|_t] = C_p \rho v A \, \Delta t \, [(T)|_z - (T)|_{z+\Delta z}]$$
$$\underset{\text{Accumulation}}{} \qquad\qquad \underset{\text{Flow in}}{} \qquad\qquad \underset{\text{Flow out}}{}$$

$$+ h A_t \, \Delta z \, \Delta t (T_s - T) \qquad\qquad 3.2\text{-}1$$
$$\underset{\text{In Through Wall}}{}$$

Dividing both sides of the equation by $\Delta z \, \Delta t$ and taking the limit as Δz and Δt both approach zero, we obtain the result

$$C_p \rho A \frac{\partial T}{\partial t} + C_p \rho v A \frac{\partial T}{\partial z} = h A_t (T_s - T) \qquad\qquad 3.2\text{-}2$$

where C_p = fluid heat capacity $(\text{Btu})(\text{lb})^{-1}(^\circ\text{F})^{-1}$
         ρ = fluid density $(\text{lb})(\text{ft}^3)^{-1}$
         A = cross-sectional area (ft^2)
         v = fluid velocity $(\text{ft})(\text{sec})^{-1}$
         A_t = heat-transfer area $(\text{ft}^2)(\text{ft})^{-1}$
         h = inside film coefficient $(\text{Btu})(\text{sec})^{-1}(\text{ft}^2)^{-1}(^\circ\text{F})^{-1}$
         T = fluid temperature
         T_s = steam temperature

This is a first-order, linear partial differential equation, which may have variable coefficients. Consequently we expect to be able to find analytical solutions for the dynamic response, at least for certain kinds of system inputs, such as, steam temperature, fluid velocity, or inlet fluid temperature variation. Of course, if the preceding assumptions are not valid for a particular system under investigation, the predictions obtained from the model might deviate widely from experimental observations

Example 3.2-2 Dynamic model of a nonisothermal tubular reactor

As a simple modification of the foregoing problem, let us suppose that the fluid reacts as it passes through the exchanger tubes. If we consider a

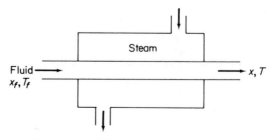

Figure 3.2-2. Nonisothermal tubular reactor.

[1] See footnote 4, p. 58.

first-order, irreversible, exothermic reaction and make the same assumptions as in the previous example, the material and energy balances for the system become

$$\frac{\partial x}{\partial t} + v\frac{\partial x}{\partial z} = -kx \qquad\qquad 3.2\text{-}3$$

$$C_p\rho A\frac{\partial T}{\partial t} + C_p\rho v A\frac{\partial T}{\partial z} = hA_t(T_s - T) + (-\Delta H)kAx \qquad\qquad 3.2\text{-}4$$

where

$$k = k_0 \exp\frac{-E}{RT} \qquad\qquad 3.2\text{-}5$$

Since this set of coupled, first-order, partial differential equations is nonlinear and may have variable coefficients, we can no longer expect to obtain analytical solutions.

Example 3.2-3 Dynamic model of a packed bed separation unit

In Example 3.1-2 we discussed the use of plate columns to separate one or more chemical components from a mixture. Exactly the same princples are applicable in a packed column, and therefore this type of unit is often used as an absorber, extractor, or still. The inert packing material makes it possible to achieve an initimate contact between the two immiscible phases and helps to keep the composition and/or temperature uniform across the bed cross section.

If we let G equal inert gas rate (lb moles)(hr)$^{-1}$(ft^2)$^{-1}$, L equal inert solvent rate (lb moles)(hr)$^{-1}$(ft^2)$^{-1}$, Y equal gas phase concentration (moles of transferable component) (moles of inert gas)$^{-1}$, X equal liquid phase concentration (moles of transferable component) (moles of inert solvent)$^{-1}$, H equal moles of inert solvent held up by the packing per unit of tower volume, h equal moles of inert gas held up by the packing per unit of tower volume, k_g equal mass transfer coefficient (lb moles)(hr)$^{-1}$(ft^2)$^{-1}$(unit of

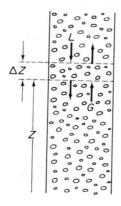

Figure 3.2-3. Packed–bed separation unit.

$Y - Y^*)^{-1}$, a equal mass transfer area per unit tower volume, Y^* equal gas phase concentration in equilibrium with a liquid phase composition X, m equal distribution coefficient, A equal cross-sectional area of the column, z equal distance, and t equal time, and assume that both phases pass countercurrently through the bed under essentially plugflowconditions, that the rate of mass transfer is controlled by the gas phase resistance, that the gas and liquid holdups in the bed remain constant, that the system is isothermal, and that the equilibrium relationship is linear

$$Y^* = mX \qquad\qquad 3.2\text{-}6$$

then a total material balance for the element Δz gives

$$H\frac{\partial X}{\partial t} + h\frac{\partial Y}{\partial t} = -G\frac{\partial Y}{\partial z} + L\frac{\partial X}{\partial z} \qquad\qquad 3.2\text{-}7$$

Similarly, a material balance for the gas phase alone gives the result

$$h\frac{\partial Y}{\partial t} = -G\frac{\partial Y}{\partial z} - k_g a(Y - Y^*) \qquad\qquad 3.2\text{-}8$$

or after using the equilibrium relationship, Eq. 3.2-6, to eliminate Y^*,

$$h\frac{\partial Y}{\partial t} = -G\frac{\partial Y}{\partial z} - k_g a(Y - mX) \qquad\qquad 3.2\text{-}9$$

Equations 3.2-7 and 3.2-9 provide sufficient information for determining $Y(z, t)$ and $X(z, t)$ once the system inputs, the inlet concentrations and flow rates, have been specified. This set of first-order partial differential equations is linear, although they may have variable coefficients. Hence we expect to be able to obtain analytical solutions for the dynamic response in some cases.

Of course, it would also be possible to write a material balance equation for just the liquid phase. However, this result can be obtained by merely subtracting Eq. 3.2-9 from Eq. 3.2-7, and therefore does not provide any new information. Similarly, it would be possibile to rederive the equations in terms of mole fractions, total molar flow rates, and the other units used in Example 3.1-2. As mentioned previously, the use of mole fractions is normally more convenient for a case where only a small amount of mass transfer takes place, so that the total flow rates of the two immiscible phases remain essentially unchanged as they pass through the column.

In some cases, especially for extraction units, the plug flow assumption does not appear valid. For these systems it is common practice to add an axial dispersion term to the material balance equations. There are important industrial systems, such as an SO_3-absorber, where large heat effects are present. In order to account for these effects, it is necessary to derive energy balances for the gas and liquid phases, and possibly for the packing material. Similarly, there are a great number of packed-bed systems, such as absorption columns, ion-exchange columns, chromatographic columns, and heat regenerators, where only one phase moves through the column and transfers mass or heat to the packing material. The models for these processes are essentially the same as those above; the gas-phase material balance remains unchanged and the flow velocity of the solid phase, L in Eq. 3.2-7, is set equal to zero. It should also be noted that the form of equations developed previously is identical with the dynamic model of a simple countercurrent heat exchanger. Hence an understanding of the solutions of this set of partial differential equations should give us considerable insight into the dynamic response of many process units.

Example 3.2-4 Dynamic model of a packed catalytic reactor

Up to now we have limited our attention to systems that were described by sets of first-order partial differential equations. This kind of result depends on the assumptions of a flat velocity profile (plug flow conditions), the absence of heat transfer by conduction, and the absence of mass transfer by a Fick's law diffusion process. In order to illustrate the way in which higher-order differential equations arise, and to give an indication of the complexity of the models obtained when a fairly detailed description of a process is developed, we will consider the operation of a packed-bed catalytic reactor. We suppose that a gas stream containing some reactant material flows into a bed packed with porous, spherical, catalyst particles. The reactant diffuses across a stagnant film surrounding each particle and then diffuses through the pores toward the center of the particle. For simplicity, we will assume that there is only a single, irreversible, exothermic reaction and that the reaction rate may be written as a first-order homogeneous expression, despite the fact that the reaction actually takes place on the surface of the catalyst material.

If we first examine the interior of a single catalyst pellet and assume that

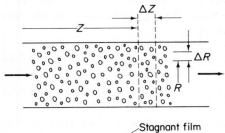

Figure 3.2-4. Catalytic reactor.

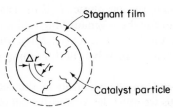

Figure 3.2-5. Single catalyst particle.

heat conduction in the radial direction, a Fick's law diffusion process in the radial direction, and chemical reaction will describe the changes occurring within the particle, then a mass and energy balance over a spherical shell of thickness Δr give the results

$$\epsilon_p \frac{\partial C_s}{\partial t} = D_p \frac{1}{r^2} \frac{\partial}{\partial r}\left(r^2 \frac{\partial C_s}{\partial r}\right) - kC_s \qquad 3.2\text{-}10$$

$$\rho_s C_{ps} \frac{\partial T_s}{\partial t} = k_p \frac{1}{r^2} \frac{\partial}{\partial r}\left(r^2 \frac{\partial T_s}{\partial r}\right) + (-\Delta H)kC_s \qquad 3.2\text{-}11$$

where

$$k = k_0 \exp \frac{-E}{RT_s} \qquad 3.2\text{-}12$$

and
$\quad \epsilon_p$ = void fraction of particle
$\quad C_s$ = reactant concentration in the particle
$\quad r$ = radius
$\quad D_p$ = diffusivity of reactant in particle
$\quad k_p$ = thermal conductivity of porous solid
$\quad \rho_s$ = density of particle
$\quad C_p$ = heat capacity of particle
$\quad T_s$ = temperature in particle
$(-\Delta H)$ = heat of reaction
$\quad k$ = reaction rate constant

It is clear that both the mass and energy flux across the center of the particle must be equal to zero at every instant of time. Thus one pair of boundary conditions for the preceding equations is

$$\frac{\partial C_s(0, t)}{\partial r} = \frac{\partial T_s(0, t)}{\partial r} = 0 \qquad 3.2\text{-}13$$

The boundary conditions at the surface of the particle, $r = r_0$, are obtained by equating the material and energy fluxes through the surface to the corresponsing fluxes through the stagnant film

$$D_p \frac{\partial C_s}{\partial r} = k_g(C_f - C_s), \qquad \text{at } r = r_0, t = t \qquad 3.2\text{-}14$$

$$k_p \frac{\partial T_s}{\partial r} = h(T_f - T_s), \qquad \text{at } r = r_0, t = t \qquad 3.2\text{-}15$$

where C_f is the reactant concentration in the gas phase, T_f is the gas temperature, k_g is the mass transfer coefficient for the stagnant film, and h represents heat-transfer coefficient for the stagnant film. Once $C_f(t)$, $T_f(t)$, and a set of initial conditions have been specified, we can solve the foregoing equations and determine $C_s(r, t)$ and $T_s(r, t)$.

Of course, C_f and T_f both depend on the position in the reactor, the values of R and z, as well as time. Thus we must derive another set of material and energy balance equations describing heat and mass transfer in the gas phase. For this purpose it is common practice to assume that the tube does not contain any packing, that there is no flow in the radial direction in the bed, and that the velocity profile is flat. With these assumptions, the transport equations become

$$\epsilon_b \frac{\partial C_f}{\partial t} = D_z \frac{\partial^2 C_f}{\partial z^2} - V \frac{\partial C_f}{\partial z} + D_R \frac{1}{R} \frac{\partial}{\partial R}\left(R \frac{\partial C_f}{\partial R}\right) - k_g(C_f - C_s) \qquad 3.2\text{-}16$$

$$\epsilon_b C_{pf} \rho_f \frac{\partial T_f}{\partial t} = k_z \frac{\partial^2 T_f}{\partial z^2} - V C_{pf} \rho_f \frac{\partial T_f}{\partial z} + k_R \frac{1}{R} \frac{\partial}{\partial R}\left(R \frac{\partial T_f}{\partial R}\right) - h(T_f - T_s) \qquad 3.2\text{-}17$$

where D_z and D_R = effective diffusion coefficients in the axial and radial directions

k_z and k_R = effective thermal conductivities in the axial and radial directions

v = fluid velocity

z = distance along the tube

R = radial distance

C_{pf} = heat capacity of gas

ρ_f = gas density

and ϵ_b = bed porosity

There is no reaction rate term in these equations, for it has been assumed that reaction occurs only at the interior of the catalyst particles.

Since there can be no mass or heat transfer across the axis of the tube, one pair of boundary conditions is

$$\frac{\partial C_f}{\partial R}(z, 0, t) = \frac{\partial T_f}{\partial R}(z, 0. t) = 0 \qquad 3.2\text{-}18$$

Similarly, there can be no mass transfer across the reactor wall, $R = R_0$, so that

$$\frac{\partial C_f}{\partial R}(z, R_0, t) = 0 \qquad 3.2\text{-}19$$

The wall temperature can also be determined by making an energy balance in the neighborhood of the wall. However, before this step can be accomplished, it is necesssary to specify whether the wall is adiabatic, whether there is a cooling jacket around the reactor, or whether there is heat loss by convection and radiation. The problem of determining the boundary conditions at the inlet and outlet of the section packed with catalyst material has been discussed in detail by Wehner and Wilhelm.[2] Their results show that it is necessary to write an additional set of material and energy balances for both the sections before and after the catalyst bed. These equations are coupled to the preceding set by recognizing that the fluxes, as well as the composition and temperature, must be continuous at the inlet and outlet of the reactor.

[2] J. F. Wehner and R. H. Wilhelm, *Chem. Eng. Sci.*, **6**, 89 (1956).

The remaining boundary conditions for the fore- and after-sections are established by recognizing that far downstream in the after-section the composition and temperature must be finite and that far upstream in the fore-section the composition and temperature must be equal to their inlet values. Of course, it is also necessary to specify a set of initial conditions throughout the entire system before an attempt is made to solve the equations, the nature of these conditions depending on the particular problem under investigation.

Therefore we find that the model describing the dynamic response of a packed catalytic reactor consists of a complicated set of second-order partial differential equations, some of which are nonlinear and which may have variable coefficients. If we had considered complex kinetic mechanisms, where it is necessary to write at least one additional material balance equation for each independent reaction, the difficulty of finding solutions would increase tremendously. In these cases, and for the problem considered, attempts are often made to simplify the equations by neglecting the radial and/or axial gradients and to approximate the reaction inside the particles by a homogeneous rate equation and an effectiveness factor.

SECTION 3.3 AN APPROXIMATE MODEL FOR DYNAMIC SYSTEMS

In the previous sections we discussed methods for developing approximate dynamic models for chemical processes. Most of the models presented consisted of sets of coupled, nonlinear, ordinary or partial differential equations, and therefore we cannot expect to find analytical solutions. Remembering that one of our primary motives for considering process dynamics was to be able to estimate how much effect the system dynamics would have on the optimum steady state control of a plant, we find that we are still a long way away from our goal. In fact, unless we can find some way of simplifying the models in order to obtain at least a crude estimate of the dynamic response, we will be faced with a mathematical problem that is even more complex than the original optimum steady state control problem.

The major difficulty in the system equations is caused by the presence of the nonlinear terms. Thus if we can find a way of linearizing the equations, and especially if we can develop a procedure for approximating a set of partial differential equations by a set of linear, ordinary differential equations, then we will be able to apply the very powerful tools of linear analysis, and it will always be possible to obtain an estimate of the dynamic response. Procedures for accomplishing this objective are described below. Initially we consider the linearization and discretization techniques, and then we discuss the solutions of coupled sets of linear differential equations. Our goal is to show that we can always estimate the response of any unit in terms of a combination of simple, first- and second-order ordinary differential equations. Of course, the resulting estimates might be very poor approximations of the actual system response because of the assumptions used in the development of the model and the linearization technique. A method that can be used to evaluate the validity of the approximation is described in Chapter 6.

Taylor Series Expansion

If we consider a plant that initially operates at some optimum steady state condition and, at time zero, one of the process inputs (a disturbance) suddenly changes to a new value, we know that the plant will move to a new steady state level. Hence one problem of interest will be the determination of the dynamic response of the system to a step input. We have also shown that it is possible to adjust the control variables in such a way that we obtain the maximum profit from the plant when it finally arrives at a different optimum steady state point. Once this optimum steady state control problem has been solved, we will know the final steady state conditions. Thus we can visualize the dynamics problem as one where the system moves from some initial state, the old steady state operating point, to an equilibrium state, the new steady state. In other words, we see that the response of a system to step changes in the inputs is essentially the same as the behavior observed when the plant moves from a set of initial conditions to a known steady state. For either kind of problem, we are interested in the dynamic response of the system in the neighborhood of a steady state operating point.

Since we can approximate the behavior of any well-behaved function in the neighborhood of some point by a Taylor series expansion, we will attempt to use this method to simplify the dynamic models discussed previously. For a system having two dependent variables, the Taylor series expansion of an arbitrary function f can be written as

$$f(x_1, x_2) = f(x_{1s}, x_{2s}) + (x_1 - x_{1s})\left(\frac{\partial f}{\partial x_1}\bigg|_s\right) + (x_2 - x_{2s})\left(\frac{\partial f}{\partial x_2}\bigg|_s\right)$$

$$+ \frac{1}{2}(x_1 - x_{1s})^2\left(\frac{\partial^2 f}{\partial x_1^2}\bigg|_s\right) + \frac{1}{2}(x_2 - x_{2s})^2\left(\frac{\partial^2 f}{\partial x_2^2}\bigg|_s\right)$$

$$+ (x_1 - x_{1s})(x_2 - x_{2s})\left(\frac{\partial^2 f}{\partial x_1 \partial x_2}\bigg|_s\right) + \cdots \qquad \text{3.3-1}$$

where the subscript s refers to the steady state operating point of interest and $(\partial f/\partial x_1|_s)$ means that the derivative of the function is evaluated at the steady state conditions. Thus we can approximately represent any nonlinear function by a multidimensional polynomial: the sum of a constant, a number of linear terms, a number of quadratic terms, cubic terms, and so on. By examining this result we find that if we are willing to restrict our attention to small deviations from steady state conditions, so that $(x_1 - x_{1s})$ and $(x_2 - x_{2s})$ are very small, there will be some region around the steady state operating point where all the quadratic, cubic, and higher-order terms will be negligible in comparison with the constant and linear terms. That is, if $(x_1 - x_{1s})$ is a very small number, then $(x_1 - x_{1s})^2$ and $(x_1 - x_{1s})^3$ will be extremely small and can be neglected. Another way of stating this result is that we can represent the characteristics of a nonlinear surface in some sufficiently small neighborhood of a point of interest by an n-dimensional plane, a hyperplane. Thus we find it is possible to linearize well-behaved, nonlinear functions in some region of a steady state operating point, although we do not know how large this region might be. The problem of determining the approximate size of this region has received relatively little attention in the literature, but a procedure for estimating its value is discussed in detail in Chapter 6. Several examples that illustrate the linearization technique are presented below.

Example 3.3-1 A linear model for a nonisothermal, continuous-stirred-tank reactor

The approximate equations describing the dynamics of a nonisothermal, continuous-stirred-tank reactor were derived in Section 3.1. After dividing each equation by the coefficient of the derivative term, the model becomes

$$\frac{dA}{dt} = \frac{q}{V}(A_f - A) - kA \qquad 3.3\text{-}2$$

$$\frac{dT}{dt} = \frac{q}{V}(T_f - T) + \frac{(-\Delta H)}{C_p \rho}kA + \frac{UA_H Kq_H}{VC_p\rho(1 + Kq_H)}(T_H - T)$$

$$3.3\text{-}3$$

where

$$k = k_0 \exp\frac{-E}{RT} \quad \text{and} \quad K = \frac{2C_{pH}\rho_H}{UA_H} \qquad 3.3\text{-}4$$

If we consider a case where q, A_f, T_f, T_H, and q_H deviate slightly from some steady state solution, which will cause A and T to vary, then we can expand the nonlinear functions on the right-hand sides of the system equations in a Taylor series that contains these deviation variables. Neglecting the second- and higher-order terms, we obtain the result

$$\frac{dA}{dt} = \left[\frac{q_s}{V}(A_{fs} - A_s) - k_s A_s\right] - (A - A_s)\left(\frac{q_s}{V} + k_s\right)$$
$$- (T - T_s)\left(\frac{Ek_s A_s}{RT_s^2}\right) + (q - q_s)\left(\frac{A_{fs} - A_s}{V}\right)$$
$$+ (A_f - A_{fs})\frac{q_s}{V} \qquad 3.3\text{-}5$$

$$\frac{dT}{dt} = \left[\frac{q_s}{V}(T_{fs} - T_s) \pm \frac{(-\Delta H)}{C_p\rho}k_sA_s \mp \frac{UA_HKq_{Hs}}{VC_p\rho(1 + Kq_{Hs})}(T_{Hs} - T_s) \right]$$

$$+ (A - A_s)\left[\frac{(-\Delta H)k_s}{C_p\rho} \right] - (T - T_s)\left[\frac{q_s}{V} - \frac{(-\Delta H)Ek_sA_s}{C_p\rho RT_s^2} \right]$$

$$+ \frac{UA_HKq_{Hs}}{VC_p\rho(1 + Kq_{Hs})} \right] + (q - q_s)\left(\frac{T_{fs} - T_s}{V} \right) + (T_f - T_{fs})\frac{q_s}{V}$$

$$+ (T_H - T_{Hs})\frac{UA_HKq_{Hs}}{VC_p\rho(1 + Kq_{Hs})}$$

$$+ (q_H - q_{Hs})\frac{UA_HK(T_{Hs} - T_s)}{VC_p\rho(1 + Kq_{Hs})^2} \qquad\qquad 3.3\text{-}6$$

where the subscript s again refers to the steady state operating point of interest. At this steady state the accumulation terms, the time derivatives in Eqs. 3.3-2 and 3.3-3, must be equal to zero. Hence the terms in brackets in Eqs. 3.3-5 and 3.3-6, which are merely the steady state equations describing the system, must vanish. This result, the fact that the constant term in the Taylor series expansion (see Eq. 3.3-1) is equal to zero, will always be obtained for the systems we will consider.

Generally it is more convenient to work with dimensionless variables rather than those obtained directly from the conservation equations. There is no unique way to make the quantities dimensionless, but one approach is to let

$$z_1 = \frac{A}{A_{fs}}, \quad z_2 = \frac{TC_p\rho}{(-\Delta H)A_{fs}}, \quad z_{2f} = \frac{T_fC_p\rho}{(-\Delta H)A_{fs}}$$

$$z_{2H} = \frac{T_HC_p\rho}{(-\Delta H)A_{fs}} \qquad \tau = \frac{q_st}{V} \qquad\qquad 3.3\text{-}7$$

and then to multiply the material balance equation by the factor (V/q_sA_{fs}) and the energy balance equation by the factor $(C_p\rho V/(-\Delta H)A_{fs}q_s)$. Substitution of the dimensionless z variables into the system equations leads to the results

$$\frac{dz_1}{d\tau} = \left(1 + \frac{k_sV}{q_s} \right)(z_1 - z_{1s}) - \frac{EC_p\rho k_sVz_{1s}}{R(-\Delta H)A_{fs}q_sz_{2s}^2}(z_2 - z_{2s})$$

$$\times (1 - z_{1s})\left(\frac{q - q_s}{q_s} \right) + \left(\frac{A_f - A_{fs}}{A_{fs}} \right) \qquad\qquad 3.3\text{-}8$$

$$\frac{dz_2}{d\tau} = \frac{k_sV}{q_s}(z_1 - z_{1s}) - \left[1 - \frac{EC_p\rho k_sVz_{1s}}{R(-\Delta H)A_{fs}q_sz_{2s}^2} + \frac{UA_HKq_{Hs}}{q_sC_p\rho(1 + Kq_{Hs})} \right]$$

$$\times (z_2 - z_{2s}) + (z_{2fs} - z_{2s})\left(\frac{q - q_s}{q_s} \right) + (z_{2f} - z_{2fs})$$

$$+ \frac{UA_HKq_{Hs}}{q_sC_p\rho(1 + Kq_{Hs})}(z_{2H} - z_{2Hs}) + \frac{UA_HK(z_{2Hs} - z_{2s})q_{Hs}}{q_sC_p\rho(1 + Kq_{Hs})^2}$$

$$\times \left(\frac{q_H - q_{Hs}}{q_{Hs}} \right) \qquad\qquad 3.3\text{-}9$$

In order to simplify the equations further, we will let

$$a_{11} = -\left(1 + \frac{k_s V}{q_s}\right), \quad a_{12} = -\frac{EC_p \rho k_s V z_{1s}}{R(-\Delta H)A_{fs}q_s z_{2s}^2}, \quad a_{21} = \frac{k_s V}{q_s}$$

$$a_{22} = -\left[1 - \frac{EC_p \rho k_s V z_{1s}}{R(-\Delta H)A_{fs}q_s z_{2s}^2} + \frac{UA_H K q_{Hs}}{q_s C_p \rho (1 + K q_{Hs})}\right],$$

$$b_{11} = 1 - z_{1s} \qquad b_{21} = z_{2fs} - z_{2s}$$

$$b_{22} = \frac{UA_H K q_{Hs}(z_{2Hs} - z_{2s})}{q_s C_p \rho (1 + K q_{Hs})^2},$$

$$c_{11} = 1 \qquad c_{22} = 1$$

$$c_{23} = \frac{UA_H K q_{Hs}}{q_s C_p \rho (1 + K q_{Hs})} \qquad\qquad\qquad\qquad 3.3\text{-}10$$

These terms are all constants, but they depend on the particular steady state operating condition of interest. Also, we will introduce new variables that represent the dimensionless deviations from steady state

$$x_1 = z_1 - z_{1s} \qquad x_2 = z_2 - z_{2s}$$

$$x_{1f} = \frac{A_f - A_{fs}}{A_{fs}} \qquad x_{2f} = z_{2f} - z_{2fs}$$

$$x_{2H} = z_{2H} - z_{2Hs}, \quad u_1 = \frac{q - q_s}{q_s}, \quad u_2 = \frac{q_H - q_{Hs}}{q_{Hs}} \qquad 3.3\text{-}11$$

Substituting these definitions into the dimensionless model equations gives

$$\frac{dx_1}{d\tau} = a_{11}x_1 + a_{12}x_2 + b_{11}u_1 + c_{11}x_{1f} \qquad\qquad 3.3\text{-}12$$

$$\frac{dx_2}{d\tau} = a_{21}x_1 + a_{22}x_2 + b_{21}u_1 + b_{22}u_2 + c_{22}x_{2f} + c_{23}x_{2H} \quad 3.3\text{-}13$$

This is a set of linear, ordinary differential equations with constant coefficients, and so we expect to be able to find analytical solutions for the dynamic response. Although it is a simple matter to solve the equations by using elementary techniques, we will follow a matrix calculus approach, which is especially advantageous for more complex systems.

Writing the equations in vector-matrix form[1] gives

$$\begin{pmatrix} \dfrac{dx_1}{d\tau} \\ \dfrac{dx_2}{d\tau} \end{pmatrix} = \begin{pmatrix} a_{11} & a_{12} \\ a_{21} & a_{22} \end{pmatrix}\begin{pmatrix} x_1 \\ x_2 \end{pmatrix} + \begin{pmatrix} b_{11} & 0 \\ b_{21} & b_{22} \end{pmatrix}\begin{pmatrix} u_1 \\ u_2 \end{pmatrix} + \begin{pmatrix} c_{11} & 0 & 0 \\ 0 & c_{22} & c_{23} \end{pmatrix}\begin{pmatrix} x_{1f} \\ x_{2f} \\ x_{2H} \end{pmatrix}$$

$$3.3\text{-}14$$

or letting

$$\mathbf{x} = \begin{pmatrix} x_1 \\ x_2 \end{pmatrix}, \quad \mathbf{u} = \begin{pmatrix} u_1 \\ u_2 \end{pmatrix}, \quad \mathbf{v} = \begin{pmatrix} x_{1f} \\ x_{2f} \\ x_{2H} \end{pmatrix}$$

[1] See Appendix A.

$$\mathbf{A} = \begin{pmatrix} a_{11} & a_{12} \\ a_{21} & a_{22} \end{pmatrix}, \quad \mathbf{B} = \begin{pmatrix} b_{11} & 0 \\ b_{21} & b_{22} \end{pmatrix}, \quad \mathbf{C} = \begin{pmatrix} c_{11} & 0 & 0 \\ 0 & c_{22} & c_{23} \end{pmatrix}$$

3.3-15

the equations become

$$\frac{d\mathbf{x}}{d\tau} = \mathbf{Ax} + \mathbf{Bu} + \mathbf{Cv}$$

3.3-16

Here $\mathbf{x}$ is called the state vector,[2] for its elements are the dependent variables of the system (A and T) and they describe the state of the system; $\mathbf{u}$ is called the control vector, for its elements are the control variables for the process (where we have implicitly assumed that we can manipulate the feed rate q and heating flow rate q_H); and $\mathbf{v}$ is the disturbance vector, for its elements are inputs, which change with time but cannot be manipulated. Before discussing the solutions of this matrix differential equation, we will show that the models describing other kinds of plants can be reduced to this same form.

Example 3.3-2 Linear model for a plate, gas absorber

The approximate equation describing the dynamic response of the nth tray of a plate absorption column was derived in Example 3.1-2.

$$H\frac{dx_n}{dt} = Lx_{n+1} - (L + mV)x_n + mVx_{n-1}$$

3.3-17

A Taylor series expansion gives

$$H\frac{dx_n}{dt} = [\overbrace{L_s x_{n+1,s} - (L_s + mV_s)x_{n,s} + mV_s x_{n-1,s}}^{0}]$$
$$+ L_s(x_{n+1} - x_{n+1,s}) - (L_s + mV_s)(x_n - x_{n,s})$$
$$+ mV_s(x_{n-1} - x_{n-1,s}) + (x_{n+1,s} - x_{n,s})(L - L_s)$$
$$- m(x_{n,s} - x_{n-1,s})(V - V_s)$$

3.3-18

where we have neglected second- and higher-order terms and have recognized that the term in brackets is simply the steady state material balance equation. Making the equation dimensionless by letting

$$z_n = \frac{x_n}{x_{N+1,s}}, \quad \tau = \frac{L_s t}{H}, \quad M = \frac{mV_s}{L_s}$$

3.3-19

we obtain

$$\frac{dz_n}{dt} = (z_{n+1} - z_{n+1,s}) - (1 + M)(z_n - z_{n,s}) + M(z_{n-1} - z_{n-1,s})$$
$$+ (z_{n+1,s} - z_{n,s})\left(\frac{L - L_s}{L_s}\right) - M(z_{n,s} - z_{n-1,s})\left(\frac{V - V_s}{V_s}\right)$$

3.3-20

[2] The term state vector appears frequently in modern control theory and has an exact mathematical definition; for example, see L. B. Koppel, *Introduction to Control Theory*, p. 56, Prentice-Hall, Englewood Cliffs, N.J., 1968. However, for sets of first-order differential equations, the dependent variables can be considered to be the state variables.

Now, if we introduce the dimensionless deviation variables

$$x_n = z_n - z_{n,s}, \quad u_1 = \frac{L - L_s}{L_s}, \quad u_2 = \frac{V - V_s}{V_s} \qquad \text{3.3-21}$$

and the constants (which depend on the steady state operating point of interest)

$$r_{1n} = z_{n+1,s} - z_{n,s} \qquad r_{2n} = -M(z_{n,s} - z_{n-1,s}) \qquad \text{3.3-22}$$

our result is

$$\frac{dx_n}{d\tau} = x_{n+1} - (1 + M)x_n + Mx_{n-1} + r_{1n}u_1 + r_{2n}u_2 \qquad \text{3.3-23}$$

(It should be noted that the same symbol x_n has been used to represent the mole fraction in the liquid phase in Eq. 3.3-17 and the dimensionless deviation from the steady state composition in Eq. 3.3-23.)

Since this result is valid for any plate, we can write

$$\frac{dx_1}{d\tau} = -(1 + M)x_1 + x_2 + 0 + \cdots + r_{11}u_1 + r_{21}u_2 + Mx_0$$

$$\frac{dx_2}{d\tau} = +Mx_1 - (1 + M)x_2 + x_3 + 0 + \cdots + r_{12}u_1 + r_{22}u_2$$

$$\frac{dx_3}{d\tau} = 0 + Mx_2 - (1 + M)x_3 + x_4 + 0 + \cdots + r_{13}u_1 + r_{23}u_2$$

$$\cdot \qquad\qquad\qquad\qquad\qquad\qquad\qquad\qquad \cdot$$
$$\cdot \qquad\qquad\qquad\qquad\qquad\qquad\qquad\qquad \cdot$$
$$\cdot \qquad\qquad\qquad\qquad\qquad\qquad\qquad\qquad \cdot$$

$$\frac{dx_N}{d\tau} = 0 + \cdots + Mx_{N-1} - (1 + M)x_N + r_{1N}u_1 + r_{2N}u_2 + x_{N+1}$$

$$\text{3.3-24}$$

where x_0 is related to the deviation in the inlet vapor composition y_0 by the equilibrium relationship and x_{N+1} refers to the deviation in inlet liquid composition. Then we define the state, control, and disturbance vectors as

$$\mathbf{x} = \begin{pmatrix} x_1 \\ x_2 \\ x_3 \\ \cdot \\ \cdot \\ \cdot \\ x_N \end{pmatrix}, \quad \mathbf{u} = \begin{pmatrix} u_1 \\ u_2 \end{pmatrix}, \quad \mathbf{v} = \begin{pmatrix} Mx_0 \\ 0 \\ 0 \\ \cdot \\ \cdot \\ \cdot \\ x_{N+1} \end{pmatrix} \qquad \text{3.3-25}$$

and a set of matrices as

$$\mathbf{A} = \begin{pmatrix} -(1+M) & 1 & 0 & 0 \cdots & & 0 \\ M & -(1+M) & 1 & 0 \cdots & & 0 \\ 0 & M & -(1+M) & 1 \cdots & & 0 \\ \cdot & & & & & \cdot \\ \cdot & & & & & \\ \cdot & & & & & \\ 0 & & \cdots & & & -(1+M) \end{pmatrix}$$

$$\mathbf{B} = \begin{pmatrix} r_{11} & r_{21} \\ r_{12} & r_{22} \\ r_{13} & r_{23} \\ \cdot & \cdot \\ \cdot & \cdot \\ \cdot & \cdot \\ r_{1N} & r_{2N} \end{pmatrix} \qquad \mathbf{C} = \begin{pmatrix} 1 & 0 & 0 \cdots & 0 \\ 0 & 1 & 0 \cdots & \\ 0 & 0 & 1 \cdots & \\ \cdot & & & \\ \cdot & & & \\ \cdot & & & \\ 0 \cdots & & & 1 \end{pmatrix} \qquad 3.3\text{-}26$$

where $\mathbf{A}$ is an $N \times N$ matrix, B is an $N \times 2$ matrix, and $\mathbf{C}$ is the $N \times N$ identity matrix. With these definitions, our system equations reduce to the form

$$\frac{d\mathbf{x}}{d\tau} = \mathbf{Ax} + \mathbf{Bu} + \mathbf{Cv} \qquad 3.3\text{-}27$$

which was our previous result. It is important to note that the coefficient matrix $\mathbf{A}$ is a tridiagonal matrix, because this form possesses special properties. The analytical solution of this set of equations will be discussed in the next chapter.

Example 3.3-3 Linear model of a nonisothermal tubular reactor

As a final example of the linearization procedure and the manipulations required to put the model into matrix form, we will consider the model for a nonisothermal tubular reactor derived in Example 3.2-2.

$$\frac{\partial x}{\partial t} + v\frac{\partial x}{\partial z} = -kx \qquad 3.3\text{-}28$$

$$C_p \rho A \frac{\partial T}{\partial t} + C_p \rho v A \frac{\partial T}{\partial z} = hA_t(T_w - T) + (-\Delta H)kAx \qquad 3.3\text{-}29$$

where

$$k = k_0 \exp \frac{-E}{RT} \qquad 3.3\text{-}30$$

and we have used the symbol T_w instead of T_s to represent the wall or steam temperature in order to avoid confusion in our subsequent analysis. Solving the equations explicitly for the time derivatives and expanding the nonlinear functions on the right-hand side in a Taylor series, we obtain

$$\frac{\partial x}{\partial t} = -\left[v_s \frac{\partial x_s}{\partial z} + k_s x_s \right]^{0} - v_s \frac{\partial(x - x_s)}{\partial z}$$
$$- (v - v_s)\frac{\partial x_s}{\partial z} - (k_s)(x - x_s) - \frac{Ek_s x_s}{RT_s^2}(T - T_s) \qquad 3.3\text{-}31$$

$$\frac{\partial T}{\partial t} = \left[-v_s \frac{\partial T_s}{\partial z} + \frac{hA_t}{C_p \rho A}(T_{ws} - T_s) + \frac{(-\Delta H)}{C_p \rho}k_s x_s \right]^{0}$$
$$- (v - v_s)\frac{\partial T_s}{\partial z} - v_s \frac{\partial(T - T_s)}{\partial z} + \frac{hA_t}{C_p \rho A}[(T_w - T_{ws}) - (T - T_s)]$$
$$+ \frac{(-\Delta H)k_s}{C_p \rho}(x - x_s) + \frac{(-\Delta H)Ek_s x_s}{C_p \rho RT_s^2}(T - T_s) \qquad 3.3\text{-}32$$

Again, the terms in brackets must be equal to zero because they are the steady state equations for the system. The linearized equations can be put into dimensionless form by letting

$$z_1 = \frac{x}{x_{fs}}, \quad z_2 = \frac{TC_p\rho}{(-\Delta H)x_{fs}}, \quad z_w = \frac{T_wC_p\rho}{(-\Delta H)x_{fs}}$$

$$\tau = \frac{v_s t}{L} \qquad Z = \frac{z}{L} \qquad \text{3.3-33}$$

where x_{fs} is the steady state feed composition and L is the total reactor length. Substituting these definitions gives the results

$$\frac{\partial z_1}{\partial \tau} + \frac{\partial(z_1 - z_{1s})}{\partial Z} = -\left(\frac{k_s L}{v_s}\right)(z_1 - z_{1s}) - \left(\frac{EC_p\rho k_s L z_{1s}}{R(-\Delta H)x_{fs}v_s z_{2s}^2}\right)$$

$$\times (z_2 - z_{2s}) - \frac{\partial z_{1s}}{\partial Z}\left(\frac{v - v_s}{v_s}\right) \qquad \text{3.3-34}$$

$$\frac{\partial z_2}{\partial \tau} + \frac{\partial(z_2 - z_{2s})}{\partial Z} = +\frac{k_s L}{v_s}(z_1 - z_{1s}) - \left[\frac{hA_t L}{C_p\rho Av_s} - \frac{EC_p\rho k_s L z_{1s}}{R(-\Delta H)x_{fs}v_s z_{2s}^2}\right]$$

$$\times (z_2 - z_{2s}) - \frac{\partial z_{2s}}{\partial Z}\left(\frac{v - v_s}{v_s}\right) + \frac{hA_t L}{C_p\rho Av_s}(z_w - z_{ws})$$

$$\text{3.3-35}$$

Introducing the dimensionless deviation variables

$$y_c = z_1 - z_{1s}, \quad y_T = z_2 - z_{2s}, \quad u_1 = \frac{v - v_s}{v_s}$$

$$y_w = z_w - z_{ws} \qquad \text{3.3-36}$$

the equations become

$$\frac{\partial y_c}{\partial \tau} + \frac{\partial y_c}{\partial Z} = -\left(\frac{k_s L}{v_s}\right)y_c - \left[\frac{EC_p\rho k_s L z_{1s}}{R(-\Delta H)x_{fs}v_s z_{2s}^2}\right]y_T - \left(\frac{\partial z_{1s}}{\partial Z}\right)u_1 \qquad \text{3.3-37}$$

$$\frac{\partial y_T}{\partial \tau} + \frac{\partial y_T}{\partial Z} = \left(\frac{k_s L}{v_s}\right)y_c - \left[\frac{hA_t L}{C_p\rho Av_s} - \frac{EC_p\rho k_s L z_{1s}}{R(-\Delta H)x_{fs}v_s z_{2s}^2}\right]y_T$$

$$-\left(\frac{\partial z_{2s}}{\partial Z}\right)u_1 + \left(\frac{hA_t L}{C_p\rho Av_s}\right)y_w \qquad \text{3.3-38}$$

Thus we find that the linearization procedure reduces a coupled set of non-linear, partial differential equations to a coupled set of linear, partial differential equations having variable coefficients; that is, the coefficients depend on the steady state composition and temperature profiles in the reactor. Since it is not possible to find analytical expressions for the manner in which the coefficients vary with length (the steady state equations are nonlinear, ordinary differential equations), we cannot expect to be able to find analytical solutions for the dynamic response. Therefore we will have to look for

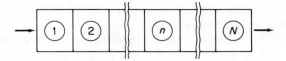

Figure 3.3-1. Series of stirred-tank reactors.

additional ways to simplify the linear model of the distributed parameter system.

One simplifying approach is to assume that we can represent the tubular reactor by a number of stirred-tank reactors placed in series. Writing a material and an energy balance for each one of these n reactors, linearizing the equations, and introducing a set of dimensionless deviation variables leads to the results

$$\frac{dy_{c1}}{d\tau} + N(y_{c1} - y_{cf}) = -\left(\frac{k_{s1}L}{v_s}\right)y_{c1} - \left[\frac{EC_p\rho Lk_{s1}z_{cs1}}{R(-\Delta H)x_{fs}v_s z_{Ts1}^2}\right]y_{T1}$$
$$- N(z_{cs1} - z_{csf})u_1 \qquad\qquad 3.3\text{-}39$$

$$\frac{dy_{T1}}{d\tau} + N(y_{T1} - y_{Tf}) = \left(\frac{k_{s1}L}{v_s}\right)y_{c1}$$
$$- \left[\frac{hA_tL}{C_p\rho Av_s} - \frac{EC_p\rho Lk_{s1}z_{cs1}}{R(-\Delta H)x_{fs}v_s z_{Ts1}^2}\right]y_{T1}$$
$$- N(z_{Ts1} - z_{Tsf})u_1 + \left(\frac{hA_tL}{C_p\rho Av_s}\right)y_w \qquad 3.3\text{-}40$$

$$\frac{dy_{c2}}{d\tau} + N(y_{c2} - y_{c1}) = -\left(\frac{k_{s2}L}{v_s}\right)y_{c2} - \left[\frac{EC_p\rho Lk_{s2}z_{cs2}}{R(-\Delta H)x_{fs}v_s z_{Ts2}^2}\right]y_{T2}$$
$$- N(z_{cs2} - z_{cs1})u_1 \qquad\qquad 3.3\text{-}41$$

$$\frac{dy_{T2}}{d\tau} + N(y_{T2} - y_{T1}) = \left(\frac{k_{s2}L}{v_s}\right)y_{c2}$$
$$- \left[\frac{hA_tL}{C_p\rho Av_s} - \frac{EC_p\rho Lk_{s2}z_{cs2}}{R(-\Delta H)x_{fs}v_s z_{Ts2}^2}\right]y_{T2}$$
$$- N(z_{Ts2} - z_{Ts1})u_1 + \left(\frac{hA_tL}{C_p\rho Av_s}\right)y_w \qquad 3.3\text{-}42$$

and there will be $2N$ of these equations. Of course, these expressions could be obained also by using a backward difference approximation for the spatial derivatives and making appropriate modifications of the coefficients in the set of linear, partial differential equations.[3] Similarly, it would be possible to use forward or central finite-difference approximations to replace the set of linear, partial differential equations by a larger set of linear, ordinary differential equations. However, the backward difference approach has the

[3] See Exercise 30, p. 111.

advantage of being equivalent to a series of stirred tank reactors; therefore it is the method we will follow.

Even though the technique for discretizing, or lumping, the parameters of the distributed parameter system is not unique, and despite the fact that we are forced to deal with a larger number of equations, we have greatly simplified the problem by being able to obtain a set of linear differential equations with constant coefficients. These coefficients will be known numerically from the solution of either the optimum steady state design or control problems. If we define two state vectors, two control vectors, and three disturbance vectors as

$$\mathbf{x}_c = \begin{pmatrix} y_{c1} \\ y_{c2} \\ \cdot \\ \cdot \\ \cdot \\ y_{cN} \end{pmatrix}, \quad \mathbf{x}_T = \begin{pmatrix} y_{T1} \\ y_{T2} \\ \cdot \\ \cdot \\ \cdot \\ y_{TN} \end{pmatrix}, \quad \mathbf{x}_{cf} = \begin{pmatrix} Ny_{cf} \\ 0 \\ \cdot \\ \cdot \\ \cdot \\ 0 \end{pmatrix}, \quad \mathbf{x}_{Tf} = \begin{pmatrix} Ny_{Tf} \\ 0 \\ \cdot \\ \cdot \\ \cdot \\ 0 \end{pmatrix} \quad \text{3.3-43}$$

$$\mathbf{x}_w = \begin{pmatrix} y_w \\ y_w \\ \cdot \\ \cdot \\ \cdot \\ y_w \end{pmatrix}, \quad \mathbf{u}_1 = -N \begin{pmatrix} z_{cs1} - z_{csf} \\ z_{cs2} - z_{cs1} \\ \cdot \\ \cdot \\ \cdot \\ z_{csN} - z_{csN-1} \end{pmatrix} u_1, \quad \mathbf{u}_2 = -N \begin{pmatrix} z_{Ts1} - z_{Tsf} \\ z_{Ts2} - z_{Ts1} \\ \cdot \\ \cdot \\ \cdot \\ z_{TsN} - z_{TsN-1} \end{pmatrix} u_2$$

and a set of constants

$$a_{n,n} = -\left(N + \frac{k_{sn}L}{v_s}\right) \qquad a_{n,n-1} = N;$$

$$b_{n,n} = -\frac{EC_p\rho L k_{sn}z_{csn}}{R(-\Delta H)x_{fs}v_s z_{Tsn}^2}$$

$$c_{n,n} = -\left[\frac{hA_tL}{C_p\rho A v_s} - \frac{EC_p\rho L k_{sn}z_{csn}}{R(-\Delta H)x_{fs}v_s z_{Tsn}^2}\right] \qquad \text{3.3-44}$$

$$c_{n,n-1} = N, \quad d_{n,n} = \frac{k_{sn}L}{v_s}, \quad g_{n,n} = \frac{hA_tL}{C_p\rho A v_s}$$

which are the elements of the matrices

$$\mathbf{A} = \begin{pmatrix} a_{11} & 0 & 0 & \cdots & 0 & 0 \\ a_{21} & a_{22} & 0 & & & \cdot \\ 0 & a_{32} & a_{33} & & & \cdot \\ \cdot & & & & & \\ \cdot & & & & & \\ 0 & \cdots & & & a_{N,N-1} & a_{N,N} \end{pmatrix}$$

$$\mathbf{B} = \begin{pmatrix} b_{11} & 0 & 0 & \cdots & 0 & 0 \\ 0 & b_{22} & 0 & & \cdot & \cdot \\ 0 & 0 & b_{33} & & \cdot & \cdot \\ \cdot & & & & & \\ \cdot & & & & & \\ \cdot & & & & & \\ 0 & & & & 0 & b_{N,N} \end{pmatrix}$$

$$\mathbf{C} = \begin{pmatrix} c_{11} & 0 & \cdots & & 0 & 0 \\ c_{21} & c_{22} & & & \cdot & \cdot \\ \cdot & & & & \cdot & \cdot \\ \cdot & & & & \cdot & \cdot \\ 0 & \cdots & & & c_{N,N-1} & c_{N,N} \end{pmatrix} \qquad 3.3\text{-}45$$

$$\mathbf{D} = \begin{pmatrix} d_{11} & 0 & \cdots & 0 & 0 \\ 0 & d_{22} & & \cdot & \cdot \\ \cdot & & & \cdot & \cdot \\ \cdot & & & & \\ 0 & \cdots & & 0 & d_{N,N} \end{pmatrix}$$

$$\mathbf{G} = \begin{pmatrix} g_{11} & 0 & \cdots & 0 & 0 \\ 0 & g_{22} & & \cdot & \cdot \\ \cdot & & & \cdot & \cdot \\ \cdot & & & & \\ 0 & \cdots & & 0 & g_{N,N} \end{pmatrix}$$

then the system equations can be put into the form

$$\frac{d\mathbf{x}_c}{d\tau} = \mathbf{A}\mathbf{x}_c + \mathbf{B}\mathbf{x}_T + \mathbf{u}_1 + \mathbf{x}_{cf} \qquad 3.3\text{-}46$$

$$\frac{d\mathbf{x}_T}{d\tau} = \mathbf{C}\mathbf{x}_c + \mathbf{D}\mathbf{x}_T + \mathbf{u}_2 + \mathbf{x}_{Tf} + \mathbf{G}\mathbf{x}_w \qquad 3.3\text{-}47$$

The preceding are simultaneous, matrix differential equations where the coefficient matrices are either diagonal or bidiagonal. Of course, it would be possible to combine the equations into a single matrix equation having more complicated coefficient matrices. It is interesting to note that the feed composition and temperature disturbances appear explicitly in the equations, rather than the boundary conditions of the set of partial differential equations.

Analytical methods for determining the dynamic response will be discussed in further detail later in the text. Our main purpose here was to show that it is possible to approximate the dynamic behavior of a distributed parameter system with a set of matrix differential equations, although the error involved in the approximation has not yet been established.

Solution of Matrix Differential Equations

We have expended a considered amount of effort to show that dynamic models of processes can be approximated by a matrix differential equation. The whole purpose of this work was to obtain mathematical equations simple enough to solve analytically so that we could estimate the dynamic response of the plant. Consequently, we need to review the methods for solving matrix differential equations. Although the following discussion is too abbreviated to serve as an introduction to a reader who is completely unfamiliar with matrix analysis, the discussions of specific examples should provide some indication of the value of this approach and the amount of effort required. Additional information is readily available,[4,5,6] and a more complete discussion is given in Appendix A.

The matrix form of the dynamic models was shown to be

$$\frac{d\mathbf{x}}{dt} = \mathbf{Ax} + \mathbf{Bu} + \mathbf{Cv} \qquad\qquad 3.3\text{-}48$$

where $\mathbf{x}$ was the state vector (the vector of dependent variables), $\mathbf{u}$ was the control vector (the plant inputs which can be manipulated), and $\mathbf{v}$ was the disturbance vector (the input variables that cannot be manipulated). These dimensionless variables represent deviations from some steady state operating condition of interest. The matrix coefficients in the equation have elements that are constants but that depend on the particular steady state under investigation. The simplest dynamic response we could investigate would be a case where both $\mathbf{u}$ and $\mathbf{v}$ were equal to zero and the system proceeded from some initial state (the initial conditions on the set of differential equations) to the steady state operating point (which is located at the origin, $\mathbf{x} = \mathbf{0}$, since we are considering only deviations from steady state). As noted previouly, this problem is essentially the same as the step response of the system, except that we are looking backward through time. Thus we are interested in finding solutions of the equation

$$\frac{d\mathbf{x}}{dt} = \mathbf{Ax} \qquad \mathbf{x}(0) = \mathbf{x}_0 \qquad\qquad 3.3\text{-}49$$

where $\mathbf{x}_0$ is the vector of the initial values of the state, or dependent, variables.

---

[4] N. R. Amundson, *Mathematical Methods in Chemical Engineering*, Prentice-Hall, Englewood Cliffs, N.J., 1966.

[5] V. G. Jenson and G. V. Jeffreys, *Mathematical Methods in Chemical Engineering*, Ch. 12, Academic Press, N.Y., 1963.

[6] C. R. Wylie, Jr., *Advanced Engineering Mathematics*, 3rd ed., Ch. 10, McGraw-Hill, N.Y., 1966.

In an analogous fashion to a single first-order differential equation, we might assume a solution like

$$\mathbf{x} = \exp(\mathbf{A}t)\mathbf{x}_0 \qquad\qquad 3.3\text{-}50$$

and define the matrix exponential by a series

$$\exp(\mathbf{A}t) = \mathbf{I} + \frac{\mathbf{A}t}{1!} + \frac{\mathbf{A}^2 t^2}{2!} + \frac{\mathbf{A}^3 t^3}{3!} + \cdots \qquad 3.3\text{-}51$$

where $\mathbf{I}$ is an identity matrix having the same dimensions as the square matrix $\mathbf{A}$. If we differentiate this solution, we find that

$$\frac{d\mathbf{x}}{dt} = \left(\mathbf{A} + \frac{\mathbf{A}^2 t}{1!} + \frac{\mathbf{A}^3 t^2}{2!} + \cdots\right)\mathbf{x}_0$$

$$= \mathbf{A}\left(I + \frac{\mathbf{A}t}{1!} + \frac{\mathbf{A}^2 t^2}{2!} + \cdots\right)\mathbf{x}_0 = \mathbf{A}\mathbf{x} \qquad 3.3\text{-}52$$

so that the assumed solution does indeed satisfy both the differential equation and the initial condition. However, even though we have found a solution, it is not very convenient to have a matrix appearing as an exponent or to have an infinite series that we must sum. It is sometimes possible to simplify the results by applying Sylvester's theorem,[7] which provides a relationship between a matrix polynomial and sums of matrices, but the algebra involved makes this approach impractical if there are more than a few differential equations in the model. Therefore we will explore the possibility of uncoupling the equations by making a diagonal transformation.

Matrix Diagonalization

Suppose that we introduce a new set of state variables $\mathbf{y}$, which are related to the original state variables $\mathbf{x}$ by a (nonsingular) linear transformation $\mathbf{P}$

$$\mathbf{x} = \mathbf{P}\mathbf{y} \qquad\qquad 3.3\text{-}53$$

where the elements of the $\mathbf{P}$ matrix will all be constants, but we will not specify their exact values as yet. If this transformation is valid at every instant of time, it also follows that

$$\frac{d\mathbf{x}}{dt} = \mathbf{P}\frac{d\mathbf{y}}{dt} \qquad\qquad 3.3\text{-}54$$

Substitution of these expressions into our original matrix differential equation, Eq. 3.3-48, gives the result

$$\mathbf{P}\frac{d\mathbf{y}}{dt} = \mathbf{A}\mathbf{P}\mathbf{y} + \mathbf{B}\mathbf{u} + \mathbf{C}\mathbf{v} \qquad\qquad 3.3\text{-}55$$

Now, if we premultiply both sides of the equation by $\mathbf{P}^{-1}$, the inverse of $\mathbf{P}$,

---

[7] See any of the references given in footnotes 4, 5, and 6 on p. 86.

we obtain

$$\frac{d\mathbf{y}}{dt} = \mathbf{P}^{-1}\mathbf{APy} + \mathbf{P}^{-1}\mathbf{Bu} + \mathbf{P}^{-1}\mathbf{Cv} \qquad 3.3\text{-}56$$

The ideal situation would occur if we could find a $\mathbf{P}$ matrix that led to a matrix product of $\mathbf{P}^{-1}\mathbf{AP}$ that would be diagonal, because then we have separated all the equations. For example, consider a situation where $\mathbf{u}$ and $\mathbf{v}$ are both zero and

$$\mathbf{P}^{-1}\mathbf{AP} = \mathbf{M} \qquad 3.3\text{-}57$$

where

$$M = \begin{pmatrix} m_{11} & 0 & \cdots & 0 \\ 0 & m_{22} & & \\ \cdot & & & \\ \cdot & & & \\ \cdot & & & \\ 0 & & \cdots & m_{nn} \end{pmatrix} \qquad 3.3\text{-}58$$

Then our new state equations would have the form

$$\begin{vmatrix} \dfrac{dy_1}{dt} \\ \dfrac{dy_2}{dt} \\ \cdot \\ \cdot \\ \cdot \\ \dfrac{dy_n}{dt} \end{vmatrix} = \begin{vmatrix} m_{11} & 0 & \cdots & 0 \\ 0 & m_{22} & & \cdot \\ \cdot & & & \cdot \\ \cdot & & & \\ \cdot & & & \\ 0 & & \cdots & m_{nn} \end{vmatrix} \begin{vmatrix} y_1 \\ \cdot \\ y_2 \\ \cdot \\ \cdot \\ y_n \end{vmatrix} \qquad 3.3\text{-}59$$

or after multiplying the matrix on the right by the column vector and equating the elements of the two column vectors,

$$\frac{dy_1}{dt} = m_{11}y_1$$

$$\frac{dy_2}{dt} = m_{22}y_2$$

$$\cdot \qquad\qquad\qquad\qquad\qquad 3.3\text{-}60$$

$$\cdot$$

$$\frac{dy_n}{dt} = m_{nn}y_n$$

Obviously this set of equations can be solved with ease, no matter how many equations are in the set.

The procedure described is practical for many engineering systems of interest. Standard texts on matrix analysis[8,9] indicate that (for the case of

---

[8] See footnotes 4, 5, and 6 on p. 86.

[9] See the discussion in Appendix A.

distinct eigenvalues) we should choose the columns of **P** as the right-hand eigenvectors (characteristic vectors) of the coefficient matrix **A**. The diagonal matrix, which we will call

$$\mathbf{P}^{-1}\mathbf{A}\mathbf{P} = \mathbf{\Lambda} \qquad\qquad 3.3\text{-}61$$

has as its elements the eigenvalues, or characteristic roots λ of the A matrix. These are obtained from the polynomial equation

$$\det |\mathbf{A} - \lambda\mathbf{I}| = 0 \qquad\qquad 3.3\text{-}62$$

Thus the new state equation becomes

$$\frac{d\mathbf{y}}{dt} = \mathbf{\Lambda}\mathbf{y} + \mathbf{P}^{-1}\mathbf{B}\mathbf{u} + \mathbf{P}^{-1}\mathbf{C}\mathbf{y} \qquad\qquad 3.3\text{-}63$$

where

$$\mathbf{\Lambda} = \begin{pmatrix} \lambda_1 & 0 & \cdots & 0 \\ 0 & \lambda_2 & & \vdots \\ \vdots & & & \vdots \\ \vdots & & & \\ 0 & & & \lambda_n \end{pmatrix} \qquad\qquad 3.3\text{-}64$$

We will refer to this set of equations as the canonical equations (standard form) of the system. The appropriate initial conditions for the canonical equations are obtained merely by transforming the original initial state vector $\mathbf{x}_0$

$$\mathbf{y}_0 = \mathbf{P}^{-1}\mathbf{x}_0 \qquad\qquad 3.3\text{-}65$$

Example 3.3-4 Diagonalization of the nonisothermal,
continuous-stirred-tank reactor model

For the case where the control variables and disturbances are both at steady state, the linearized, stirred-tank reactor equations (see Eqs. 3.3-12 and 3.3-13) can be written as

$$\frac{dx_1}{dt} = a_{11}x_1 + a_{12}x_2 \qquad x_1(0) = x_{1_0} \qquad 3.3\text{-}66$$

$$\frac{dx_2}{dt} = a_{21}x_1 + a_{22}x_2 \qquad x_2(0) = x_{2_0} \qquad 3.3\text{-}67$$

or

$$\frac{d\mathbf{x}}{dt} = \mathbf{A}\mathbf{x} \qquad \mathbf{x}(0) = \mathbf{x}_0 \qquad\qquad 3.3\text{-}68$$

The characteristic roots for the system are determined from the equation

$$\det |\mathbf{A} - \lambda\mathbf{I}| = 0 \qquad\qquad 3.3\text{-}62$$

or

$$\begin{vmatrix} a_{11} - \lambda & a_{12} \\ a_{21} & a_{22} - \lambda \end{vmatrix} = 0 = \lambda^2 - (a_{11} + a_{22})\lambda + a_{11}a_{22} - a_{12}a_{21}$$

$$3.3\text{-}69$$

Hence

$$\lambda = \tfrac{1}{2}[(a_{11} + a_{22}) \pm \sqrt{(a_{11} + a_{22})^2 - 4(a_{11}a_{22} - a_{12}a_{21})}]$$

3.3-70

and we will let λ_1 be the root with the positive sign and λ_2 be the root with the negative sign.

The transformation matrix **P** is determined by using the right-hand eigenvectors of **A** as its columns. We know that the eigenvectors are found by substituting each value of the characteristic roots into the equation

$$\mathbf{Ax} - \lambda \mathbf{Ix} = 0 \tag{3.3-71}$$

Substitution of the first root yields two equations

$$a_{11}x_1 + a_{12}x_2 = \lambda_1 x_1 \tag{3.3-72}$$

$$a_{21}x_1 + a_{22}x_2 = \lambda_1 x_2 \tag{3.3-73}$$

or after some rearranging

$$(a_{11} - \lambda_1)x_1 = -a_{12}x_2 \tag{3.3-74}$$

$$-a_{21}x_1 = (a_{22} - \lambda_1)x_2 \tag{3.3-75}$$

Similarly, we obtain two additional equations by substituting the second root

$$a_{11}x_1 + a_{12}x_2 = \lambda_2 x_1 \tag{3.3-76}$$

$$a_{21}x_1 + a_{22}x_2 = \lambda_2 x_2 \tag{3.3-77}$$

or

$$(a_{11} - \lambda_2)x_1 = -a_{12}x_2 \tag{3.3-78}$$

$$-a_{21}x_1 = (a_{22} - \lambda_2)x_2 \tag{3.3-79}$$

However, only one equation in each set is independent. For example, if we consider the first pair and solve Eq. 3.3-74 for x_1,

$$x_1 = -\frac{a_{12}x_2}{a_{11} - \lambda_1} \tag{3.3-80}$$

Substitution of this result into Eq. 3.3-75 gives

$$\frac{a_{12}a_{21}}{a_{11} - \lambda_1}x_2 = (a_{22} - \lambda_1)x_2 \tag{3.3-81}$$

or

$$(a_{11} - \lambda_1)(a_{22} - \lambda_1) - a_{12}a_{21} = 0 \tag{3.3-82}$$

which is simply the characteristic equation of the system, Eq. 3.3-69. Thus we can fix the eigenvector only to within an arbitrary constant. For simplicity we will choose $x_1 = 1$. Then from Eqs. 3.3-74 and 3.3-78 we find that the appropriate values of x_2 are

$$x_2 = -\frac{a_{11} - \lambda_1}{a_{12}} \qquad x_2 = -\frac{a_{11} - \lambda_2}{a_{12}} \tag{3.3-83}$$

so that the $\mathbf{P}$ matrix becomes

$$\mathbf{P} = \begin{pmatrix} x_1(\lambda_1) & x_1(\lambda_2) \\ x_2(\lambda_1) & x_2(\lambda_2) \end{pmatrix} = \begin{pmatrix} 1 & 1 \\ -\dfrac{a_{11} - \lambda_1}{a_{12}} & -\dfrac{a_{11} - \lambda_2}{a_{12}} \end{pmatrix} \qquad 3.3\text{-}84$$

The determinant of $\mathbf{P}$ is

$$\det |\mathbf{P}| = -\frac{a_{11} - \lambda_2}{a_{12}} + \frac{a_{11} - \lambda_1}{a_{12}} = -\frac{\lambda_1 - \lambda_2}{a_{12}} \qquad 3.3\text{-}85$$

The cofactor of an element in $\mathbf{P}$ is obtained by crossing out the row and column containing the element, finding the determinant of the remaining terms, and multiplying this value by $(-1)^{i+j}$, where i is the row and j is the column in which the element is located. Hence the matrix of cofactors is

$$\begin{pmatrix} -\dfrac{a_{11} - \lambda_2}{a_{12}} & \dfrac{a_{11} - \lambda_1}{a_{12}} \\ -1 & 1 \end{pmatrix} \qquad 3.3\text{-}86$$

The transpose of this matrix, obtained by interchanging the rows and columns, is called the adjoint, and this can be written as

$$\text{adj } \mathbf{P} = \begin{pmatrix} -\dfrac{a_{11} - \lambda_2}{a_{12}} & -1 \\ \dfrac{a_{11} - \lambda_1}{a_{12}} & 1 \end{pmatrix} \qquad 3.3\text{-}87$$

The inverse of $\mathbf{P}$ is merely the adjoint divided by the determinant,

$$\mathbf{P}^{-1} = -\frac{a_{12}}{\lambda_1 - \lambda_2} \begin{pmatrix} -\dfrac{a_{11} - \lambda_2}{a_{12}} & -1 \\ \dfrac{a_{11} - \lambda_1}{a_{12}} & 1 \end{pmatrix} \qquad 3.3\text{-}88$$

In order to make certain that we have not made an error, and the $\mathbf{P}^{-1}\mathbf{AP}$ actually is a diagonal matrix, we can evaluate the product

$\mathbf{P}^{-1}\mathbf{AP} = \boldsymbol{\Lambda}$

$$= -\frac{a_{12}}{\lambda_1 - \lambda_2} \begin{pmatrix} -\dfrac{a_{11} - \lambda_2}{a_{12}} & -1 \\ \dfrac{a_{11} - \lambda_1}{a_{12}} & 1 \end{pmatrix} \begin{pmatrix} a_{11} & a_{12} \\ a_{21} & a_{22} \end{pmatrix} \begin{pmatrix} 1 & 1 \\ -\dfrac{a_{11} - \lambda_1}{a_{12}} & -\dfrac{a_{11} - \lambda_2}{a_{12}} \end{pmatrix}$$

$$= -\frac{a_{12}}{\lambda_1 - \lambda_2} \begin{pmatrix} -\lambda_1\dfrac{a_{11} - \lambda_2}{a_{12}} - a_{21} + \dfrac{a_{22}}{a_{12}}(a_{11} - \lambda_1) \\ \\ -\lambda_2\dfrac{a_{11} - \lambda_2}{a_{12}} - a_{21} + \dfrac{a_{22}}{a_{12}}(a_{11} - \lambda_2) \\ \\ \lambda_1\dfrac{a_{11} - \lambda_1}{a_{12}} + a_{21} - \dfrac{a_{22}}{a_{12}}(a_{11} - \lambda_1) \\ \\ \lambda_2\dfrac{a_{11} - \lambda_1}{a_{12}} + a_{21} - \dfrac{a_{22}}{a_{12}}(a_{11} - \lambda_2) \end{pmatrix}$$

The element in the first row and first column can be written as

$$\frac{1}{\lambda_1 - \lambda_2}[\lambda_1(a_{11} - \lambda_2) + a_{12}a_{21} - a_{22}(a_{11} - \lambda_1)]$$

$$= \frac{1}{\lambda_1 - \lambda_2}\{\lambda_1^2 - \lambda_1\lambda_2 - [\lambda_1^2 - (a_{11} + a_{22})\lambda_2 + a_{11}a_{22} - a_{12}a_{21}]\}$$

$$= \frac{\lambda_1(\lambda_1 - \lambda_2)}{\lambda_1 - \lambda_2} = \lambda_1$$

where the term in brackets is equal to zero because it is the characteristic equation. The fact that the other diagonal element is equal to λ_2 can be shown in a similar manner. A slight rearrangement of the off-diagonal elements reveals that these terms are identical to the characteristic equation and therefore are equal to zero. Thus our analysis is correct and

$$\mathbf{P}^{-1}\mathbf{AP} = = \begin{pmatrix} \lambda_1 & 0 \\ 0 & \lambda_2 \end{pmatrix} \qquad \text{3.3-89}$$

so that the canonical equations become (for the case where $\mathbf{u} = \mathbf{v} = 0$)

$$\frac{dy_1}{dt} = \lambda_1 y_1 \qquad y_1(0) = y_{1_0} \qquad \text{3.3-90}$$

$$\frac{dy_2}{dt} = \lambda_2 y_2 \qquad y_2(0) = y_{2_0} \qquad \text{3.3-91}$$

The solutions of these equations are

$$y_1 = y_{1_0}e^{\lambda_1 t} \qquad y_2 = y_{2_0}e^{\lambda_2 t} \qquad \text{3.3-92}$$

This result shows that if λ_1 and λ_2 are negative quantities, or have negative real parts, y_1 and y_2 will approach zero as time progresses. Thus the system will return to steady state. However, if either λ_1 or λ_2 is positive, or has a positive real part, y_1 or y_2 will tend to grow without a bound as time increases. Of course, what we really mean by this statement is that the steady state operating point under consideration is unstable, so that the system will tend to move away from that point. This motion will continue until we pass out of the region where the quadratic and higher-order terms in the deviation variables are negligible in comparison with the linear terms, and therefore our linear analysis becomes invalid. Although many textbooks talk about the system blowing up or approaching infinity as time becomes large, it is apparent that such statements are misleading.[10] In fact, we will present an example in a later chapter where the system always remains in a relatively small neighborhood of the steady state solution, even though it never approaches that value.

An additional point of difficulty that should be recognized is the significance of the canonical state variables for a case where λ_1 and λ_2 are complex

[10] Of course, the system temperature may exceed the melting point of the reactor wall or the pressure may become so high that the vessel ruptures, so that we do indeed have an explosion.

conjugate roots. Thus if $4(a_{11}a_{22} - a_{12}a_{21}) > (a_{11} + a_{22})^2$, the roots will have the form

$$\lambda_1 = \alpha + j\omega \qquad \lambda_2 = \alpha - j\omega \qquad\qquad 3.3\text{-}93$$

where $j = \sqrt{-1}$. The only way Eqs. 3.3-90 and 3.3-91 can be valid for this case is if y_1 and y_2 are complex variables. Since it is generally not convenient to work with complex state equations, we introduce an additional transformation

$$y_1 = z_1 + jz_2 \qquad y_2 = z_1 - jz_2 \qquad\qquad 3.3\text{-}94$$

Substituting these definitions into the state equations and manipulating the results somewhat gives

$$\dot{z}_1 = \alpha z_1 - \omega z_2 \qquad \dot{z}_2 = \omega z_1 + \alpha z_2 \qquad\qquad 3.3\text{-}95$$

This pair of first-order differential equations is still coupled and therefore, as we will demonstrate later, is equivalent to a second-order system. However, the equations contain only real variables and thus are normally considered canonical. Of course, these results can be obtained directly from the original state equations, Eq. 3.3-68, by making the transformation

$$x_1 = 2(\alpha - a_{11})z_1 + 2\omega z_2 \qquad x_2 = -2a_{21}z_2 \qquad\qquad 3.3\text{-}96$$

where

$$\alpha = \tfrac{1}{2}(a_{11} + a_{22}) \qquad \omega = \tfrac{1}{2}\sqrt{4(a_{11}a_{22} - a_{12}a_{21}) - (a_{11} + a_{22})^2} \qquad 3.3\text{-}97$$

Analysis of Higher-Order Systems

The procedure for uncoupling the state equations by the matrix diagonalization technique just discussed indicates that any complex system can be analyzed in terms of simple first- and second-order systems, providing that the appropriate transformation matrix can be found. For complicated systems it will be necessary to use a digital computer to find the eigenvalues (characteristic roots) and eigenvectors. Programs for this purpose are readily available. Even though eventually we are forced to obtain numerically some of the quantities required in the analysis, the foregoing procedure certainly minimizes the computing effort and will provide a relatively quick estimate of the dynamic response. It should also be noted that the method is applicable only for systems having distinct characteristic roots. An alternate procedure for the special case of repeated roots will be considered in the next chapter.

SECTION 3.4 THE USE OF PROCESS CORRELATIONS AS DYNAMIC MODELS

In the previous discussions of process dynamics, we restricted our attention to the theoretical development of dynamic models using the conservation equations. However, in some cases we assumed that empirical quantities, such as heat-transfer coefficients, could be used to represent the transport processes, even during periods of dynamic operation. Before considering the quantitative aspects of the system response, it might be wise to make certain that we understand the nature of this kind of assumption. Also, we should give additional consideration to other kinds of process correlations and assess the validity of using them in dynamic models.

Heat Transfer and Other Transport Coefficients

When a fluid is in contact with a solid surface, we often observe that the temperature gradient is confined to a relatively thin region quite close to the wall. Thus we find it convenient to approximate the temperature profile

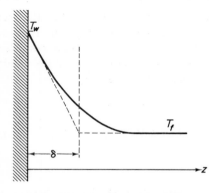

Figure 3.4-1. Temperature profile near a wall.

by two straight lines, or we assume that there is a thin film of fluid (of thickness δ) which adheres to the wall and that the total temperature drop in the system occurs across this film. With this assumption, we will have a linear temperature gradient; and from the elementary principles of conduction we know that

$$q = -k\frac{dT}{dz}\bigg|_{z=0} = k\frac{T_w - T_f}{\delta} \qquad \text{3.4-1}$$

where q represents the heat flux and k is the thermal conductivity of the fluid. The difficulty with this approach is that the film thickness normally varies a great deal with the flow conditions, so that the thickness δ cannot be determined in any simple way. Despite this obstacle, it is common practice to replace the term k/δ with a film-heat-transfer coefficient h[1] and to write the equation in the form

$$q = \frac{k}{\delta}(T_w - T_f) = h(T_w - T_f) \qquad \text{3.4-2}$$

---

[1] Actually, we define the film-heat-transfer coefficient as the surface heat flux divided by some characteristic temperature difference $h = [-k(dT/dz)|_{z=0}]/(T_w - T_f)$, but for our simple picture this is equal to k/δ.

where

$$h = \frac{k}{\delta} \qquad\qquad 3.4\text{-}3$$

For very simple steady state problems, it is possible to evaluate h in an exact manner. If we consider a heated sphere having a constant surface temperature while immersed in an infinite body of motionless fluid having a

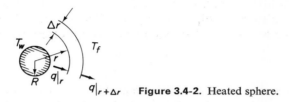

$q|_{r+\Delta r}$ **Figure 3.4-2.** Heated sphere.

constant thermal conductivity, we can determine the temperature profiles within the system analytically and thus calculate the heat flux from Eq. 3.4-1. Making a simple steady state energy balance on a spherical shell of fluid of thickness Δr gives

$$4\pi r^2 q|_r - 4\pi(r + \Delta r)^2 q|_{r+\Delta r} = 0$$

or after dividing by Δr and taking the limit as Δr approaches zero,

$$\frac{d}{dr}(r^2 q) = 0 = \frac{d}{dr}\left(r^2 k \frac{dT}{dr}\right) \qquad\qquad 3.4\text{-}4$$

Integration of this simple second-order differential equation gives the result

$$r^2 \frac{dT}{dr} = c_1 \qquad\qquad 3.4\text{-}5$$

Then, dividing by r^2 and integrating again, we obtain

$$T = -\frac{c_1}{r} + c_2 \qquad\qquad 3.4\text{-}6$$

where c_1 and c_2 are integration constants. These can be evaluated by using the boundary conditions

$$\text{at } r = R, \quad T = T_w; \qquad \text{at } r = \infty, \quad T = T_f \qquad\qquad 3.4\text{-}7$$

so that the equation for the temperature profile becomes

$$T = T_f + (T_w - T_f)\frac{R}{r} \qquad\qquad 3.4\text{-}8$$

The heat-transfer coefficient is defined by the equation

$$q\bigg|_{r=R} = -k\frac{dT}{dr}\bigg|_{r=R} = h(T_w - T_f) \qquad\qquad 3.4\text{-}9$$

or after differentiating Eq. 3.4-8 and eliminating the temperature gradient at the surface,

$$\frac{k(T_w - T_f)R}{r^2}\bigg|_{r=R} = h(T_w - T_f)$$

Hence we find that

$$h = \frac{k}{R} = \frac{2k}{D} \qquad\qquad 3.4\text{-}10$$

where D is the sphere diameter. The heat-transfer coefficient can be made dimensionless by multiplying it by D/k. The resulting dimensionless group is called the Nusselt number, Nu, and for our problem this becomes

$$\text{Nu} = \frac{hD}{k} = 2 \qquad\qquad 3.4\text{-}11$$

Thus even though the temperature profile is not concentrated in a thin film next to the sphere surface, for this problem we can determine exactly what is meant by the film-heat-transfer coefficient, or Nusselt number. Also, we find that the apparent film thickness for a linear gradient is equal to the sphere radius.

The problem becomes much more complex for flow systems. In a few cases it is possible to use boundary layer equations to develop approximate solutions for the heat-transfer coefficient; but for most problems of engineering interest, the analysis is too complex to undertake. Hence the normal practice is to conduct a series of steady state experiments on the flow system of interest or an equivalent piece of small-scale equipment, and then attempt to correlate the Nusselt number against the Reynolds number, the Prandtl number, and other dimensionless groups that are expected to be of importance. These correlations are used to design the process and to describe its steady state operating characteristics.

Our interest, however, is in the dynamic response of systems, and therefore we must find out if these concepts can be extended to the dynamic operation of a plant; or at least we must recognize which additional simplifying assumptions will have to be introduced in order to use the correlations

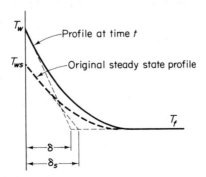

Figure 3.4-3. Step change in the wall temperature.

in estimating the dynamic behavior of the system. The difficulty involved in this extension becomes immediately clear if we examine the response to a step change in the surface temperature. Initially there will be an infinite temperature gradient in the fluid immediately adjacent to the wall. This gradient will eventually smooth out, however, and the system will approach some other steady state profile. Since the instantaneous film thickness δ is obtained by approximating the instantaneous profile by two straight lines, it is clear that this thickness will change with time. It follows that the heat-transfer coefficient h, which is inversely proportional to δ, will also be a time-dependent quantity.

An analytical expression for the time dependence of the heat-transfer coefficient can be derived for the sphere problem. This result is obtained by including an accumulation term in the energy balance equation developed earlier, Eq. 3.4-4. Following our standard procedure, we write

$$\rho C_p 4\pi r^2 \Delta r(T|_{t+\Delta t} - T|_t) = [4\pi r^2 q|_r - 4\pi(r + \Delta r)^2 q|_{r+\Delta r}]\Delta t$$

or after taking the appropriate limits and writing the heat flux in terms of the temperature gradient,

$$\rho C_p r^2 \frac{\partial T}{\partial t} = k \frac{\partial}{\partial r}\left(r^2 \frac{\partial T}{\partial r}\right) \qquad \text{3.4-12}$$

where ρ is the fluid density and C_p its heat capacity. We will assume that the system initially is at the steady state profile we developed earlier

$$T_s = T_f + (T_{ws} - T_f)\frac{R}{r} \qquad \text{at } t < 0 \qquad \text{3.4-8}$$

and then at time $t = 0$ we introduce a step change in the wall temperature

$$T_w = T_{ws} + a = T_{ws} + A(T_{ws} - T_f), \qquad \text{at } t = 0, \quad r = R$$
$$\text{3.4-13}$$

where a is the temperature increase and A is the change in dimensionless form. Also, if we consider a case where a large amount of fluid is present, we would not expect to observe any temperature variations at very large distances from the surface of the sphere, so that

$$\text{at } r = \infty, \qquad T = T_f \qquad \text{3.4-14}$$

The system equation, Eq. 3.4-12, and the three boundary conditions, Eqs. 3.4-8, 3.4-13, and 3.4-14, provide enough information to determine the temperature profiles in the fluid as a function of time. Next, the instantaneous heat flux at the surface can be calculated, and the heat-transfer coefficient evaluated, using the equation

$$q\Big|_{R,t} = -k\frac{dT}{dr}\Big|_{R,t} = h(T_{ws} - T_f)(1 + A) = h(T_{ws} + a - T_f)$$
$$\text{3.4-15}$$

This procedure leads to the result

$$\mathrm{Nu} = \frac{hD}{k} = 2\left[1 + \frac{A}{(1+A)\sqrt{\pi\tau}}\right]$$ 3.4-16

where τ is a dimensionless time variable

$$\tau = \frac{kt}{C_p \rho R^2}$$ 3.4-17

A sketch of this equation for the case where $A = 0.1$ is given in Figure 3.4-4.

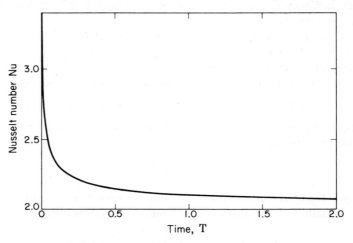

Figure 3.4-4. Time dependence of heat-transfer coefficient.

Thus we find that the Nusselt number, or heat-transfer coefficient, is infinite at time zero and decreases to its original value as time increases; and the apparent film thickness will increase from zero to its original value. The fact that the final film thickness must be independent of the temperature difference agrees with the previous steady state analysis.

It is also possible to show that if the wall temperature is varied sinusoidally, the heat-transfer coefficient will be a periodic function of time. Similarly, mass transfer coefficients and friction factors can be expected to be time-dependent quantities. This behavior occurs because of the definition of the transport coefficient

$$q\Big|_{R,t} = -k\frac{dT}{dr}\Big|_{R,t} = h[T_w(t) - T_f]$$ 3.4-18

since the actual heat flux at the surface is determined by both resistance and capacitance effects in the fluid, whereas the approximate expression for the flux, $h(T_w - T_f)$, implicitly assumes that only resistance to heat transfer is important and that the capacity effects are negligible in the film. In other words, we have assumed that a distributed parameter system can be approximated by a lumped parameter model. It is apparent that this assumption will be valid only if there is a much larger capacitance somewhere else in the system so that the time it takes the heat-transfer coefficient to decay to its steady state value is negligible in comparison with the response time of the remainder of the system. Normally we will have to rely on the steady state correlations for transport coefficients because it is seldom possible to solve even the steady state equations describing heat transfer in complex flow systems; as a result, we cannot expect to be able to develop rigorous dynamic models.

A similar situation is encountered when we consider other types of correlations, such as the plate efficiency of a distillation column or other separation unit. If the flow rates to a plate vary with time, it is difficult to predict how the froth will change or to predict the effect that this variation might have on the plate efficiency. Furthermore, we are not certain that the plate efficiency would have the same meaning if we carried out a chemical reaction in the column. Hence the available correlations might not be applicable even for steady state design purposes. This kind of uncertainty is present for most of the two-phase flow correlations. Although it will often be necessary to use the steady state correlations in dynamic models in order to estimate the dynamic response of the system, being aware of the limitations of this kind of an approximation is important.

Correlations for Process Units

Steady state correlations are frequently used to describe the behavior of process units. Extensive use is made of this approach in reactor design, particularly in the case of complex kinetic mechanisms, such as catalytic cracking, butane alkylation, and certain polymerization reactions. Since a process correlation has limitations with respect to dynamic operation and the determination of the optimum reactor design, we will discuss the basic assumptions involved in some detail.

The normal approach used to develop a steady state correlation is to conduct a set of experiments either on the existing unit or on a small-scale system believed similar to the plant equipment. All the system inputs and design parameters are varied, either one at a time or according to some pattern established by a statistical analysis, over the entire range of interest; and the corresponding values of the system outputs are measured. Then attempts are made to correlate these steady state experimental data—that is, to develop a relationship that can be used to predict the results of new experiments.

Of course, the quality of the correlation will largely depend on the form selected for the model. An approach commonly followed by statisticians is to construct an approximate response surface for the system by using a multivariable regression analysis to find the constants in a polynominal equation that minimize the square of the errors between the predicted and experimental points.

$$C = A_0 + A_1 x + A_2 y + A_3 T + A_4 x^2 + A_5 xy + \cdots + A_n xyT + \cdots$$

$$\text{3.4-19}$$

where C might be the conversion or some other output variable, x might represent a system input, y might be an equipment-size parameter, and T might be temperature or another dependent variable. With this type of model, data obtained from a statistically designed set of experiments, and certain assumptions concerning the random nature of the errors, it is possible to develop a quantitative description of the quality of the fit of the model to to the process. Even if the system variables are combined into dimensionless groups, which will minimize the number of constants that must be considered, a large number of constants must still be established by the regression analysis for a complex process. Therefore it is common practice to limit the model to the linear terms and a few quadratic terms. In other words, a multidimensional quadratic polynomial is used to approximate the unknown nonlinear surface describing the steady state characteristics of the system.

An alternate method is to assume that the model can be represented by the equation

$$C = A_0 (N_1)^{A_1} (N_2)^{A_2} (N_3)^{A_3} \ldots \qquad \text{3.4-20}$$

where $N_1, N_2, \ldots$ are the dimensionless groups obtained from a dimensional analysis of the general conservation equations or the application of the Buckingham π theorem and $A_0, A_1, \ldots$ are constants that are determined by a regression analysis. Obvisualy this method is essentially the same as the one just described except that a different form is chosen for the model. This kind of correlation is widely used in chemical engineering.

Neither of the preceding approaches gives any consideration to the fundamental conservation principles governing the system behavior. Therefore attempts sometimes are made to account, in a crude manner, for some of the phenomena that are expected to take place. For example, it is often assumed that a catalytic cracking reaction can be approximated by a first-order rate expression and that the simple batch reactor equation can be used to describe the unit

$$\ln \left(\frac{1}{1 - C} \right) = k_0 e^{-E/RT} t \qquad \text{3.4-21}$$

where C is the conversion of gas oil, t is the residence time in the reactor, and k_0 is assumed to depend on all the other design parameters. The functional dependence of k_0 is then established, using a regression analysis similar to one of those discussed earlier. In this way, a certain amount of structure can be built in to the correlation.

Of course, it is possible to extend this last method and develop approximate, steady state, models based only on a set of conservation equations. As a specific example of this procedure, we can consider the model of the catalytic cracking unit given by Eqs. 3.1-38 to 3.1-43. We know that a dynamic model must reduce to a steady state description of the system when all the accumulation terms are set equal to zero. Thus an approximate set of steady state equations is

Reactor

$$0 = V_0(y_0 - y_1) + \frac{P_1 H_1}{RT_1}(1 - y_1)A_1 e^{-E/RT_1} \qquad 3.4\text{-}22$$

$$0 = W(x_2 - x_1) + 0.1\frac{P_1 H_1}{RT_1}(1 - y_1)A_1 e^{-E/RT_1} \qquad 3.4\text{-}23$$

$$0 = C_{ps}W(T_2 - T_1) + C_{p1}V_0(T_0 - T_1) - \frac{(\Delta H_1)P_1 H_1}{RT_1}(1 - y_1)A_1 e^{-E/RT_1}$$

$$3.4\text{-}24$$

Regenerator

$$0 = V_a(y_a - y_2) - \frac{mP_2 H_2}{RT_2}y_2 A_2 e^{-E_2/RT_2} \qquad 3.4\text{-}25$$

$$0 = W(x_1 - x_2) - \frac{P_2 H_2}{RT_2}y_2 A_2 e^{-E_2/RT_2} \qquad 3.4\text{-}26$$

$$0 = C_{ps}W(T_1 - T_2) + C_{p2}V_a(T_a - T_2) - \frac{(\Delta H_2)P_2 H_2}{RT_2}y_2 A_2 e^{-E_2/RT_2} + Q$$

$$3.4\text{-}27$$

The preceding is a set of coupled, nonlinear, algebraic equations, which we must find some way of solving for the system outputs. One simple approach is to solve for y_1 in terms of T_1 and y_2 in terms of T_2, using the first material balance expressions for the reactor and regenerator

$$y_1 = \frac{V_0 y_0 + (P_1 H_1/RT_1)A_1 e^{-E_1/RT_1}}{V_0 + (P_1 H_1/RT_1)A_1 e^{-E_1/RT_1}} \qquad 3.4\text{-}28$$

$$y_2 = \frac{V_a y_a}{V_a + (mP_2 H_2/RT_2)A_2 e^{-E_2/RT_2}} \qquad 3.4\text{-}29$$

When the coke equations for the two units are added, the result is

$$0.1\frac{P_1 H_1}{RT_1}(1 - y_1)A_1 e^{-E_1/RT_1} = \frac{P_2 H_2}{RT_2}y_2 A_2 e^{-E_2/RT_2} \qquad 3.4\text{-}30$$

which essentially says that the rate of coke formation equals the rate of coke burning at steady state. Hence the steady state values of x_1 and x_2 can only be determined within some arbitrary constant. This relationship can be used to simplify the reactor energy equation so that it becomes possible to solve for T_1 in terms of T_2.

$$T_1 = \left(\frac{1}{C_{ps}W + C_{p1}V_0}\right)\left\{C_{ps}WT_2 + C_{p1}V_0T_0 - \frac{(\Delta H_2)P_2H_2}{0.1RT_2}\right.$$

$$\left.\times \left[\frac{V_a y_a A_2 e^{-E_2/RT_2}}{V_a + (mP_2H_2/RT_2)A_2 e^{-E_2/RT_2}}\right]\right\} \qquad 3.4\text{-}31$$

If we now substitute this expression into the regenerator energy equation, we obtain

$$\left(C_{ps}W + C_{p2}V_a - \frac{C_{ps}^2 W^2}{C_{ps}W + C_{p1}V_0}\right)T_2 - \left(C_{p2}V_aT_a + \frac{C_{ps}WC_{p1}V_0T_0}{C_{ps}W + C_{p1}V_0}\right) - Q$$

$$= -\left\{\frac{(\Delta H_2)P_2H_2}{RT_2}\left[\frac{V_a y_a A_2 e^{-E_2/RT_2}}{V_a + (mP_2H_2/RT_2)A_2 e^{-E_2/RT_2}}\right] + \frac{C_{ps}W(\Delta H_1)P_2H_2}{0.1RT_2}\right.$$

$$\left.\times \left[\frac{V_a y_a A_2 e^{-E_2/RT_2}}{V_a + (mP_2H_2/RT_2)A_2 e^{-E_2/RT_2}}\right]\right\} \qquad 3.4\text{-}32$$

This result is a transcendental equation, but it contains only one unknown variable, T_2. A solution can be obtained simply by plotting the right- and left-hand sides of the equation against T_2 and finding the point where the two curves intersect. Because of the presence of the exponential terms in the right-hand side of the equation, however, it is sometimes possible to predict multiple steady state solutions (a nonunique design problem). Similar behavior can be observed with the simple stirred-tank reactor model. This problem is discussed in additional detail in Section 7.1. Once the regenerator temperature has been calculated, the remaining variables can be determined, using Eqs. 3.4-28 through 3.4-31.

This final type of steady state model is much more complex than those described earlier, both because of the nonlinear terms appearing in the equations and because it is actually a multiple correlation—that is, it can be used to predict six (really five) output variables. However, it contains many less unknown constants. In fact, if kinetic parameters are known from laboratory studies, it should be possible to predict the results of plant experiments without any data-fitting procedure (provided, of course, that the assumptions used to derive the model are correct).

It would be possible to use a polynomial or other form of model to construct this type of a multiple correlation. Alternatively, it would be possible to expand Eqs. 3.4-22 through 3.4-27 in a Taylor series around some operating point of interest and in this way develop an approximate multidimensional set of polynomial surfaces for the five output variables. In this last case, all the constants in the polynomial equations will be known; whereas in the former case, they must all be determined by using a regression analysis. The theoretical model should also be applicable to other catalytic cracking units and can be used to predict the dynamic response of the system, whereas the earlier types of correlations are valid only for interpolating steady state results.

SECTION 3.5 SUMMARY

Theoretical models describing the dynamic operation of process units are derived via the fundamental principles of the conservation of mass, energy, and momentum. The best engineering judgment available is used to decide which dynamic effects are expected to be the most significant, and then simplifying assumptions are introduced to remove the extraneous terms from the conservation equations. Of course, these assumptions might not be correct; therefore it is necessary to verify the models experimentally.

One immediate test of the validity of the model is based on its steady state behavior. After the accumulation terms in the conservation equations have been set equal to zero, and the resulting set of equations solved for the system outputs in terms of the system inputs and the design parameters, it is possible to compare the theoretical predictions either directly with experimental data or with the best set of process correlations available. If an agreement is not obtained, it is necessary to reexamine the assumptions and to change the parameters in the model and/or the form of the equations until the results correspond. Thus steady state data obtained in laboratory studies, pilot plant investigations, and plant test runs can be used for an initial evaluation of a dynamic model. Additional confidence can be placed in the results if different kinds of equipment have been used for the various scales of experiments. For example, a batch reactor has been used for the laboratory study and a series of stirred tank reactors was used in the pilot plant investigation, so that different theoretical forms of the conservation equations, but the same kinetic scheme, describe all the data equally well.

Once the model has been shown to predict steady state operating data, possibly after several modifications, it is still necessary to show that it also properly describes the dynamic behavior. From the examples discussed previously, we know that most process units are described by coupled sets of ordinary or partial differential equations, and that in many instances these sets of equations are nonlinear and have variable coefficients. Consequently, it will usually be necessary to resort to numerical techniques in order to solve the equations for the dynamic response.

Since obtaining numerical solutions of the system equations is often expensive, it is common practice to develop an approximate analytical solution first. This approach is particularly useful when we are attempting to design a set of meaningful dynamic experiments that can be used to verify the model. The system equations are linearized by expanding all the terms in a Taylor series around a steady state operating point of interest and neglecting all second- and higher-order terms. All spatial derivatives are replaced by backward finite-difference approximations. This procedure results in a set of linear differential equations having constant coefficients, which can always be solved analytically. In fact, a matrix diagonalization procedure can be used to uncouple the equations so that the dynamic response of a unit can be approximately expressed in terms of the response of simple first- and second-order systems. Hence an understanding of the dynamic characteristics of very simple systems will make it possible, in many cases, to estimate the dynamic characteristics of complex process units. The analysis of these simple systems will be discussed in the next chapter.

QUESTIONS FOR DISCUSSION

When a chemist undertakes a laboratory study of a new reaction, he often takes data in a batch reactor at conditions corresponding to complete conversion or thermodynamic equilibrium. If the process economics appear promising, a pilot plant might be constructed to study the reaction in a continuous system. Normally data are gathered at several steady state operating conditions in an attempt to ascertain the most profitable operating region.

Are any of these basic laboratory data useful if we are interested in establishing the dynamic characteristics of a process? If you were in charge of the whole project, what kind of experiments would you recommend? How would you try to sell top management on your approach? How would you expect the cost of your experimental program to compare to the conventional approach?

EXERCISES

1. (A) Show that the equilibrium relationship $y = mx + b$ reduces to the equation $Y = mX$ if we let $Y = y - b/(1 - m)$ and $X = x - b/(1 - m)$.

2. (A*) A dynamic model for a plate absorber was developed in Example 3.1-2 for the case of unit-plate efficiency. In order to make the model more realistic, we could assume that the plate efficiency

$$E = \frac{y_n - y_{n-1}}{y_n^* - y_{n-1}}$$

is constant throughout the column. In this expression y_n^* is the vapor composition in equilibrium with liquid mixture of composition x_n, so that the equilibrium relationship becomes

$$y_n^* = mx_n + b$$

Develop a dynamic model of a plate absorption column, using these assumptions.

3. (A*) Liquid is pumped into a cylindrical storage tank at a rate $q_i(t)$, ft³/min. The exit pipe from the bottom of the tank is parallel to the ground and is L ft long. Assuming laminar flow through the exit pipe, and no dynamic effects in this part of the system, derive a dynamic model that describes the height of the liquid in the tank. Carefully define your terms and list any additional assumptions you make.

4. (A*) Suppose that we install a valve in the exit line of the system described in the previous problem, so that the volumetric flow rate leaving the tank is proportional to the square root of the liquid level in the tank. Derive a dynamic model for this case. Also, evaluate the steady state solution of the model and then linearize the dynamic equation around this steady state solution.

5. (A) Assuming that all the resistance to heat transfer is in a thin-fluid film surrounding the bulb and that all the capacitance is in the mercury, derive an expression for the dynamic response of a thermometer. Also, derive a dynamic model for the liquid level in a tank, assuming that the effluent flow rate is proportional to the level. For each problem, identify the stored quantity, the driving force, the flow, the resistance, and the capacitance.

6. (A‡) It is apparent that the treatment of the reaction kinetics in the dynamic model of a catalytic cracking unit described in Example 3.1-4 was greatly oversimplified. Postulate an alternate kinetic mechanism that includes parallel and/or consecutive reactions and then incorporate this kinetic scheme into a dynamic model.

7. (A) Bilous and Amundson[1] considered the dynamics of an isothermal, continuous-stirred-tank reactor followed by a simple (single-stage) separating unit— for example, an extractor, crystallizer, or settler. The reaction was reversible, $A \rightleftharpoons B$, and the effluent stream from the separator, which was rich in unreacted material, was recycled. It was assumed that the reaction rate was first order, the equilibrium relationship for the separator was linear, and the rate of mass transfer between phases in the separator could be written in terms of a mass transfer coefficient and a linear driving force. Making certain that you define all your terms carefully, show that the dynamic model for the plant can be put into the form

Reactor

$$V\frac{dA}{dt} = qA_f + QA_2 - (q + Q)A - V(k_1A - k_2B)$$

$$V\frac{dB}{dt} = QB_2 - (q + Q)B + V(k_1A - k_2B)$$

Lean Phase in Separator

$$H\frac{dA_2}{dt} = (q + Q)A - QA_2 - (k_{gA}a_v)(A_2 - m_1A_1)$$

$$H\frac{dB_2}{dt} = (q + Q)B - QB_2 - (k_{gB}a_v)(B_2 - m_2B_1)$$

Rich Phase in Separator

$$h\frac{dA_1}{dt} = (k_{gA}a_v)(A_2 - m_1A_1) - qA_1$$

[1] O. Bilous and N. R. Amundson, *AIChE Journal*, **1**, 513 (1955).

$$h\frac{dB_1}{dt} = (k_{gB}a_v)(B_2 - m_2B_1) - qB_1$$

8. (B*) In Example 2.1-1 we determined the optimum design of a second-order reaction in a single, isothermal, continuous-stirred-tank reactor. Using the values given in that example—$A_f = 1.0$ lb mole/ft^3, $k = 1.2$ ft^3/(lb mole)(hr), and $q = 100$ ft^3/hr—find the composition, production rate, and total cost of two reactors in series, each having a volume of 500 ft^3. Then write the dynamic equations for the two reactors in series. Next, linearize the equations around the steady state operating point. Finally, find the characteristic roots of the equations and determine the transformation matrix that can be used to diagonalize the equations.

9. (B*) If we consider a plate, gas absorption unit containing only two trays for a case where the equilibrium relationship is linear, we can write a dynamic model for the unit as[2]

$$H\frac{dx_1}{dt} = -(L_1 + Vm)x_1 + L_2x_2 + Vmx_0$$

$$H\frac{dx_2}{dt} = Vmx_1 - (L_2 + Vm)x_2$$

where x_0 is the liquid composition in equilibrium with the vapor stream entering the bottom of the column. For the simplest case, we might hope that the liquid flow rate was constant, so that $L_1 = L_2 = L$. However, there are situations where the liquid rate is observed to be time dependent, and we often model the tray hydraulics with the simple equations

$$\tau\frac{dL_1}{dt} = -L_1 + L_2$$

$$\tau\frac{dL_2}{dt} = -L_2 + L_3$$

where L_3 is the liquid rate entering the top plate.

First consider the case of constant flow rates. Find the steady state solutions of the material balance expressions, linearize the equations around this steady state operating point, determine the characteristic roots of the linearized equations, and then find the transformation matrix that can be used to diagonalize the equations. Next, try to reproduce this procedure for the case of variable liquid flow rates. At what point does the procedure fail? Why?

10. (B*) A simple model for a pair of exothermic parallel reactions, $A \longrightarrow B$ and $A \longrightarrow C$, in a continuous-stirred-tank reactor containing a cooling coil might be written as

$$V\frac{dA}{dt} = q(A_f - A) - k_1VA^2 - k_2VA$$

$$V\frac{dB}{dt} = -qB + k_1VA^2$$

[2] A derivation of this model is presented in D. R. Coughanowr and L. B. Koppel, *Process Systems Analysis and Control*, p. 338, McGraw-Hill, N.Y., 1965, and a modification of the model was discussed by F. P. Lees, *Ind. Eng. Chem. Fundamentals*, **9**, 512 (1970).

$$VC_p\rho\frac{dT}{dt} = qC_p\rho(T_f - T) + (-\Delta H_1)k_1 VA^2 + (-\Delta H_2)k_2 VA$$

$$- \frac{UA_cKq_c}{1 + Kq_c}(T - T_c)$$

See if you can list the assumptions implied by these equations.

Describe a procedure for calculating the steady state compositions and temperature in the reactor, linearize these equations around the steady state operating point, calculate the characteristic roots of the linearized equations, and then determine a transformation matrix that can be used to put the equations into a diagonal form.

11. (B) In some cases it is advantageous to carry out chemical reactions in separating units. For example, Jenson and Jeffreys[3] discuss the design of a staged column where animal fat is being hydrolyzed and extracted by a countercurrent stream of water. They assume that the hydrolyzable glycerine in the fat undergoes a first-order reaction in the fat phase and that there is no reaction in the sweetwater phase. Letting x = weight fraction of glycerine in the fat phase (raffinate), y = weight fraction of glycerine in the sweetwater phase (extract), and z = weight fraction of unreacted fat in the raffinate, derive a dynamic model for the column. Carefully define your terms and list your assumptions.

12. (B‡) Derive a dynamic model that can be used to describe an adsorption or ion-exchange column. If you desire, consider a particular case where benzene is being removed from an air stream by passing it through a bed packed with silica gel. Carefully define your terms and list your assumptions.

What is the meaning of steady state operation in this kind of unit? How does the model change as you change one or more of the assumptions?

13. (B‡) Show that the dynamic model for a double-pipe, countercurrent heat exchanger can have the same form as the model for a packed absorber given in Example 3.2-3. Discuss the assumptions inherent in both the heat exchanger and absorber models which might lead to significant differences in the kinds of equations used to describe each system.

14. (B) Derive a dynamic model for a packed-bed extraction unit that includes an axial dispersion term for the dispersed phase. Give an appropriate set of boundary conditions for the model.

15. (B*) Linearize the dynamic model of a catalytic cracking unit described in Example 3.1-4 and write the linearized equations in matrix notation. Justify your choice of control variables and disturbances.

16. (B) It would be possible to use a double-pipe heat exchanger as a reactor. For single, irreversible, exothermic reactions, this unit would have the great advantage that the heat generated by the reaction could be used to raise the temperature of the reacting material, which would eliminate the need for any heating fluid. Derive a dynamic model for the system.

17. (B) Determine the steady state solution of Bilous and Amundson's reactor–separator system, see Exercise 7, p. 105, linearize the equations, and write them in matrix form.

18. (B) Replace the spatial derivatives in the model of a packed absorption column described in Example 3.2-3 with a backward finite-difference approximation,

[3] See p. 328 of the reference given in footnote 5 on p. 86.

linearize the equations, and write them in matrix form. Compare your results to the model of a plate column described in Examples 3.1-2 and 3.3-2.

19. (B*) The dynamic model of the double–tube-pass, single–shell-pass heat exchanger (Figure 3.5-1) would involve a complicated set of partial differential equations. Hence Williams and Lauher[4] assumed that there was perfect mixing in each segment of the shell separated by the baffles, where the volume of the compartments next to the inlet and outlet pipes were one-half of the remaining three, and, similarly, that there was perfect mixing in the five tube sections defined by the baffles. If, in addition, it is assumed that the heat-transfer coefficient is constant and that both the capacitance and the resistance of the metal walls to heat transfer are neglible, derive a dynamic model for the exchanger. Linearize your model and write it in matrix notation.

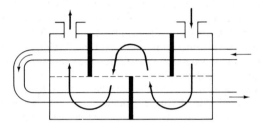

Figure 3.5-1. Multipass exchanger.

20. (B*) The analysis of laminar flow through a pipe is quite simple and is used to introduce the concept of friction factor. However, if the pressure gradient across the pipe is forced to oscillate according to the expression $\Delta P = \Delta P_0 \cos \omega t$, where ΔP is the negative of the pressure gradient and ω is the frequency of the oscillations, then at low frequencies and long times it can be shown that the average velocity and the wall shear stress in the pipe are

$$v_{\mathrm{av}} = \frac{\Delta P R^2}{8\eta L}\left[1 + A\left(\cos \omega t + \frac{\omega \rho R^2}{6} \sin \omega t\right)\right]$$

$$\tau_w = \frac{\Delta P R}{2L}\left[1 + A\left(\cos \omega t + \frac{\omega \rho R^2}{6} \sin \omega t\right)\right]$$

Use these results to find an expression for the friction factor.

21. (B) If you assume plug flow both in tube passes and in the shell of the multipass exchanger pictured in Figure 3.5-1, what kind of a dynamic model do you obtain? How does the model change if you consider accumulation effects in the metal tube walls and the exchanger shell?

22. (B*) A dynamic technique is often used to measure the axial dispersion coefficient in a packed bed. Derive dynamic models both for the case of an inert tracer material and for the case where there can be adsorption of the tracer on the packing. Carefully define your terms and list your assumptions.

[4] T. J. Williams and V. A. Lauher, *Automatic Control of Chemical and Petroleum Processes*, Gulf Publishing Co., Houston, Texas, 1963.

23. (B) Bruley and Prados[5] experimentally studied the dynamic response of a wetted-wall column. They introduced sinusoidal variations in the air temperature entering the column and then measured the outlet air temperature. By assuming adiabatic humidification conditions and that the heat capacity and transfer rates would be much higher in the liquid phase than in the gas phase, they could assume that the liquid phase temperature and the interfacial temperature were constant. Using these assumptions, derive a dynamic model for the system. Consider both the laminar and turbulent flow cases.

24. (B‡) Thermal pollution problems have recently attracted great interest from engineers. Suppose that we attempt to determine the effects of diurnal temperature changes in the air above a river on the temperature profile in the river, so that we can calculate the time-dependent behavior of the equilibrium oxygen concentration. Derive a dynamic model for the system. Carefully define your terms and list your assumptions.

25. (B‡) In the contact process for manufacturing sulfuric acid, sulfur is burned to SO_2, the SO_2 is reacted with oxygen over a catalyst to produce SO_3, and the SO_3 gas is absorbed in a packed tower by concentrated sulfuric acid. The particular concentration of acid fed to the absorber is a critical design variable, for the heat effects can be so large that there might be a significant amount of mist formation. Derive a dynamic model describing an SO_3–absorber. Carefully define your terms and list your assumptions. How would you decide whether you could neglect the accumulation of energy in the packing?

26. (B‡) Heat regenerators are encountered in a number of large-scale industrial processes, such as open-hearth furnaces, liquefaction of a vapor and the separation of its components in the liquid state. A hot gas, possibly leaving a reactor, is passed through a checkerwork of bricks, or even a bed of stones, so that the heat is removed from the gas and stored in the solid. Then a cold gas is passed through this bed, normally in the opposite direction, so that the gas is preheated before it enters the reactor. Derive a dynamic model for the system. Carefully define your terms and list your assumptions.

27. (C‡) The catalytic dehydrogenation of butane is important both in the manufacture of butadiene and as one step in the synthetic manufacture of gasoline. Dodd and Watson[6] found that the principal reaction was $C_4H_{10} \longrightarrow C_4H_8 + H_2$ and that an important secondary reaction was the production of butadiene, $C_4H_8 \longrightarrow C_4H_6 + H_2$. In the high-temperature range of interest, an appreciable amount of the reactions occurred by pyrolysis in the homogeneous phase, as well as by the catalytic path. Also, it was observed that a large number of other reactions were taking place, including.

(a) The dealkylation or cracking of butane to form methane, ethane, ethylene, and propylene.

(b) The dealkylation of butenes to form methane, ethane, propane, ethylene, propylene, and coke.

(c) The dimerization of butadiene to form 4–vinyl cyclohexene–1.

[5] D. F. Bruley and J. W. Prados, *AIChE Journal*, **10**, 612 (1964).
[6] R. H. Dodd and K. M. Watson, *Trans. Am. Inst. Chem Engrs.*, **42**, 263 (1946).

(d) The decomposition of butadiene to form hydrogen, methane, ethylene, acetylene, and coke.

Obviously a complete determination of the reaction kinetics, both homogeneous and catalytic, of this process would be extremely difficult. Hence, for gasoline manufacture, Dodd and Watson suggested using the simplified reaction model

$$C_4H_{10} \longrightarrow C_4H_8 + H_2$$
$$C_4H_{10} \longrightarrow 0.1C_4H_8 + 0.1H_2 + 1.8 \quad \text{(dealkylation products)}$$
$$C_4H_8 \longrightarrow 0.1H_2 + 1.8 \quad \text{(dealkylation products)}$$

with the rate equations

$$r_A = \frac{C(P_A - P_R P_s/K)}{(1 + K_A P_A + K_{Rs} P_{Rs})^2}$$

$$r_B = kP_A \qquad r_C = kP_R$$

where r_A = rate of reaction (moles) (mass catalyst)$^{-1}$(hr)$^{-1}$, C = overall rate constant (moles)(mass catalyst)$^{-1}$(hr)$^{-1}$(atm)$^{-1}$, K = overall gas phase equilibrium constant (atm), K_A = effective adsorption equilibrium constant of butane (atm)$^{-1}$, K_{Rs} = effective average adsorption constant of hydrogen and butenes, P_{Rs} = average partial pressure of hydrogen and butene = $(P_R + P_s)/2$, r_B and r_C = rate of cracking reactions (moles) (mass catalyst)$^{-1}$(hr)$^{-1}$, P_A and P_R = partial pressures of butane and butene, P_S = partial pressure of hydrogen, and k = cracking reaction rate constant.

Using this simplified kinetic model, develop a dynamic model for a nonisothermal, fixed-bed, catalytic reactor. Include the possibility of feeding steam to the bed to act as a diluent. Carefully define your terms and list your assumptions.

28. (C‡) Stults, Moulton, and McCarthy[7] studied some of the condensation reactions that take place when formaldehyde (F) is added to sodium paraphenolsulfonate (M) in an alkaline-aqueous solution. They found that they could represent the reactions by the equations

$$F + M \longrightarrow MA \qquad\qquad k_1 = 0.16 \text{ (liter)(g mole)}^{-1}\text{(min)}^{-1}$$
$$F + MA \longrightarrow MDA \qquad\quad k_2 = 0.50$$
$$MA + MDA \longrightarrow DDA \qquad k_3 = 0.15$$
$$M + MDA \longrightarrow DA \qquad\quad k_4 = 0.14$$
$$MA + MA \longrightarrow DA \qquad\quad k_5 = 0.03$$
$$MA + M \longrightarrow D \qquad\qquad k_6 = 0.058$$
$$F + D \longrightarrow DA \qquad\qquad k_7 = 0.50$$
$$F + DA \longrightarrow DDA \qquad\quad k_8 = 0.50$$

where M, MA, and MDA represent monomers; D, DA, and DDA represent dimers; and the process continues to form trimers. The rate constants were evaluated using the assumption that the molecularity of each reaction was identical to its stoichiometry.

Derive a dynamic model for these reactions taking place in a single, isothermal, continuous-stirred-tank reactor. Carefully define your terms and list your assumptions.

[7] F. C. Stults, R. W. Moulton, and J. L. McCarthy, *Chem. Eng. Prog. Symposium Ser.*, **48**, No. 4, 38 (1952).

29. (C‡) The reaction rate expressions for heterogeneous, catalytic reactions are based on the assumption that the rates of the individual steps taking place in the overall reaction are equal; that is, an equality exists between the rate of diffusion of reactants through a stagnant film of gas surrounding the catalyst particle; the rate of adsorption of reactants on the surface of the catalyst; the rate of surface reaction; the rate of desorption of the products from the surface; and the rate of diffusion of products across the stagnant film back into the bulk of the gas stream flowing through the catalyst bed. A further modification must be made if the catalyst particle is porous and if the rate of diffusion into the pores is important.

Normally the equations we obtain based on this assumption are so complicated that they are unmanageable. Hence we often attempt to simplify the approach by assuming that one of the steps is rate controlling and that the others are at equilibrium. The final form of the rate expression, as well as the values of the unknown constants, is usually obtained by comparing the model predictions with experimental data from isothermal reactors operated at steady state conditions.

Discuss in detail the applicability of these rate expressions in dynamic models.

30. (C) The dynamic models we use to describe systems are comprised of sets of differential equations. The derivative terms arise because it is necessary to employ a limiting process in order to define precisely the terms in the conservation equation. For example, if we consider the material balance for a second-order, irreversible reaction in an isothermal, continuous-stirred-tank reactor, we write

$$(VA)|_{t+\Delta t} - (VA)|_t = qA_f\Delta t - qA_{av}\,\Delta t - kVA_{av}^2\,\Delta t$$

where the terms on the left-hand side of the equation represent the accumulation of unreacted material and the terms on the right-hand side represent the input by convective flow, the output by convective flow, and the disappearance by chemical reaction, respectively. We expect the composition in the reactor to change with time, and therefore we must use some kind of an average concentration during the time interval Δt for the output terms. We do not know how to define this average value, but we can avoid any difficulty by dividing both sides of the equation by Δt and then taking the limit as Δt approaches zero. For a system with constant volume, we obtain

$$\lim_{\Delta t \to 0} V\left(\frac{A|_{t+\Delta t} - A|_t}{\Delta t}\right) = \lim_{\Delta t \to 0} [qA_f - qA_{av} - kVA_{av}^2]$$

The expression on the left-hand side of the equation becomes the derivative, and the average values on the right-hand side of the equation approach the instantaneous value of A at time t. Thus the final result is

$$V\frac{dA}{dt} = q(A_f - A) - kVA^2$$

Now suppose that we attempt to approximate this differential equation by a set of finite-difference equations. We could use any of the expressions below:

$$V\left(\frac{A|_{t+\Delta t} - A|_t}{\Delta t}\right) = qA_f - q(A|_t) - kV(A|_t)^2 \qquad \text{Forward Difference}$$

$$V\left(\frac{A|_{t+1/2\Delta t} - A|_{t-1/2\Delta t}}{\Delta t}\right) = qA_f - q(A|_t) - kV(A|_t)^2 \qquad \text{Central Difference}$$

$$V\left(\frac{A|_t - A|_{t-\Delta t}}{\Delta t}\right) = qA_f - q(A|_t) - kV(A|_t)^2 \qquad \text{Backward Difference}$$

All these equations are slightly different, and none agrees with our original material balance equation written for a small increment of time. However, if the time interval Δt is chosen sufficiently small, all the equations will give essentially the same results.

Apply the three definitions of finite differences to the *spatial derivative* in Example 3.3-3, and show that you obtain three different sets of linear, ordinary differential equations. Write each set of equations in matrix form. Also, show that the results for the backward-difference approximation are equivalent to the model for a series of continuous-stirred-tank reactors.

Response of
Lumped Parameter
Systems

4

The procedure for linearizing dynamic models, approximating partial differential equations by ordinary differential equations, putting the equations in matrix form, and then diagonalizing the matrix indicates that it is always possible to estimate the dynamic response of any complicated system as a linear combination of the responses of simple first- and second-order systems. In this chapter we will develop quantitative methods for determining the time-varying outputs of these simple systems and then show how the solutions can be combined to provide descriptions of complex units. Since the theory of Laplace transforms often simplifies the manipulations required in the analysis, it will be used extensively throughout the remainder of the text. The fundamental definitions and useful theorems of transform theory are introduced in the next section, and a fairly complete table of transforms is given in Appendix B.

SECTION 4.1   SYSTEM INPUTS

Before the response of any kind of plant can be determined, the inputs to the plant must be specfied. This is a problem we talked around in our previous disucssions, for little is known about the types of disturbances that enter industrial processes. Consequently, it is common practice to study the response of the system to step, impulse, and sinusoidal inputs, and sometimes to ramp and stochastic inputs. In addition to mathematical simplicity, there are other reasons for considering the manner in which the plant reacts to these input signals. For example, we know that an arbitrary forcing function can be approximated by a series of impulses, a series of step functions, or as a Fourier series of sinusoidal and cosinusoidal waves, so that sometimes it will be possible to add together the outputs caused by a number of simple input functions to find the response of a complicated disturbance signal. Additional reasons are given below.

Step Inputs

Making a sudden change in a valve setting is the easiest way of purposely introducing a disturbance into a system. Although a finite time is required to accomplish this change, normally it will be negligible in comparison to the time it takes the process to reach a new steady state operating condition. Therefore we visualize a step change as an instantaneous discontinuity in

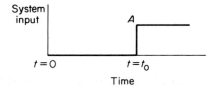

Time   **Figure 4.1-1.** Step change.

one of the input variables. Since the variables we prefer to work with in dynamics problems are dimensionless deviations from some steady state operating condition of interest, the system input will change from an initial value of zero to some value A when the step change is introduced. We often choose the initial time, $t = 0$, for the problem as the time when the step change is made, although in some problems we are interested in the response to step inputs occurring at $t = t_0$. The mathematical definition of a unit step function is

$$u(t - t_0) = \begin{cases} 1, & \text{for } t > t_0 \\ 0, & \text{for } t < t_0 \end{cases} \qquad 4.1\text{-}1$$

and a step function having a magnitude A is written as $Au(t - t_0)$.

In addition to the step change being the simplest forcing function, it also represents one of the most severe type of disturbance that we could impose on the system. It is necessary to estimate the final steady state output of a system before the step is introduced in order to ensure that it will remain in a safe operating region; for example, that the final reactor temperature will not exceed the melting point of the wall material or the sintering temperature of the catalyst. A practical application of the step response is the prediction of the plant behavior when a pump fails.

Impulse Inputs

Most disturbances entering a plant fluctuate around some average value. Hence the plant outputs normally remain in the vicinity of the original steady state design conditions. Since a step input will always force the system to some new operating level, we also must consider inputs that return to zero after some time or that have a zero mean value. A simple example of the first type would be to make a sudden change in the valve setting and, a short time later, reset it to its original value—that is, introduce a rectangular pulse. Examples of forcing functions having a zero mean are sinusoidal and stochastic disturbances, which will be considered later.

Of course, the response to a rectangular pulse will depend on ϵ, the duration of the pulse, as well as its amplitude, A/ϵ. In order to avoid specifying

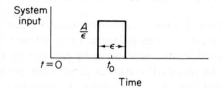

Figure 4.1-2. Rectangular pulse.

this quantity, ϵ, we often consider the response to a very quick and sharp disturbance—that is, an impulse. Obviously, as we let ϵ approach zero, the width of the pulse will approach zero and the height will become infinitely large. However, the area under the curve will always remain finite and, in fact, is always equal to A. Other functions having this kind of behavior are

$$\delta(t - t_0) = \frac{A}{\epsilon\pi} \exp\left[-\frac{(t - t_0)^2}{\epsilon^2}\right], \qquad \text{let } \epsilon \longrightarrow 0 \qquad 4.1\text{-}2$$

$$\delta(t - t_0) = \frac{\epsilon A}{\pi} \frac{\sin^2\left[(t - t_0)/\epsilon\right]}{(t - t_0)^2}, \qquad \text{let } \epsilon \longrightarrow 0 \qquad 4.1\text{-}3$$

Even though some mathematical difficulities are assoicated with this procedure, we will define an impulse function as one that has a zero width, an infinite height, and where the area under the curve is equal to the magnitude of the impulse, A. Thus

$$\delta(t - t_0) = 0, \qquad \text{if } t \neq t_0$$

and

$$\int_{t_1}^{t_2} \delta(t - t_0)\, dt = \begin{cases} 0, & \text{if } t_1 > t_0 \quad \text{or} \quad t_2 < t_0 \\ A, & \text{if } t_1 < t_0 < t_2 \end{cases} \qquad 4.1\text{-}4$$

Of course, it will not be possible to introduce experimentally an actual impulse into any real system, for it is a discontinuous function. However, if ϵ is made very small in comparison with the response time of the plant, an impulse function will be a good approximation of the actual disturbance.

Sinusoidal Inputs

For the disturbances described above, the system input is changed only at one instant of time. A simple forcing function that varies continually and that has a zero mean value is a sinusoid. This function somewhat resembles

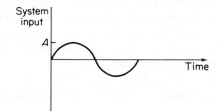

Figure 4.1-3. Sinusoidal input.

the variations in ambient temperature during a day's operation. It can be synthesized by using special equipment to manipulate a valve setting. Of course, a cosine wave can be obtained from a sine wave by translating the time axis; therefore it is important to specify the correct phase relationship at the initial time, $t = 0$, of the problem. A trigonometric identity that is useful for this purpose is

$$c_1 \cos \omega t + c_2 \sin \omega t = A \sin (\omega t + \phi) \qquad 4.1\text{-}5$$

where

$$A = \sqrt{c_1^2 + c_2^2} \qquad \tan \phi = \frac{c_1}{c_2} \qquad 4.1\text{-}6$$

Stochastic Inputs

Stochastic inputs are those that change in a random manner. Thus, even if we know the complete past history of the input signal, it is never possible to make a precise prediction of the next value. However, we can generally make certain statements concerning the statistics of this kind of a forcing function. For example, it might have a constant average value and a certain average spread (variance) around this value. Since we cannot provide a unique description of the input, we cannot expect to determine the exact value of the system output at some future time. Instead we attempt to compute the expected value of the output or attempt to ascertain the statistical characteristics of this signal.

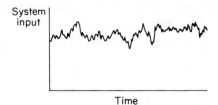

Figure 4.1-4. Stochastic input.

Although it would seem as if the disturbances entering industrial plants could be treated as stochastic signals, this approach is seldom followed. Therefore stochastic processes are not discussed in this book. Nevertheless, an extensive literature does exist,[1,2] and undoubtedly there will be many more applications of the theory in the relatively near future.

SECTION 4.2   LAPLACE TRANSFORMS

The great advantage of the Laplace transform is that it makes it possible to find the solutions of some ordinary differential equations through the manipulation of algebraic quantities. In practice, the method is used to transform an ordinary differential equation into an algebraic equation, then the algebraic equation is solved explicitly for the transform of the dependent variable, and, finally, the transform is inverted to give an expression for the system output. It is clear that this procedure is analogous to the use of logarithms to simplify the arithmetic operations of multiplication and division. The transform technique can also be used to reduce partial differential equations to ordinary differential equations, which greatly simplifies the task of finding solutions. Although a complete treatment of the theory of Laplace transforms requires an extensive knowledge of complex variables, we will present some of the important results of the theory without rigorous proofs.

The Laplace transform of an arbitrary function of time, $f(t)$, is denoted by the symbol $\mathscr{L}[f(t)]$ or $\bar{f}(s)$ and is defined by the equation

$$\mathscr{L}[f(t)] = \bar{f}(s) = \int_0^\infty e^{-st} f(t)\, dt \qquad\qquad 4.2\text{-}1$$

where s is called the Laplace parameter and is some unknown complex variable. Of course, this definition is valid only if the integral exists, which will occur for most functions of engineering interest. Thus we find that the transformation replaces some function of time by a different function of the Laplace parameter s.

[1] J. S. Bendat, *Principles and Applications of Random Noise Theory*, Wiley, N.Y., 1958.
[2] J. H. Lanning Jr. and R. H. Battin, *Random Processes in Automatic Control*, McGraw-Hill, N.Y., 1956.

This definition can be used to show that the Laplace transformation is a linear operation,

$$\mathcal{L}[c_1 f_1(t) + c_2 f_2(t)] = \int_0^\infty [c_1 f_1(t) + c_2 f_2(t)] e^{-st} dt$$

$$= c_1 \mathcal{L}[f_1(t)] + c_2 \mathcal{L}[f_2(t)] \qquad 4.2\text{-}2$$

Laplace Transforms of Simple Functions

Similarly, we can use the definition to find the Laplace transform of a great number of functions. We will be particularly interested in the transforms of the simple forcing functions described in the previous section, although we will include a few more-complicated examples. Since we normally assume that the system is initially operating at some steady state condition and then at time equal to zero one or more of the inputs changes, we will be able to set $f(t) = 0$ for all $t < 0$, providing that our variables are taken as deviations from an initial steady state.

The Laplace transform of a *step function* occurring at $t = 0$ is

$$\mathcal{L}[Au(t)] = \int_0^\infty A e^{-st} dt = -\frac{Ae^{-st}}{s}\Big|_0^\infty = \frac{A}{s} \qquad 4.2\text{-}3$$

Similarly, the Laplace transform of a sinusoid is

$$\mathcal{L}[\sin \omega t] = \int_0^\infty \sin \omega t \, e^{-st} dt = \left[\left(-\frac{e^{-st}}{s^2 + \omega^2} \right)(s \sin \omega t + \omega \cos \omega t) \right]\Big|_0^\infty$$

$$= \frac{\omega}{s^2 + \omega^2} \qquad 4.2\text{-}4$$

and the Laplace transform of a cosinusoid is

$$\mathcal{L}[\cos \omega t] = \int_0^\infty \cos \omega t \, e^{-st} dt = \left[\left(-\frac{e^{-st}}{s^2 + \omega^2} \right)(s \cos \omega t - \omega \sin \omega t) \right]\Big|_0^\infty$$

$$= \frac{s}{s^2 + \omega^2} \qquad 4.2\text{-}5$$

From the linear nature of the transform, Eq. 4.2-2, we can write

$$\mathcal{L}[c_1 \sin \omega t + c_2 \cos \omega t] = \frac{c_1 \omega + c_2 s}{s^2 + \omega^2} \qquad 4.2\text{-}6$$

The Laplace transform of a *ramp* function—that is, a function which changes linearly with time, $f(t) = At$—is

$$\mathcal{L}[At] = \int_0^\infty At e^{-st} dt = \left[\left(-\frac{Ae^{-st}}{s^2} \right)(st + 1) \right]\Big|_0^\infty = \frac{A}{s^2} \qquad 4.2\text{-}7$$

It is a simple matter to generalize this last result to functions having the form $f(t) = At^n$. Thus

$$\mathcal{L}[At^n] = A \int_0^\infty t^n e^{-st} dt$$

is determined by letting

$$t = \frac{\tau}{s} \qquad dt = \frac{d\tau}{s}$$

so that

$$\int_0^\infty t^n e^{-st} \, dt = \int_0^\infty \left(\frac{\tau}{s}\right)^n e^{-\tau} \frac{d\tau}{s} = \frac{1}{s^{n+1}} \int_0^\infty \tau^n e^{-\tau} \, d\tau$$

or

$$\mathscr{L}[At^n] = \frac{\Gamma(n+1)}{s^{n+1}} = \frac{n!}{s^{n+1}} \qquad \text{4.2-8}$$

where $\Gamma(n+1)$ is the gamma function of the argument $n+1$.

Another function of some interest is the *exponential* function, $f(t) = e^{+at}$. For this case

$$\mathscr{L}[e^{+at}] = \int_0^\infty e^{at} e^{-st} \, dt = -\frac{e^{-(s-a)t}}{s-a} \bigg|_0^\infty = \frac{1}{s-a} \qquad \text{4.2-9}$$

Clearly, we could follow this procedure and determine the Laplace transforms of a great number of elementary functions. We are spared this effort, however, for the transforms of most functions of interest are available in various mathematical tables. Some of these are included in Appendix B.

Translated Functions

The preceding discussion, and the transforms given in the appendix, is limited to functions that are equal to zero for $t < 0$. Since the initial time chosen for many problems is arbitrary, we would like to be able to extend our results to functions that are zero for $t < t_0$. In other words, we would like to be able to determine the transform of functions that have been translated along the time axis some distance t_0. From the definition we find that

$$\mathscr{L}[f(t - t_0)] = \int_0^\infty f(t - t_0) e^{-st} \, dt = e^{-st_0} \int_0^\infty f(t - t_0) e^{-s(t-t_0)} \, d(t - t_0)$$

Then letting $\tau = t - t_0$

$$\mathscr{L}[f(t - t_0)] = e^{-st_0} \int_{-t_0}^\infty f(\tau) e^{-s\tau} \, d\tau$$

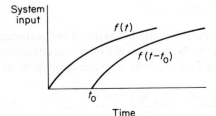

Figure 4.2-1. Translated function.

However, since $f(\tau) = 0$ for $\tau < 0$, we can replace the lower limit in the integral by zero and the dummy variable of integration by t. Hence the *shift theorem* becomes

$$\mathcal{L}[f(t - t_0)] = e^{-st_0}\mathcal{L}[f(t)] = e^{-st_0}\bar{f}(s) \qquad 4.2\text{-}10$$

so that we merely multiply the normal Laplace transform of the function by the factor e^{-st_0} in order to find the transform of the translated function.

A similar result is obtained if we consider the transform of an exponential multiplied by an arbitrary function $f(t)$

$$\mathcal{L}[e^{-at}f(t)] = \int_0^\infty f(t)e^{-at}e^{-st}\,dt = \int_0^\infty f(t)e^{-(s+a)t}\,dt$$

Now, if we introduced a new Laplace parameter, $p = s + a$, it is clear that the last integral is simply the Laplace transform with respect to p. Thus

$$\mathcal{L}[e^{-at}f(t)] = \bar{f}(p) = \bar{f}(s + a) \qquad 4.2\text{-}11$$

is an additional *shift theorem*, where we merely replace s in the normal transform by the quantity $(s + a)$.

If we apply the first shift theorem to the translated step function defined by Eq. 4.1-1, we obtain

$$\mathcal{L}[Au(t - t_0)] = \frac{Ae^{-st_0}}{s} \qquad 4.2\text{-}12$$

Similarly, a rectangular *pulse* can always be considered as the difference between two step functions, and therefore its transform is

$$\mathcal{L}\left[\frac{Au}{\epsilon}(t - t_1) - \frac{Au}{\epsilon}(t - t_2)\right] = \frac{A}{s\epsilon}e^{-st_1} - \frac{A}{s\epsilon}e^{-st_2}$$

$$= \frac{A}{s\epsilon}e^{-st_1}[1 - e^{-s(t_2-t_1)}] \qquad 4.2\text{-}13$$

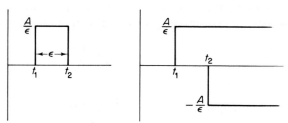

Figure 4.2-2. Square pulse.

For an *impulse* input, we are interested in the behavior of the rectangular pulse as ϵ approaches zero. Ignoring the mathematical difficulties encountered in an attempt to take this limit, we find that the Laplace transform of an impulse is

$$\mathcal{L}A[\delta(t - t_1)] = \lim_{\epsilon \to 0} \frac{Ae^{-st_1}[1 - e^{-\epsilon s}]}{s\epsilon} = \lim_{\epsilon \to 0} \frac{Ase^{-st_1}e^{-\epsilon s}}{s} = Ae^{-st_1} \qquad 4.2\text{-}14$$

where L'Hôpital's rule has been used to find the limit. Of course, if $t_1 = 0$, the Laplace transform of an impulse becomes

$$\mathscr{L}A[\delta(t)] = A \qquad\qquad 4.2\text{-}15$$

Periodic Functions

Although we have developed the Laplace transforms of some simple periodic functions, sines and cosines, in a straightforward manner, we also would like to obtain the transforms of general periodic functions. If we con-

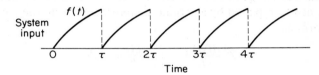

Figure 4.2-3. Periodic input.

sider some arbitrary function $f(t)$ having a period τ—that is, $f(t) = f(t + \tau)$—then by definition

$$\mathscr{L}[f(t)] = \int_0^\infty f(t)e^{-st}\,dt$$

$$= \int_0^\tau f(t)e^{-st}\,dt + \int_\tau^{2\tau} f(t)e^{-st}\,dt + \int_{2\tau}^{3\tau} f(t)e^{-st}\,dt + \cdots$$

Now, in the second integral we let $t = T + \tau$, in the third integral we let $t = T + 2\tau$, and, in general, we let $t = T + n\tau$ in the $(n + 1)$st integral. Since τ is a constant, then $dt = dT$ in each case. Also, the limits for each integral can be written as zero and τ. Thus

$$\mathscr{L}[f(t)] = \int_0^\tau f(T)e^{-sT}\,dT + \int_0^\tau f(T + \tau)e^{-s(T+\tau)}\,dT$$

$$+ \int_0^\tau f(T + 2\tau)e^{-s(T+2\tau)}\,dT + \cdots$$

$$= \int_0^\tau f(T)e^{-sT}\,dT + e^{-\tau s}\int_0^\tau f(T + \tau)e^{-sT}\,dT$$

$$+ e^{-2\tau s}\int_0^\tau f(T + 2\tau)e^{-sT}\,dT + \cdots$$

However, because of the periodic nature of the function, $f(T) = f(T + \tau) = f(T + n\tau)$, and therefore

$$\mathscr{L}[f(t)] = (1 + e^{-\tau s} + e^{-2\tau s} + \cdots)\int_0^\tau f(T)e^{-sT}\,dT$$

Recognizing that the infinite series appearing as the coefficient is equal to $1/(1 - e^{-\tau s})$, our result for the Laplace transform of a *periodic function* becomes

$$\tilde{f}(s) = \frac{\int_0^\tau f(t)e^{-st}\,dt}{1 - e^{-\tau s}} \qquad\qquad 4.2\text{-}16$$

Hence it is a simple matter to determine the transform of a sawtooth function, a square wave, or other elementary periodic function.

Laplace Transforms of Derivatives

The great advantage of the Laplace transform technique is that ordinary derivatives can be replaced by algebraic quantities. Writing the definition of this transform as

$$\mathscr{L}\!\left[\frac{df(t)}{dt}\right] = \int_0^\infty \frac{df(t)}{dt}e^{-st}\,dt$$

and integrating by parts gives

$$\int_0^\infty e^{-st}\frac{df(t)}{dt}\,dt = [e^{-st}f(t)]\Big|_0^\infty + s\int_0^\infty e^{-st}f(t)\,dt$$

For most functions of engineering interest, $e^{-st}f(t)$ will approach zero as t approaches infinity, and therefore the transform becomes

$$\mathscr{L}\!\left[\frac{df(t)}{dt}\right] = s\mathscr{L}[f(t)] - f(0^+) = s\tilde{f}(s) - f(0^+) \qquad 4.2\text{-}17$$

Hence the Laplace transform of a derivative is merely the Laplace parameter s multiplied by the Laplace transform of the function itself, minus the initial value of the function. This initial value is written in the form $f(0^+)$ to denote that we need the value just after time equal to zero in case there is a discontinuity in the dependent variable. However, if we choose our state variables as deviations from some steady state operating condition, in most cases $f(0^+)$ will be equal to zero.

The Laplace transform of a second derivative can be obtained similarly. Writing the definition

$$\mathscr{L}\!\left[\frac{d^2f(t)}{dt^2}\right] = \int_0^\infty e^{-st}\frac{d^2f(t)}{dt^2}\,dt$$

and integrating by parts, we obtain

$$\int_0^\infty e^{-st}\frac{d^2f(t)}{dt^2}\,dt = \left[e^{-st}\frac{df(t)}{dt}\right]\Big|_0^\infty + s\int_0^\infty e^{-st}\frac{df(t)}{dt}\,dt$$

Again, for most functions of interest, $e^{-st}(df(t)/dt)$ approaches zero as t approaches infinity, so that

$$\mathscr{L}\!\left[\frac{d^2f(t)}{dt^2}\right] = s\mathscr{L}\!\left[\frac{df(t)}{dt}\right] - \frac{df(0^+)}{dt} \qquad 4.2\text{-}18$$

or after substituting our previous result,

$$\mathscr{L}\left[\frac{d^2f(t)}{dt^2}\right] = s^2\tilde{f}(s) - sf(0^+) - \frac{df(0^+)}{dt} \qquad 4.2\text{-}19$$

If our dependent variables represent deviations from steady state, the Laplace transform of a second deviative is simply s^2 times the transform of the variable.

This procedure can be generalized, and it can be shown that the Laplace transform of an nth derivative is

$$\mathscr{L}\left[\frac{d^nf(t)}{dt^n}\right] = s\mathscr{L}\left[\frac{d^{n-1}f(t)}{dt}\right] - \frac{d^{n-1}f(0^+)}{dt}$$
$$= s^n\tilde{f}(s) - [s^{n-1}f(0^+) + s^{n-2}f'(0^+) + \cdots \qquad 4.2\text{-}20$$
$$+ sf^{n-2}(0^+) + f^{n-1}(0^+)]$$

or it is equal to s^n times the transform of the function for the case where the initial value of the function and its first $(n-1)$ derivatives are equal to zero.

Laplace Transforms of Integrals

The method of Laplace transforms also makes it possible to replace integrals by algebraic quantities. According to the definition,

$$\mathscr{L}\left[\int_a^t f(t)\,dt\right] = \int_0^\infty \left[\int_a^t f(\lambda)\,d\lambda\right] e^{-st}\,dt$$

where the dummy variable λ has been introduced for convenience. Integrating by parts, where

$$u = \int_a^t f(\lambda)\,d\lambda \qquad dv = e^{-st}\,dt$$
$$du = f(t)\,dt \qquad v = \frac{-e^{-st}}{s}$$

we obtain

$$\int_0^\infty \left[\int_a^t f(\lambda)\,d\lambda\right] e^{-st}\,dt = \left[\frac{-e^{-st}}{s}\int_a^t f(\lambda)\,d\lambda\right]\Big|_0^\infty + \frac{1}{s}\int_0^\infty f(t)e^{-st}\,dt$$

As t approaches infinity the first term normally approaches zero, and therefore

$$\mathscr{L}\left[\int_a^t f(t)\,dt\right] = \frac{1}{s}\tilde{f}(s) + \frac{1}{s}\int_a^0 f(\lambda)\,d\lambda \qquad 4.2\text{-}21$$

If we consider variables that represent deviations from steady state, then the second integral will be equal to zero and the Laplace transform of an integral is merely the transform of the function divided by s. It should be apparent that this procedure can be extended in a straightfoward manner to find the transform of multiple integrals.

SECTION 4.3 RESPONSE OF FIRST-ORDER SYSTEMS

In Section 3.3 we showed that it was always possible to approximate the dynamic behavior of any system by the set of linear differential equations

$$\frac{d\mathbf{x}}{dt} = \mathbf{A}\mathbf{x} + \mathbf{B}\mathbf{u} + \mathbf{C}\mathbf{v} \qquad 4.3\text{-}1$$

where $\mathbf{x}$ was the state vector, $\mathbf{u}$ the control vector, $\mathbf{v}$ the disturbance vector, and $\mathbf{A}$, $\mathbf{B}$, and $\mathbf{C}$ were constant matrices. Of course, this approximation might not give valid predictions over the total range of dependent variables of interest, but it often provides considerable information about the dynamic characteristics of the process. Furthermore, it is usually possible to uncouple the preceding set of equations by making a canonical transformation, so that the equations can be written

$$\frac{d\mathbf{y}}{dt} = \mathbf{\Lambda}\mathbf{y} + \mathbf{P}^{-1}\mathbf{B}\mathbf{u} + \mathbf{P}^{-1}\mathbf{C}\mathbf{v} \qquad 4.3\text{-}2$$

Providing that all the characteristic roots are distinct and real, this set of equations is just a set of uncoupled, first-order differential equations.[1] For example, the ith equation in the set can be written as

$$\frac{dy_i}{dt} = \lambda_i y_i + f_i \qquad 4.3\text{-}3$$

where λ_i is the ith characteristic root and f_i is the ith component of the vector $(\mathbf{P}^{-1}\mathbf{B}\mathbf{u} + \mathbf{P}^{-1}\mathbf{C}\mathbf{v})$. Since the forcing function f_i is, in general, a linear combination of all the control and disturbance variables, we can always relate the system response for a particular form of $f_i(t)$ to the response for some actual system input.

Although the simple first-order differential equation, Eq. 4.3-3, arises naturally from elementary matrix manipulations, it has been traditional in the literature on process dynamics and control theory to write the equation for a first-order system in the form

$$\tau\frac{dy}{dt} + y = g(t) \qquad 4.3\text{-}4$$

where $\tau = -1/\lambda$ and $g(t) = -f(t)/\lambda$. The quantity τ, called the time constant of the process, has received considerable emphasis in the literature, for it obviously has a major affect on the system response. Thus we will usually write the system equation in terms of a time constant, although it is apparent that the two forms are equivalent.

---

[1] Similar equations are obtained for complex roots, but then the canonical variables must also be complex. Therefore we use a different transformation for that case; see Eq. 3.3-96.

Transfer Function

If we take the Laplace transform of Eq. 4.3-4 for a case where $y = 0$ and $g(t) = 0$ for $t < 0$, we obtain the result

$$(\tau s + 1)\tilde{y} = \tilde{g} \qquad\qquad 4.3\text{-}5$$

where $\tilde{y}$ is the transform of the state variable and $\tilde{g}$ is the transform of the forcing function. Some rearrangement gives

$$\frac{\tilde{y}}{\tilde{g}} = H(s) = \frac{1}{\tau s + 1} \qquad\qquad 4.3\text{-}6$$

This quantity, the ratio of the Laplace transform of the system output to the system input, is called the *transfer function* of the process and is often denoted as $H(s)$. The transfer function completely describes, in some way, the dynamic characteristics of the process, for it is valid for any system input $g(t)$. However, the actual response of the system depends on the particular form chosen for the forcing function. The dynamic response for several simple forcing functions is given below.

Impulse Response

For an impulse input,

$$\tilde{g} = A \qquad\qquad 4.3\text{-}7$$

according to Eq. 4.2-15, so that the equation for the Laplace transform of the system output becomes

$$\tilde{y} = \frac{A}{\tau s + 1} = \frac{A/\tau}{s + 1/\tau} \qquad\qquad 4.3\text{-}8$$

From Eq. 4.2-9 we know that the function having this Laplace transform is just

$$y(t) = \frac{A}{\tau} e^{-t/\tau} \qquad\qquad 4.3\text{-}9$$

so that this expression must be the impulse response of the system. Often it is convenient to write the result in dimensionless form

$$\frac{y(t)}{A/\tau} = e^{-t/\tau} \qquad\qquad 4.3\text{-}10$$

A sketch of this solution is given in Figure 4.3-1. The graph is applicable to cases where τ is a positive constant or where

$$\lambda = -\frac{1}{\tau} \qquad\qquad 4.3\text{-}11$$

is negative and real. If this is not the case, the output would tend to increase

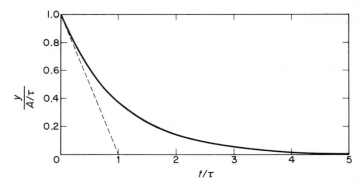

Figure 4.3-1. Impulse response of a first-order system.

exponentially until the approximate system equation was no longer valid—
that is, we know that physical systems cannot increase in an unbounded
manner.

As shown in the graph, the solution of the system equation predicts that
the system output $y(t)$ instantaneously increases to a value of unity (or the
magnitude of the impulse) and then takes an infinite length of time to return
to zero. We know that it is not physically realistic to expect an abrupt rise
of this nature and that at large values of t/τ the value of y will be indistinguish-
ible from its steady state value because of the limited accuracy of our measuring instruments. Thus if we speak of some characteristic time, called the time
constant of the process, which is the time it takes for the system to return
63.2 percent of the way toward its steady state value—that is,

$$\text{at } \frac{t}{\tau} = 1 \quad \text{or} \quad t = \tau, \qquad \frac{y}{A/\tau} = e^{-1.0} = 0.368 \qquad 4.3\text{-}12$$

we know that the system approximately will be at its original steady state
value after three (95 percent of the way) to five (99 percent of the way) time
constants. Obviously the particular value selected is somewhat arbitrary
because of the asymptotic nature of the solution.

An impulse response experiment can be used to determine the time
constant of a first-order system or to determine the validity of a first-order
model. Thus for any arbitrary value of A chosen for the experiment, y can
be measured as a function of time. Then from the equation describing the
system response we know that the time it takes for y/A to decay to 63.2
percent of its initial value will be equal to the time constant, the initial slope
of a plot of y/A versus t will be minus $1/\tau^2$, and the slope of the straight line
obtained when log (y/A) is plotted against t is minus $1/2.303\tau$. Since any
actual input is only an approximation of an impulse function, the initial
measurements cannot be expected to be accurate; therefore the first two
methods just described will not give as good results as the third. In addition,

the time constant obtained from the semilogarithmic plot used in the third method is based on all the experimental data, and no undue emphasis is placed on the initial values. However, if the time it takes to introduce an arbitrary pulse is much smaller than the time constant of the system, there will be little error in the impulse approximation and all three methods will give essentially the same results.

Step Response

The step response of the system, often called the *transient response*, is determined in a similar manner. The Laplace transform of the step input is (see Eq. 4.2-3)

$$\tilde{g} = \frac{A}{s} \qquad\qquad 4.3\text{-}13$$

so that the transform of the system output as obtained from the transfer function, Eq. 4.3-6, becomes

$$\tilde{y} = \frac{A}{s}\left(\frac{1}{\tau s + 1}\right) \qquad\qquad 4.3\text{-}14$$

This result is equivalent to the expression

$$\tilde{y} = A\left[\frac{1}{s} - \frac{1}{s + (1/\tau)}\right]$$

Thus from the linear nature of the Laplace transform and the specific results obtained in Eqs. 4.2-3 and 4.2-9, the function of time that corresponds to this Laplace transform must be

$$y(t) = A(1 - e^{-t/\tau}) \qquad\qquad 4.3\text{-}15$$

which is the step response of the system.

A dimensionless plot of the system response is given in Figure 4.3-2. For positive values of τ (negative values of λ), the output approaches A

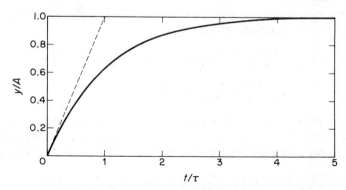

Figure 4.3-2. Step response of a first-order system.

asymptotically, whereas it decreases without a bound (actually, until the linear equation is no longer valid) if τ is negative. Thus, as we know from our previous discussions, a step input forces the system to go to a new steady state condition.

The time constant of the system can best be determined from experimental data by plotting the logarithm of the fractional incomplete response, $\log [1 - (y/A)]$, against t and measuring the slope. However, the time it takes to reach 63.2 percent of the final value or the initial slope method can also be used to estimate the time constant. All three methods will lead to essentially the same result if the time actually required to introduce the step (e.g., the time required to change a valve setting) is very small in comparison with the time constant of the system.

It is interesting to note that a step function is the integral of an impulse and that the integral of the impulse response is identical to the step response. One might expect, therefore, that the response to a ramp input, which is merely the integral of a step function, would be equal to the time integral of the step response. A proof of this spectulation is left as an exercise for the reader.

Frequency Response

If the forcing function is sinusoidal, $g(t) = A \sin \omega t$, the Laplace transform of the input is (see Eq. 4.2-4)

$$\tilde{g} = \frac{A\omega}{s^2 + \omega^2} \qquad \text{4.3-16}$$

and from the system transfer function, Eq. 4.3-6, we find that the transform of the output signal must be

$$\tilde{y} = \left(\frac{1}{\tau s + 1}\right)\left(\frac{A\omega}{s^2 + \omega^2}\right) \qquad \text{4.3-17}$$

After some manipulations, which will be described in additional detail in the next section, this equation can be written

$$\tilde{y} = \left(\frac{A\omega\tau}{1 + \tau^2\omega^2}\right)\left[\frac{1}{s + (1/\tau)}\right] + \left(\frac{A}{1 + \tau^2\omega^2}\right)\left(\frac{\omega}{s^2 + \omega^2}\right)$$
$$- \left(\frac{A\tau\omega}{1 + \tau^2\omega^2}\right)\left(\frac{s}{s^2 + \omega^2}\right) \qquad \text{4.3-18}$$

Once this expression has been obtained, the linearity property of transforms, Eq. 4.2-2, and some of our previous results, Eqs. 4.2-9, 4.2-4, and 4.2-5, can be used to find the function of time having this Laplace transform. Thus

$$y = \left(\frac{A\tau\omega}{1 + \tau^2\omega^2}\right)e^{-t/\tau} + \left(\frac{A}{1 + \tau^2\omega^2}\right)\sin \omega t - \left(\frac{A\tau\omega}{1 + \tau^2\omega^2}\right)\cos \omega t \qquad \text{4.3-19}$$

Again, τ must be positive, and therefore λ negative, in order to have a bounded output and a valid result.

Sometimes the solution is simplified by substituting the trigonometric identity given by Eqs. 4.1-5 and 4.1-6.

$$c_1 \cos \omega t + c_2 \sin \omega t = c_3 \sin (\omega t + \phi) \qquad 4.1\text{-}5$$

where

$$c_3 = \sqrt{c_1^2 + c_2^2} \qquad \tan \phi = \frac{c_1}{c_2} \qquad 4.1\text{-}6$$

Applying this identity to our result, we obtain

$$y = \left(\frac{A\tau\omega}{1 + \tau^2\omega^2}\right)e^{-t/\tau} + \frac{A}{\sqrt{1 + \tau^2\omega^2}} \sin (\omega t + \phi) \qquad 4.3\text{-}20$$

where

$$\phi = \tan^{-1}(-\tau\omega) \qquad 4.3\text{-}21$$

As t becomes very large and approaches infinity, the first term in the solution approaches zero, while the remainder of the solution is a periodic function of time. Hence the first term is often referred to as the transient portion of the solution and the periodic terms are called the steady state or pseudo-steady steady state solution. Some care must be taken not to confuse this nomenclature with the normal steady state output of the system (the result obtained when there is no accumulation in the system) and the transient response of the plant (the response to a step input).

If we compare the steady state portion of the solution with the input signal

$$\text{Input:} \quad g = A \sin \omega t$$

$$\text{Output:} \quad y = \frac{A}{\sqrt{1 + \tau^2\omega^2}} \sin (\omega t + \phi) \qquad 4.3\text{-}22$$

where

$$\phi = \tan^{-1}(-\tau\omega) \qquad 4.3\text{-}23$$

we find that the output is a sine wave having the same frequency ω as the

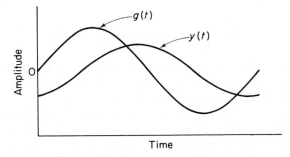

Figure 4.3-3. Frequency response of a first-order system.

input signal, the ratio of the amplitudes of the output to input signal, $(1 + \tau^2\omega^2)^{-1/2}$, is independent of the input amplitude A and is always less than unity, the output signal will always lag behind the input signal, for the phase angle ϕ is always negative (see Figure 4.3-3), and the amplitude of the output signal approaches zero as the frequency of the forcing function increases for any given value of τ. It is interesting to note that this last result is in qualitative agreement with our knowledge that rapid variations of the accelerator position of an automobile essentially have no affect on the car velocity.

The dependence of the ratio of the output to input amplitudes, called the amplitude ratio or the system gain, $|G|$, and the phase angle, ϕ, on frequency can be neatly represented on a Bode plot. Actually, a Bode diagram consists of two graphs: the phase angle and the logarithm of the gain plotted against the logarithm of the frequency, or the dimensionless group $\tau\omega$. Since the gain $|G|$ is

$$|G| = \frac{1}{\sqrt{1 + \tau^2\omega^2}} \qquad 4.3\text{-}24$$

we can write

$$\log|G| = -\tfrac{1}{2}\log(1 + \tau^2\omega^2) \qquad 4.3\text{-}25$$

This equation shows that as $\tau\omega$ approaches zero, G must approach unity and the slope of the curve must be equal to zero. Also, as $\tau\omega$ becomes very large, $\log|G|$ approximately is equal to $-\log\tau\omega$, which is a line having a slope of -1 passing through the point where $|G| = 1$ when $\tau\omega = 1$. Thus the gain curve can be approximated by two straight lines, which are asymptotes for the actual curve and which intersect at the frequency $\omega_c = 1/\tau$, called the *corner frequency*.

A Bode plot for a first-order system is given below. In some cases the gain $|G|$, or amplitude ratio, is given in decibels, where

$$\text{decibels} = 20\log|G| \qquad 4.3\text{-}26$$

This scale is plotted as the right-hand ordinate on the diagram. The graph shows that the maximum deviation of the gain curve from the two straight-line asymptotic approximations occurs at the corner frequency, and at this point

$$|G| = \frac{1}{\sqrt{1 + \tau^2\omega^2}} = \frac{1}{\sqrt{2}} = 0.707 \qquad 4.3\text{-}27$$

Since this maximum deviation only corresponds to a 30 percent error, the asymptotic approximations often provide sufficient accuracy for making engineering estimates of the frequency response.

The graph also indicates the importance of making a theoretical prediction of the process time constant, or corner frequency, before undertaking an experimental study of the frequency response. If the frequencies selected for the experimental investigation are much larger than the corner frequency, $\omega_c = 1/\tau$, then no fluctuations in the process output will be observed and no

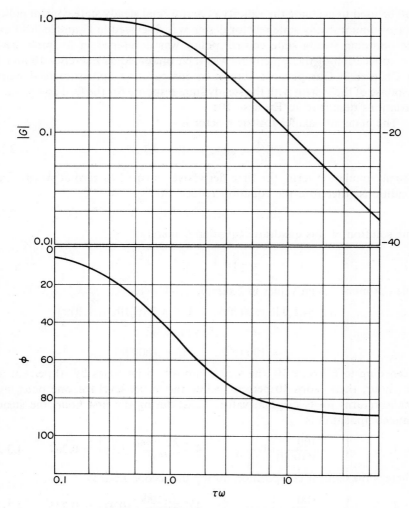

Figure 4.3-4. Bode plot for a first-order system.

information will be obtained. Similarly, if the values for the experiment are too low, the output signal will be identical with the input and again no information will be obtained. In addition, the graph shows that it is necessary to design test equipment capable of operating over a wide frequency range before initiating any experimental work.

Example 4.3-1 Response of an isothermal reactor

A dynamic model for an isothermal second-order reaction in a continuous-stirred-tank reactor was derived in Example 3.1-1. This model, along with the mathematical techniques for determining the dynamic response of the system, can be used to evaluate the validity of an optimal steady state control policy. The optimum steady state reactor design was discussed in Example 2.1-1 and the optimum steady state control policy was developed in Example 2.3-1. An application of these methods was then presented as Exercises 10 and 11 in Chapter 2. Our purpose here is to compare the approximate dynamic response of the system with the steady state response for the feed composition variations described in Exercise 10:

The material balance for the reactor is

$$V\frac{dA}{dt} = q(A_f - A) - kVA^2 \qquad 3.1\text{-}5$$

The accumulation term, the time derivative, is equal to zero at steady state conditions; therefore the equation reduces to

$$q(A_{fs} - A_s) - kVA_s^2 = 0 \qquad 2.1\text{-}4$$

The solution of this quadratic equation is merely

$$A_s = -\frac{1}{2}\frac{q}{kV}\left[1 - \sqrt{1 + 4\frac{kV}{q}A_{fs}}\right]$$

The optimum design values in Example 2.1-1 were

$$A_{fs} = 1.0 \text{ lb mole/ft}^3 \qquad k = 1.2 \text{ ft}^3/(\text{lb mole})(\text{hr})$$

$$A = 0.25 \text{ lb mole/ft}^3$$

$$q = 100 \text{ ft}^3/\text{hr} \qquad V = 1000 \text{ ft}^3$$

According to Exercise 10, the feed composition increases by 10 percent for one hour, then drops 10 percent below the initial level for one hour, and finally returns to its original value. Thus, during the first hour, the steady state composition is

$$A_s = -\frac{1}{2}\frac{100}{(1.2)(1000)}\left[1 - \sqrt{1 + 4\frac{(1.2)(1000)}{100}(1.1)}\right] = 0.264 \qquad 4.3\text{-}28$$

whereas the effluent composition during the second hour is

$$A_s = -\frac{1}{2}\frac{100}{(1.2)(1000)}\left[1 - \sqrt{1 + 4\frac{(1.2)(1000)}{100}(0.9)}\right] = 0.235 \qquad 4.3\text{-}29$$

The steady state analysis predicts that these changes occur instantaneously in response to the input changes—that is, the system has no accumulation. It is interesting to note that the average reactant composition during the period is slightly lower than the optimum steady state value, which means that the conversion is higher.

In order to evaluate the actual system response, we must include the accumulation term in the material balance and find the solution of Eq. 3.1-5. If we let

$$x = A - A_s \quad \text{and} \quad x_f = A_f - A_{fs} \qquad 4.3\text{-}30$$

the equation becomes

$$V\frac{dx}{dt} = [q(A_{fs} - A_s) - kVA_s^2] + qx_f - (q + 2kVA_s)x - kVx^2 \quad \text{4.3-31}$$

where the term in brackets is equal to zero, for it is the steady state equation (see Eq. 2.1-4). The transformation defined by Eqs. 4.3-30 merely translates the steady state solution to the origin; that is, we replace the actual system input and output variables by their deviations from steady state conditions. In order to make the state and disturbance variables dimensionless, and to eliminate some of the constants, we let

$$y = \frac{x}{A_{fs}} = \frac{A - A_s}{A_{fs}} \qquad y_f = \frac{x_f}{A_{fs}} = \frac{A_f - A_{fs}}{A_{fs}} \qquad \text{4.3-32}$$

and

$$\tau = \frac{1}{2kA_s + (q/V)}, \quad \beta = \frac{kA_{fs}}{2kA_s + (q/V)}, \quad \gamma = \frac{q/V}{2kA_s + (q/V)} \qquad \text{4.3-33}$$

so that the state equation becomes

$$\tau\frac{dy}{dt} + y + \beta y^2 = \gamma y_f \qquad \text{4.3-34}$$

Although it is possible to solve this equation analytically for certain kinds of forcing functions, y_f, we will first linearize the equation and use the methods described above to obtain an approximation of the dynamic response. Then we will compare the approximate solution with the exact response in order to develop some appreciation for the error introduced by the linearization procedure.

If we expand the preceding state equation in a Taylor series about the origin, which is the steady state solution for the deviation variables, we find that the equation remains unchanged. (This result will always be obtained when the nonlinear terms appear as polynomials.) If we neglect all second- and higher-order terms, the result is

$$\tau\frac{dy}{dt} + y = \gamma y_f \qquad \text{4.3-35}$$

This is the general form we considered for first-order systems, and there-

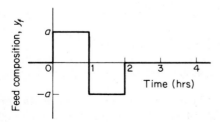

Figure 4.3-5. Feed-composition fluctuations.

fore we can use the results developed previously to estimate the response to simple forcing functions. However, since the disturbance described in Exercise 10 of Chapter 2 is slightly different than those we have treated in detail, it will be necessary to modify the analysis somewhat.

The forcing function can be considered a combination of step inputs, where the size of the step is taken as $\pm a$. Following the normal procedure used to derive Eq. 4.2-13, we find that the Laplace transform of the input signal is

$$\mathscr{L}[au(t-0) - 2au(t-1) + au(t-2)] = \frac{a}{s} - \frac{2a}{s}e^{-s} + \frac{a}{s}e^{-2s} \quad 4.3\text{-}36$$

Hence the Laplace transform of the state equation gives

$$(\tau s + 1)\tilde{y} = \gamma\left(\frac{a}{s} - \frac{2a}{s}e^{-s} + \frac{a}{s}e^{-2s}\right)$$

or

$$\tilde{y} = \frac{\gamma a}{s(\tau s + 1)}(1 - 2e^{-s} + e^{-2s}) \quad 4.3\text{-}37$$

Expanding the denominator into its individual factors and separating the three terms in the numerator gives

$$y = \gamma a\left[\frac{1}{s} - \frac{1}{(s + 1/\tau)}\right] - 2\gamma ae^{-s}\left[\frac{1}{s} - \frac{1}{(s + 1)/\tau}\right] \quad 4.3\text{-}38$$
$$+ \gamma ae^{-2s}\left[\frac{1}{s} - \frac{1}{(s + 1/\tau)}\right]$$

Now, inverting term by term and using the shift theorem, we obtain

$$y(t) = \gamma a(1 - e^{-t/\tau}), \qquad 0 < t < 1$$
$$y(t) = \gamma a(1 - e^{-t/\tau}) - 2\gamma a(1 - e^{-(t-1)/\tau}), \qquad 1 < t < 2 \quad 4.3\text{-}39$$
$$y(t) = \gamma a(1 - e^{-t/\tau}) - 2\gamma a(1 - e^{-(t-1)/\tau})$$
$$+ \gamma a(1 - e^{-(t-2)/\tau}), \qquad 2 < t < \infty$$

or in terms of a unit step function and delays,

$$y(t) = \gamma a[1 - e^{-t/\tau}] - 2\gamma au(t-1)[1 - e^{-t/\tau}] \quad 4.3\text{-}40$$
$$+ \gamma au(t-2)[1 - e^{-t/\tau}]$$

For the problem of interest,

$$A = A_{fs}y + A_s = y + 0.25 \qquad a = \frac{A_f - A_{fs}}{A_{fs}} = 0.1$$

$$\tau = \frac{1}{2kA_s + (q/V)} = \frac{1}{2(1.2)0.25 + (100/1000)} = 1.429$$

$$\gamma = \frac{q/V}{2kA_s + (q/V)} = 0.1(1.429) = 0.1429$$

The solution of Eq. 4.3-40 is plotted in Figure 4.3-6, along with the predictions of the steady state model, Eqs. 4.3-28 and 4.3-29. It is apparent from the graph

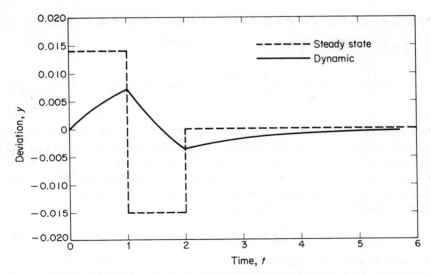

Figure 4.3-6. Approximate response of an isothermal reactor.

that accumulation effects are significant in this problem and that the optimum steady state control policy will not be valid during the period when the upsets enter the system, for the process never operates at steady state during this period. Again, it is interesting to note that the time average conversion[2] is slightly higher than the optimum steady state value.

In order to determine the exact response of the system described by Eq. 4.3-34 for the forcing function given in Figure 4.3-5, we have to consider each time interval individually. Letting

$$\theta = \frac{t}{\tau} \qquad\qquad 4.3\text{-}41$$

the equation for the first time interval becomes

$$\frac{dy}{d\theta} = \gamma a - y - \beta y^2 \qquad\qquad 4.3\text{-}42$$

Separating the variables and integrating gives the response for the first interval

$$\theta = \frac{1}{\sqrt{1 + 4a\beta\gamma}} \ln\left(\frac{-2\beta y - 1 - \sqrt{1 + 4a\beta\gamma}}{-2\beta y - 1 + \sqrt{1 + 4a\beta\gamma}}\right) + c_1 \qquad 4.3\text{-}43$$

where c_1 is an integration constant that can be evaluated from the initial condition

$$\text{at } t = 0 \quad \text{or} \quad \theta = 0; \qquad y = 0 \qquad\qquad 4.3\text{-}44$$

---

[2] The time-average composition over some time interval of interest is defined as $A_{av} = \frac{1}{T}\int_0^T A\,dt$

After solving for c_1 and manipulating the results somewhat, we obtain

$$y = \left(\frac{\sqrt{1 + 4a\beta\gamma} - 1}{2\beta} \right)$$

$$\times \left\{ \frac{1 - \exp(-\sqrt{1 + 4a\beta\gamma}\,\theta)}{1 - [(1 - \sqrt{1 + 4a\beta\gamma})/(1 + \sqrt{1 + 4a\beta\gamma})] \exp(-\sqrt{1 + 4a\beta\gamma}\,\theta)} \right\}$$

4.3-45

As β approaches zero, so that the quadratic term in Eq. 4.3-42 becomes negligible in comparison to the linear term, the foregoing equation reduces to the first term in Eq. 4.3-39. This result is valid in the interval $0 < t < 1$, or $0 < \theta < 0.7$. The final value of y when $\theta = 0.7$ becomes the initial condition for the next interval; that is, we shift the time axis, and the system equation for this next period becomes

$$\frac{dy}{d\theta} = -\gamma a - y - \beta y^2 \qquad\qquad 4.3\text{-}46$$

A solution of this equation can be determined by separating the variables, integrating, and using the initial condition given above to eliminate the integration constant. The solution will describe the system response for the interval, $1 < t < 2$ or $0.7 < \theta < 1.4$. The procedure can then be repeated to find the response for the third interval, $2 < t < \infty$. A plot of the results is given in Figure 4.3-7. A comparison of the graphs shows that the approximate solution provides a good estimate of the actual response because the nonlinear term (a quadratic) is not very significant for the process parameters under consideration.

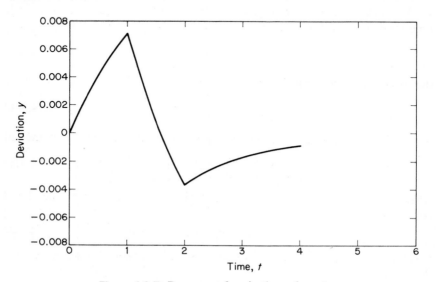

Figure 4.3-7. Response of an isothermal reactor.

Although it was possible to determine the exact step response of this system analytically, a simple solution for the frequency response is not available. In fact, for most forcing functions it is necessary to rely on the linearized equation for a first estimate of the dynamic response. Since a step change in some process input only changes its level from one constant value to another, there is always a better chance of finding an exact solution for this forcing function than any other. Then a comparison between the exact solution and the results from the linear analysis will give some indication of the error introduced by the linearization procedure. Perturbation theory, which will be discussed in a later chapter, can also be used for this purpose. It should be clearly recognized that the linear analysis does not give a valid predication of the steady state gain of the system—that is, the change in the final steady state output corresponding to a step change in one of the process inputs is not correct. For example, if the disturbance in the preceding problem had merely been a single step input, the approximate solution

$$y(t) = \gamma a(1 - e^{-t/\tau}) \qquad\qquad 4.3\text{-}39$$

would apply for all time. The corresponding exact solution is

$$y(t) = \frac{(\sqrt{1 + 4a\beta\gamma} - 1)}{2\beta}$$

$$\times \left\{ \frac{1 - \exp(-\sqrt{1 + 4a\beta\gamma}\,\theta)}{1 - [(1 - \sqrt{1 + 4a\beta\gamma})/(1 + \sqrt{1 + 4a\beta\gamma})] \exp(-\sqrt{1 + 4a\beta\gamma}\,\theta)} \right\}$$

$$4.3\text{-}45$$

As time approaches infinity, the final steady state values predicted by these equations are

$$\gamma a \quad \text{and} \quad \frac{\sqrt{1 + 4a\beta\gamma} - 1}{2\beta} \qquad\qquad 4.3\text{-}47$$

Of course, if β is very small so that the quadratic term in the state equation is negligible in comparison to the linear term, the two results will be essentially the same; that is, L'Hôpital's rule can be used to show that the second expression approaches the first as β approaches zero.

Example 4.3-2 Approximation of impulse response

Since the results of the preceding example indicate that the optimum steady state control policy is not applicable for the type of disturbance under consideration, it will be necessary to consider some kind of a dynamic control system. However, before attempting to develop a new kind of controller, we might decide to perform a simple dynamic experiment that can be used to test the validity of the dynamic model. One possible experiment would be to add an additional amount of reactant for a short time and measure the time dependence of the effluent concentration. As a specific problem consider the case where the feed concentration is increased by 10 percent over a 15-sec interval and is then returned to normal. Determine the appropriate response of the model for this input, and compare the results with the response of an impulse input.

Solution

This problem is merely a simplified version of the preceding example. Hence the Laplace transform of the input is

$$\mathscr{L}[au(t - 0) - au(t - t_0)] = \frac{a}{s} - \frac{a}{s}e^{-t_0 s} \qquad 4.3\text{-}48$$

and the expression for the transform of the output becomes

$$y = \frac{\gamma a}{s(\tau s + 1)}(1 - e^{-t_0 s}) \qquad 4.3\text{-}49$$

The inverse transform gives the solution

$$y(t) = \gamma a(1 - e^{-t/\tau}), \qquad 0 < t < t_0 \qquad 4.3\text{-}50$$

$$y(t) = \gamma a(1 - e^{-t/\tau}) - \gamma a(1 - e^{-(t-t_0)/\tau}), \qquad t_0 < t < \infty \qquad 4.3\text{-}51$$

The response curve for a case where $a = 0.1$, $\tau = 1.429$, $\gamma = 0.1429$, and $t_0 = 15/3600 = 0.00417$ is plotted in Figure 4.3-8.

If we assume that the forcing function is an impulse, its transform is $\mathscr{L}at_0[\delta(t)] = (at_0)$ and the transform of the output signal is

$$y = \frac{\gamma(at_0)}{(\tau s + 1)} \qquad 4.3\text{-}52$$

Inversion gives the expression (see Eq. 4.3-9)

$$y(t) = \frac{\gamma(at_0)}{\tau}e^{-t/\tau} \qquad 4.3\text{-}53$$

This solution is compared with the previous result in Figure 4.3-8. It is ap-

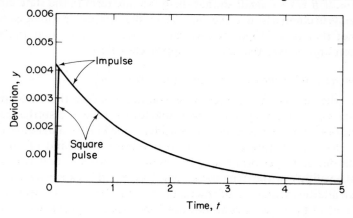

Figure 4.3-8. Impulse response of an isothermal reactor.

parent from the graph that the impulse approximation is valid in this case, as would be expected because t_0, the duration of the pulse, is very small compared with τ, the time constant of the system. Normally we anticipate a fairly good agreement if t_0 is less than 10 percent of τ.

Example 4.3-3 A problem on cultural regression

Recently we received a letter from a former student who is presently employed by the Peace Corps and working to improve the lives of the Abominable Snowmen high in the Himalaya Mountains. Since present government policies have left the people somewhat cold, a dissident group is planning a "cultural revolution." Being a typical student, our young hero naturally sides with the good guys and wants to help them build up their cultural equipment—that is, guns. Although the revolutionaries have approached both the East and the West, no one has figured out how to ship arms to this inaccessible spot in the Himalayas. Besides, the Snowmen have no vote in the United Nations and, unfortunately for them, are not being exploited by any outside power. Thus they must do the best that they can with their present arsenal, which consists of a one-year stockpile of snowballs and a monstrous old canon that has a barrel 4.36 ft in diameter.

The current method for producing cannon balls is to drop globs of molten iron from a shot tower into a large pool of water. Then the cannon ball is allowed to cool for a day, and the next day another load is dropped. The limiting step in the process is the cooling period, and our young hero is asked if this part of the process can be speeded up.

Solution

He reasons that each cannon ball will fall like a ton of bricks, except that there will be less air resistance, and therefore not much cooling will take place during the time it takes the ball to hit the water. Also, he believes that he will be able to obtain a conservative (although he prefers the word crude) estimate of the cooling time if he neglects all the radiation and convection effects, ignores the fact that the water will boil when it contacts the molten iron, disregards any phase changes that might occur in the iron during the cooling period, and assumes that the conductivity of the iron is so high that there will be no temperature gradients inside the cannon ball. In other words, he decides to make an initial estimate by treating the problem as a sphere losing heat to an infinite body of stagnant water. He writes the energy balance as

$$\rho C_p V \frac{dT}{dt} = -hA(T - T_w) \qquad\qquad 4.3\text{-}54$$

and lets

$$\theta = \frac{T - T_w}{T_0 - T_w} \qquad\qquad 4.3\text{-}55$$

where T_0 is the initial temperature of the cannon ball, which he considers to be the melting point of iron. Thus

$$\frac{d\theta}{dt} = -\frac{hA\theta}{\rho C_p V} \qquad\qquad 4.3\text{-}56$$

He solves this equation by separating the variables and integrating, thereby obtaining

$$\theta = \exp\left(-\frac{hA}{\rho C_p V}t\right) \qquad\qquad 4.3\text{-}57$$

He uses the result

$$\text{Nu} = \frac{hD}{k} = 2.0$$

and assumes that the cannon balls are cool enough to handle when their temperature has decreased to $35°C$. Then, for the values given below,

$$\rho = 450 \text{ lb/ft}^3 \qquad k = 0.40 \text{ Btu/(hr)(ft}^2)(°F \text{ per foot}),$$

$$T_0 = 1275°C \qquad T_w = 10°C,$$

$$C_p = 0.1178 \text{ Btu/(lb)(}°F) \qquad r = 2.18 \text{ ft}$$

he finds that

$$\theta = \frac{35 - 10}{1275 - 10} = 1.978 \times 10^{-2}$$

$$\frac{hA}{\rho C_p V} = \frac{k 4\pi r^2}{r \rho C_p \frac{4}{3}\pi r^3} = \frac{3k}{\rho C_p r^2} = \frac{3(0.40)}{(450)(0.1178)(2.18)^2} = 0.00266 \text{ (hr)}^{-1}$$

so that the required cooling time is

$$t = \frac{\ln (1.978 \times 10^{-2})}{-2.66 \times 10^{-3}} = 1472 \text{ hours}$$

Although convinced that this answer must be at the lunatic fringe, he still is not certain that the Abominable Snowmen are not giving him an abominable snow job. Thus he recognizes that his simplifying assumptions concerning the mechanism of heat loss must be changed in order to obtain a more realistic estimate of the cooling time. However, he does not know which assumptions should be modified, which is why he wrote the letter. Could you help him develop a better solution?

At first glance the preceding problem—that is, heat transfer from a solid body at a uniform temperature into a fluid—does not fit into our general analytical approach. However, if we change our outlook on the problem somewhat, this apparent discrepancy can be resolved quite easily. Thus we can write the energy equation as

$$\rho C_p V \frac{dT}{dt} = -hA(T - T_a) \qquad\qquad 4.3\text{-}58$$

where T_a represents the ambient temperature. Now if we consider that the system is at steady state conditions initially and that the ambient temperature is at the melting point of iron—at $t = 0$, $T_a = T_0$—the steady state solution of the energy equation is merely $T = T_0$. Letting

$$\tau = \frac{\rho C_p V}{hA}, \quad y = T - T_0, \quad y_f = T_a - T_0 \qquad 4.3\text{-}59$$

the equation becomes

$$\tau \frac{dy}{dt} + y = y_f \qquad 4.3\text{-}60$$

and the Laplace transform of the equation is

$$(\tau s + 1)\tilde{y} = \tilde{y}_f \qquad 4.3\text{-}61$$

For a step change in the ambient temperature, where we suddenly decrease T_a to a value T_w, we have

$$y_f = T_w - T_0 \qquad \tilde{y}_f = \frac{T_w - T_0}{s} \qquad 4.3\text{-}62$$

and the equation for the transform of the system output becomes

$$\tilde{y} = \frac{T_w - T_0}{s(\tau s + 1)} \qquad 4.3\text{-}63$$

Using our previous results for the inverse transform of this expression, Eqs. 4.3-14 and 4.3-15, we find that

$$y(t) = T - T_0 = (T_w - T_0)(1 - e^{-t/\tau}) \qquad 4.3\text{-}64$$

or after some manipulation,

$$\frac{T - T_w}{T_0 - T_w} = \theta = e^{-t/\tau} \qquad 4.3\text{-}65$$

which is identical to our previous result. Thus we have been able to show that the step response of a process is equivalent to a problem where some initial condition is imposed on a dynamic system, an assertion we put forth in Chapter 2. Also, it should be recognized that the simple analysis given here is often useful for estimating the response characteristics of thermometers, thermocouples, and so on.

Example 4.3-4 Frequency response of the liquid level in a tank

Suppose that we have a positive displacement pump producing a flow that is approximately given by the expression

$$q_i = q_{is} + a \sin \omega t \qquad 4.3\text{-}66$$

where the fluctuations are caused by the motion of the piston in the pump. In order to smooth out the fluctuations, we introduce the stream into a tank having a free surface. Then the liquid flows out of the bottom of the tank through a horizontal pipe having an inside radius R and a length L. For the

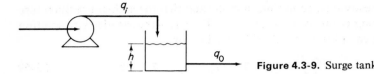

Figure 4.3-9. Surge tank.

sake of simplicity, we will consider a case where the flow in the pipe is laminar[3] and we are required to deliver some amount of material q_{os} to another unit. The problem is to choose an appropriate size for the surge tank so that the output fluctuations are only 10 percent as large as the input variations.

Solution

Poiseuille's equations for laminar flow in a pipe is

$$q_o = \frac{\pi R^4 \, \Delta P}{8\eta L} \qquad\qquad 4.3\text{-}67$$

Providing that the end of the pipe is at atmospheric pressure,

$$\Delta P = \rho g h \qquad\qquad 4.3\text{-}68$$

where h is the liquid level in the tank. Hence

$$q_0 = \frac{\pi R^4 \rho g}{8\eta L} h = Kh \qquad\qquad 4.3\text{-}69$$

where

$$K = \frac{\pi R^4 \rho g}{8\eta L} \qquad\qquad 4.3\text{-}70$$

Now if we make a material balance on the tank contents, we obtain

$$\frac{d\rho A h}{dt} = \rho q_i(t) - \rho q_0(t) \qquad\qquad 4.3\text{-}71$$

or after substituting the expressions given above,

$$A\frac{dh}{dt} = q_{is} + a \sin \omega t - Kh \qquad\qquad 4.3\text{-}72$$

For a process operating at steady state up until time equal to zero, the accumulation term in the material balance must be equal to zero; therefore

$$q_{is} = q_{os} = Kh_s \qquad\qquad 4.3\text{-}73$$

Letting

$$y = h - h_s, \quad \tau = \frac{A}{K}, \quad \gamma = \frac{a}{K} \qquad\qquad 4.3\text{-}74$$

the material balance becomes

$$\tau\frac{dy}{dt} + y = \gamma \sin \omega t \qquad\qquad 4.3\text{-}75$$

---

[3] We are also assuming that the acceleration effects in the pipe are negligible.

The Laplace transform of the system output can be determined

$$\tilde{y} = \left(\frac{1}{\tau s + 1}\right)\left(\frac{\gamma \omega}{s^2 + \omega^2}\right) \qquad 4.3\text{-}76$$

and from our previous results, Eqs. 4.3-22 and 4.3-23, we know that the steady state frequency response must be

$$y = \frac{\gamma}{\sqrt{1 + \tau^2\omega^2}}\sin(\omega t + \phi) \qquad \text{where } \phi = \tan^{-1}(-\tau\omega) \qquad 4.3\text{-}77$$

If we want to fix the size of the tank so that the amplitude of the output oscillations is only 10 percent of the input amplitude, we require that

$$\frac{1}{\gamma}\left[\frac{\gamma}{\sqrt{1 + \tau^2\omega^2}}\right] = \frac{1}{\sqrt{1 + \tau^2\omega^2}} = 0.1 \qquad 4.3\text{-}78$$

or

$$\tau = \frac{\sqrt{99}}{\omega} \qquad 4.3\text{-}79$$

Since

$$\tau = \frac{A}{K} = \left(\frac{8\eta L}{\pi R^4 \rho g}\right)A = \frac{\sqrt{99}}{\omega}$$

we can solve for the tank area, once the frequency of the fluctuations has been specified. The tank height corresponding to steady state operation can be determined by using Eqs. 4.3-73 and 4.3-69, since the process flow rate q_{os} is known from the basic design information. Of course, the actual tank height must be at least an amount (a/A) higher than this value in order to accomodate the fluctuations, and an additional safety factor should be added to ensure that the tank will not overflow. Moreover, we should check the final design conditions to make certain that the flow in the outlet pipe actually is laminar.

SECTION 4.4   INVERSION OF LAPLACE TRANSFORMS, AND LIMIT THEOREMS

The results of the previous section show that it is a relatively simple matter to derive an expression for the Laplace transform of the system output once the forcing function has been specified. However, in almost every case it was necessary to manipulate the results and put them into a form where the inverse transform became obvious; that is, the function of time corresponding to this Laplace transform could be recognized. Since it is not immediately apparent how to accomplish these manipulations in order always to arrive at the proper form, we need to develop a procedure for this purpose. Such a procedure does exist and is called the method of partial fractions.

Method of Partial Fractions

If we return to the analysis of the step response of a first-order system, we find that it was necessary to replace a complicated transform expression by one involving simple factors

$$\frac{A}{s}\left(\frac{1}{\tau s + 1}\right) = A\left[\frac{1}{s} - \frac{1}{(s + 1/\tau)}\right] \qquad \text{4.4-1}$$

A similar operation was required for the case of the frequency response

$$\left(\frac{1}{\tau s + 1}\right)\left(\frac{A\omega}{s^2 + \omega^2}\right) = \left(\frac{A\omega\tau}{1 + \tau^2\omega^2}\right)\left[\frac{1}{(s + 1/\tau)}\right] + \left(\frac{A}{1 + \tau^2\omega^2}\right)\left(\frac{\omega}{s^2 + \omega^2}\right)$$
$$- \left(\frac{A\tau\omega}{1 + \tau^2\omega^2}\right)\left(\frac{s}{s^2 + \omega^2}\right) \qquad \text{4.4-2}$$

The reason this procedure was successful was that we had previously determined the Laplace transforms of elementary functions of time so that each of the simple factors could be inverted term by term. Of course, we could determine the Laplace transforms of a great number of more complicated functions of time and hope that our list of results would include all functions of interest to engineers. It seems more reasonable, however, to look for a technique that can be used to break down complex expressions into their simple factors. In order to develop a general method, we need to consider a fairly general form for the system equations.

We already know that we can approximate the response of most systems by the set of equations

$$\frac{d\mathbf{x}}{dt} = \mathbf{Ax} + \mathbf{Bu} + \mathbf{Cv} \qquad \text{4.4-3}$$

Assuming that the state vector $\mathbf{x}$ represents deviations from some steady state condition of interest and that the system is initially at steady state, taking the Laplace transform of this set of equations gives the result

$$s\mathbf{I}\tilde{\mathbf{x}} = \mathbf{A}\tilde{\mathbf{x}} + \mathbf{B}\tilde{\mathbf{u}} + \mathbf{C}\tilde{\mathbf{v}} \qquad \text{4.4-4}$$

where s is the Laplace parameter, $\mathbf{I}$ is the identity matrix, $\tilde{\mathbf{x}}$, $\tilde{\mathbf{u}}$, and $\tilde{\mathbf{v}}$ are the Laplace transforms of the state, control, and disturbance vectors, $\mathbf{A}$, $\mathbf{B}$, and $\mathbf{C}$, are constant matrices, and the initial conditions for $\mathbf{x}$ have been set equal to zero. Following our normal procedure, we can solve for the transform of the output

$$\tilde{\mathbf{x}} = (s\mathbf{I} - \mathbf{A})^{-1}(\mathbf{B}\tilde{\mathbf{u}} + \mathbf{C}\tilde{\mathbf{v}}) \qquad \text{4.4-5}$$

where the superscript (-1) means the inverse of the matrix. Once the transforms of $\tilde{\mathbf{u}}$ and $\tilde{\mathbf{v}}$ have been specified, the right-hand side of this matrix equation will depend only on s and the system parameters contained in $\mathbf{A}$, $\mathbf{B}$, and $\mathbf{C}$. For the forcing functions of interest (e.g., impulses, steps, and periodic signals), the transforms of $\tilde{\mathbf{u}}$ and $\tilde{\mathbf{v}}$ usually are ratios of polynomials in s having higher powers of s in the denominator than in the numerator. Hence any row in the column vector $(\mathbf{B\tilde{u}} + \mathbf{C\tilde{v}})$, which is merely a linear combination of these terms, will also be a ratio of polynomials in s where the denominator again has a higher order. The matrix quantity $(s\mathbf{I} - \mathbf{A})^{-1}$ is the transpose of the cofactor matrix of $(s\mathbf{I} - \mathbf{A})$ divided by the determinant. Since the determinant is an nth order polynomial in s, where n is the number of state equations, and since each cofactor is at most an $(n - 1)$th order polynomial in s, every term in the matrix is a ratio of polynomials in s where the denominator is of higher order. Thus multiplication of the matrix by the column vector will lead to a new column vector containing ratios of polynomials in s, where the highest order polynomial always appears in the denominator.

Of course, if the canonical form of the equations is used, generally we only need to consider first- and second-order systems. However, to illustrate the generality of the method, we will consider Laplace transforms having the form

$$\tilde{x}(s) = \frac{p(s)}{q(s)} = \frac{b_m s^m + b_{m-1} s^{m-1} + \cdots + b_1 s + b_0}{s^n + a_{n-1} s^{n-1} + \cdots + a_1 s + a_0} \qquad 4.4\text{-}6$$

where $p(s)$ and $q(s)$ are polynomials in s and the order of the denominator is greater than that of the numerator—that is, $m < n$. Once this ratio of polynomials has been determined, it becomes necessary to find the roots of the denominator. This problem is identical to the determination of the roots of the characteristic equation, which was discussed in Section 3.3. Then the equation can be written

$$\tilde{x}(s) = \frac{p(s)}{q(s)} = \frac{b_m s^m + b_{m-1} s^{m-1} + \cdots + b_1 s + b_0}{(s - \lambda_1)(s - \lambda_2) \cdots (s - \lambda_n)}$$

or

$$\tilde{x}(s) = \frac{p(s)}{q(s)} = \frac{b_m(s - \gamma_1)(s - \gamma_2) \cdots (s - \gamma_m)}{(s - \lambda_1)(s - \lambda_2) \cdots (s - \lambda_n)} \qquad 4.4\text{-}7$$

where γ_i represents one of the m roots of the numerator and λ_i is one of the characteristic roots.

Our goal is to find an equivalent expression in terms of simple factors,

$$\tilde{x}(s) = \frac{b_m(s - \gamma_1)(s - \gamma_2) \cdots (s - \gamma_m)}{(s - \lambda_1)(s - \lambda_2) \cdots (s - \lambda_n)}$$

$$= \frac{K_1}{s - \lambda_1} + \frac{K_2}{s - \lambda_2} + \cdots + \frac{K_n}{s - \lambda_n} \qquad 4.4\text{-}8$$

where K_i are unknown constants that must be determined. Obviously, if we put the terms on the right-hand side of this equation over a common denominator, the numerators of the two expressions must be equal and we could find the value of K_i by comparing coefficients of various powers of s. Despite the fact that this procedure does, in fact, lead to the correct results, it requires a considerable amount of algebraic manipulation, which we would like to avoid. A much simpler approach, providing that the characteristic roots are distinct, is to multiply both sides of the equation by one of the factors, say $(s - \lambda_1)$,

$$\left[\frac{b_m(s - \gamma_1)(s - \gamma_2) \cdots (s - \gamma_m)}{(s - \lambda_1)(s - \lambda_2) \cdots (s - \lambda_n)}\right](s - \lambda_1)$$

$$= \frac{K_1(s - \lambda_1)}{s - \lambda_1} + \frac{K_2(s - \lambda_1)}{s - \lambda_2} + \cdots + \frac{K_n(s - \lambda_1)}{s - \lambda_n} \qquad 4.4\text{-}9$$

Now, if we let $s = \lambda_1$, all the terms on the right-hand side of the equation are equal to zero except the first, and we obtain

$$K_1 = \frac{b_m(\lambda_1 - \gamma_1)(\lambda_1 - \gamma_2) \cdots (\lambda_1 - \gamma_m)}{(\lambda_1 - \lambda_2)(\lambda_1 - \lambda_3) \cdots (\lambda_1 - \lambda_n)} \qquad 4.4\text{-}10$$

Repeating this procedure for each one of the factors in turn, we can easily evaluate $K_1, K_2, \ldots, K_n$. This result is valid both for real roots and complex conjugate roots but fails if there are multiple roots.

Example 4.4-1 Partial fraction expansion for step response

If we apply the foregoing procedure to the transform expression for the step response of a first-order system (see Eq. 4.3-14)

$$\tilde{y} = \frac{A}{s}\left(\frac{1}{\tau s + 1}\right) \qquad 4.4\text{-}11$$

We write

$$\tilde{y} = \frac{A}{\tau}\left(\frac{1}{s}\right)\left[\frac{1}{(s + 1/\tau)}\right] = \frac{A}{\tau}\left[\frac{K_1}{s} + \frac{K_2}{(s + 1/\tau)}\right] \qquad 4.4\text{-}12$$

Multiplying both sides of the equation by s, we obtain

$$\frac{A}{\tau}\left[\frac{1}{(s + 1/\tau)}\right] = \frac{A}{\tau}\left[K_1 + \frac{K_2 s}{(s + 1/\tau)}\right]$$

Next, setting $s = 0$, we find that

$$\frac{A}{\tau}\left(\frac{1}{1/\tau}\right) = \frac{A}{\tau}K_1$$

or

$$K_1 = \tau \qquad 4.4\text{-}13$$

Repeating this procedure for the other factor, we first obtain

$$\frac{A}{\tau}\left(\frac{1}{s}\right) = \frac{A}{\tau}\left\{\frac{K_1[(s + 1/\tau)]}{s} + K_2\right\}$$

and after setting s equal to $(-1/\tau)$ we see that

$$\frac{A}{\tau}\left(-\frac{1}{1/\tau}\right) = \frac{A}{\tau}K_2$$

or

$$K_2 = -\tau \qquad \text{4.4-14}$$

Substituting these expressions for K_1 and K_2 into Eq. 4.4-12, we obtain

$$\tilde{y} = \frac{A}{\tau}\left(\frac{1}{s}\right)\left[\frac{1}{(s + 1/\tau)}\right] = A\left[\frac{1}{s} - \frac{1}{(s + 1/\tau)}\right] \qquad \text{4.4-15}$$

which is identical to our previous result, Eq. 4.3-15, and can be verified by putting the terms over a common denominator.

Example 4.4-2 Partial fraction expansion for frequency response

Exactly the same procedure can be applied to the transform equation for the frequency response (see Eq. 4.3-17).

$$\tilde{y} = \left(\frac{1}{\tau s + 1}\right)\left(\frac{A\omega}{s^2 + \omega^2}\right) \qquad \text{4.4-16}$$

The roots of the quadratic term in the denominator are $s = \pm j\omega$, where $j = \sqrt{-1}$, so that when we write the equation in terms of its simple factors we obtain

$$\frac{A\omega}{\tau}\left[\frac{1}{(s + 1/\tau)}\right]\left(\frac{1}{s + j\omega}\right)\left(\frac{1}{s - j\omega}\right) = \frac{A\omega}{\tau}\left[\frac{K_1}{(s + 1/\tau)} + \frac{K_2}{s + j\omega} + \frac{K_3}{s - j\omega}\right]$$

$$\text{4.4-17}$$

Multiplying both sides of this equation by the factor $(s + 1/\tau)$ and letting $s = -1/\tau$ leads to the result

$$K_1 = \frac{1}{(-1/\tau)^2 + \omega^2} = \frac{\tau^2}{1 + \tau^2\omega^2} \qquad \text{4.4-18}$$

Also, multiplying by the factor $(s + j\omega)$ and letting $s = -j\omega$ gives

$$K_2 = \left[\frac{1}{(-j\omega + 1/\tau)}\right]\left(\frac{1}{-2j\omega}\right) = \frac{\tau}{2\omega}\left(-\frac{1}{\tau\omega + j}\right)\left(\frac{\tau\omega - j}{\tau\omega - j}\right)$$

$$= \left(-\frac{\tau}{2\omega}\right)\left(\frac{\tau\omega - j}{1 + \tau^2\omega^2}\right) \qquad \text{4.4-19}$$

Similarly, the result for K_3 is

$$K_3 = \left[\frac{1}{(j\omega + 1/\tau)}\right]\left(\frac{1}{2j\omega}\right) = \left(-\frac{\tau}{2\omega}\right)\left(\frac{\tau\omega + j}{1 + \tau^2\omega^2}\right) \qquad \text{4.4-20}$$

which is just the complex conjugate of the expression for K_2. Hence the partial fraction expansion of Eq. 4.4-17 becomes

$$\frac{A\omega}{\tau}\left[\frac{1}{(s + 1/\tau)}\right]\left(\frac{1}{s + j\omega}\right)\left(\frac{1}{s - j\omega}\right)$$

$$= \frac{A\omega\tau}{1 + \tau^2\omega^2}\left[\frac{1}{(s + 1/\tau)}\right] - \frac{A}{2}\left(\frac{\tau\omega - j}{1 + \tau^2\omega^2}\right)\left(\frac{1}{s + j\omega}\right)$$

$$- \frac{A}{2}\left(\frac{\tau\omega + j}{1 + \tau^2\omega^2}\right)\left(\frac{1}{s - j\omega}\right) \qquad \text{4.4-21}$$

After putting the last two terms over a common denominator, we get

$$\frac{A\omega}{\tau}\left[\frac{1}{(s+1/\tau)}\right]\left(\frac{1}{s^2+\omega^2}\right)=\left(\frac{A\omega\tau}{1+\tau^2\omega^2}\right)\left[\frac{1}{(s+1/\tau)}\right]$$
$$+\frac{A(\omega-\tau\omega s)}{1+\tau^2\omega^2}\left(\frac{1}{s^2+\omega^2}\right) \qquad 4.4\text{-}22$$

and if we break the last term up into two parts, we obtain the desired result.

It should be noted that we can apply the general method to the separate factors in Eq. 4.4-21. The solution for $y(t)$ becomes

$$y=\frac{A\tau\omega}{1+\tau^2\omega^2}e^{-t/\tau}-\frac{A}{2}\left(\frac{\tau\omega-j}{1+\tau^2\omega^2}\right)e^{-j\omega t}-\frac{A}{2}\left(\frac{\tau\omega+j}{1+\tau^2\omega^2}\right)e^{j\omega t} \qquad 4.4\text{-}23$$

If we substitute the identity

$$e^{(\alpha+j\omega)t}=e^{\alpha t}(\cos\omega t+j\sin\omega t) \qquad 4.4\text{-}24$$

the imaginary terms cancel out, and the solution is

$$y=\left(\frac{A\tau\omega}{1+\tau^2\omega^2}\right)e^{-t/\tau}+\left(\frac{A}{1+\tau^2\omega^2}\right)\sin\omega t-\left(\frac{A\tau\omega}{1+\tau^2\omega^2}\right)\cos\omega t \qquad 4.4\text{-}25$$

which again is identical to our previous result. Thus we find that the method of simple factors is valid for complex conjugate roots as well as real roots.

Example 4.4-3 Generalization of frequency response

The transfer function of a first-order process is given by the expression

$$\frac{\tilde{y}}{\tilde{g}}=H(s)=\frac{1}{\tau s+1} \qquad 4.3\text{-}6$$

For a sinusoidal forcing function $\tilde{g}=A\omega/(s^2+\omega^2)$, and the transform of the output becomes

$$\tilde{y}=\left(\frac{1}{\tau s+1}\right)\left(\frac{A\omega}{s^2+\omega^2}\right) \qquad 4.3\text{-}17$$

After a considerable amount of manipulation, we found that the steady state frequency response of the system is

$$y(t)=\frac{A}{\sqrt{1+\tau^2\omega^2}}\sin(\omega t+\phi) \qquad 4.3\text{-}20$$

where

$$\phi=\tan^{-1}(-\tau\omega) \qquad 4.3\text{-}21$$

It is interesting to note that if we merely substitute $j\omega$ for s into the transfer function, we can determine the system gain (the ratio of the amplitude of the output signal to the input signal) and the phase angle directly. Thus

$$H(s=j\omega)=\frac{1}{1+j\tau\omega}\left(\frac{1-j\tau\omega}{1-j\tau\omega}\right)=\frac{1-j\tau\omega}{1+\tau^2\omega^2}$$

Since we define the magnitude of a complex number as the square root of the sums of the squares of the real and imaginary parts and we define the tangent of the phase angle as the ratio of the imaginary part to the real part, we have

$$|G| = \left[\frac{1 + \tau^2\omega^2}{(1 + \tau^2\omega^2)^2}\right]^{1/2} = \frac{1}{\sqrt{1 + \tau^2\omega^2}}$$

$$\tan\phi = -\tau\omega \quad \text{or} \quad \phi = \tan^{-1}(-\tau\omega)$$

which are identical to the results above.

Show that this procedure of substituting $j\omega$ for s will also give the correct frequency response for a case where the transfer function is given by a ratio of polynomials

$$H(s) = \frac{p(s)}{q(s)} = \frac{b_m(s - \gamma_1)(s - \gamma_2)\cdots(s - \gamma_m)}{(s - \lambda_1)(s - \lambda_2)\cdots(s - \lambda_n)}, \qquad n > m \qquad 4.4\text{-}8$$

Solution

If the transfer function is

$$\frac{\tilde{y}(s)}{\tilde{g}(s)} = H(s) = \frac{b_m(s - \gamma_1)(s - \gamma_2)\cdots(s - \gamma_m)}{(s - \lambda_1)(s - \lambda_2)\cdots(s - \lambda_n)}$$

for a sinusodial forcing function, we have

$$\tilde{y}(s) = H(s)\left(\frac{A\omega}{s^2 + \omega^2}\right) = \frac{b_m(s - \gamma_1)(s - \gamma_2)\cdots(s - \gamma_m)}{(s - \lambda_1)(s - \lambda_2)\cdots(s - \lambda_n)}\left(\frac{A\omega}{s^2 + \omega^2}\right) \qquad 4.4\text{-}26$$

Since $n > m$, we can expand this expression in terms of its factors[1]

$$y(s) = \frac{K_1}{s - \lambda_1} + \frac{K_2}{s - \lambda_2} + \cdots + \frac{K_n}{s - \lambda_n} + \frac{K_\alpha}{s + j\omega} + \frac{K_\beta}{s - j\omega} \qquad 4.4\text{-}27$$

Then we know that the solution will have the form

$$y(t) = K_1 e^{\lambda_1 t} + K_2 e^{\lambda_2 t} + \cdots + K_n e^{\lambda_n t} + K_\alpha e^{-j\omega t} + K_\beta e^{j\omega t}$$

If the real part of all the characteristic roots λ_i are negative, the first n terms in the preceding equation approach zero as t approaches infinity (even if the roots are complex numbers), so that the steady state portion of the solution becomes

$$y(t) = K_\alpha e^{-j\omega t} + K_\beta e^{j\omega t} \qquad 4.4\text{-}28$$

In order to evaluate the coefficients K_α and K_β, we follow our normal procedure and mutiply both sides of Eq. 4.4-27 by the factor $(s + j\omega)$ and then let $s = -j\omega$. Thus

$$K_\alpha = \left[(s + j\omega)H(s)\frac{A\omega}{(s + j\omega)(s - j\omega)}\right]\bigg|_{s=-j\omega} = \frac{AH(-j\omega)}{-2j} \qquad 4.4\text{-}29$$

Similarly, we find that

$$K_\beta = \frac{AH(j\omega)}{2j} \qquad 4.4\text{-}30$$

---

[1] It is a simple matter to extend the analysis to include the case of repeated roots.

Substituting these results into the steady state solution, Eq. 4.4-28, gives

$$y(t) = A \left[\frac{H(j\omega)e^{j\omega t} - H(-j\omega)e^{-j\omega t}}{2j} \right] \qquad 4.4\text{-}31$$

However, we know that $H(j\omega)$ is simply some complex number; therefore we can write

$$H(j\omega) = \alpha + j\beta = |H(j\omega)|e^{j\phi} \qquad 4.4\text{-}32$$

$$H(-j\omega) = \alpha - j\beta = |H(-j\omega)|e^{-j\phi} \qquad 4.4\text{-}33$$

where α and β are the real and imaginary parts of the complex number and

$$|H(j\omega)| = \alpha^2 + \beta^2 \qquad \phi = \tan^{-1}\frac{\beta}{\alpha} \qquad 4.4\text{-}34$$

Hence our steady state solution can be written as

$$y(t) = A|H(j\omega)| \left\{ \frac{\exp[j(\omega t + \phi)] - \exp[-j(\omega t + \phi)]}{2j} \right\} \qquad 4.4\text{-}35$$

or recognizing that the term in brackets is merely the complex representation of the sine function, as

$$y(t) = A|H(j\omega)| \sin(\omega t + \phi) \qquad 4.4\text{-}36$$

Thus the steady state frequency response of any linear system is always a sine wave having the same frequency as the input signal. The ratio of the amplitude of the output signal to the input signal can be found by substituting $j\omega$ for s in the transfer function, rationalizing the result, and determining the magnitude of the final complex number. Similarly, the phase angle is the arc tangent of the ratio of the imaginary part of the number to the real part.

The only case we have not discussed is one where repeated roots are encountered. For example, if we consider the function

$$\tilde{y} = \frac{A}{s} \frac{1}{(\tau s + 1)^2} \qquad 4.4\text{-}37$$

which later we will show corresponds to the step response of two vessels placed in series, and attempt to use our previous method, we would write

$$\frac{A}{s} \frac{1}{(\tau s + 1)^2} = \frac{A}{\tau^2} \left[\frac{K_1}{s} + \frac{K_2}{(s + 1/\tau)} + \frac{K_3}{(s + 1/\tau)} \right] \qquad 4.4\text{-}38$$

Multiplying both sides of the equation by s and setting $s = 0$ gives

$$K_1 = \tau^2 \qquad 4.4\text{-}39$$

However, when we multiply both sides of the equation by the factor $(s + 1)/\tau$ to obtain

$$\frac{A}{\tau s} \left[\frac{1}{(s + 1)/\tau} \right] = \frac{A}{\tau^2} \left\{ \frac{K_1[(s + 1/\tau)]}{s} + K_2 + K_3 \right\}$$

and let $s = -1/\tau$, we run into trouble. The term involving K_1 disappears, but the term on the left-hand side becomes unbounded. Also, we cannot separate the factors K_2 and K_3. Hence the proposed form for the expansion, Eq. 4.4-38, cannot be correct.

Another possibility would be to write the expansion in the form

$$\frac{A}{s}\frac{1}{(\tau s + 1)^2} = \frac{A}{\tau^2}\left\{\frac{K_1}{s} + \frac{K_2}{[(s + 1/\tau)]^2} + \frac{K_3}{(s + 1/\tau)}\right\} \qquad 4.4\text{-}40$$

If we multiply both sides of this expression by s and set $s = 0$, we obtain the same result for K_1. Similarly, if we multiply both sides by the factor $[(s + 1/\tau)]^2$, we find that

$$\frac{A}{\tau^2 s} = \frac{A}{\tau^2}\left\{\frac{K_1[(s + 1/\tau)]^2}{s} + K_2 + K_3(s + 1/\tau)\right\} \qquad 4.4\text{-}41$$

Setting $s = -1/\tau$ gives

$$K_2 = -\tau \qquad 4.4\text{-}42$$

However, this procedure cannot be used to find K_3, for if we multiply both sides of Eq. 4.4-40 by $(s + 1/\tau)$ and set $s = -1/\tau$, we again find that the left-hand side of the equation becomes unbounded. This difficulty is resolved by differentiating both sides of Eq. 4.4-41 with respect to s

$$-\frac{A}{\tau^2 s^2} = \frac{A}{\tau^2}\left\{\frac{2K_1 s[(s + 1/\tau)] - K_1[(s + 1/\tau)]^2}{s^2} + K_3\right\}$$

$$= \frac{A}{\tau^2}\left\{\frac{[(s + 1/\tau)][(s - 1/\tau)]}{s^2} + K_3\right\} \qquad 4.4\text{-}43$$

Now if we let $s = -1/\tau$ and solve for K_3, we get

$$K_3 = -\tau^2 \qquad 4.4\text{-}44$$

Substituting the expressions for K_1, K_2, and K_3 into Eq. 4.4-40 gives

$$y = \frac{A}{s}\left[\frac{1}{(\tau s + 1)^2}\right] = \frac{A}{s} - \frac{A}{\tau}\frac{1}{[(s + 1/\tau)]^2} - \frac{A}{(s + 1/\tau)} \qquad 4.4\text{-}45$$

A transform leading to a quadratic factor can be established by combining Eqs. 4.2-8 and 4-2-11.

$$\mathscr{L}[te^{-at}] = \frac{1}{(s + a)^2} \qquad 4.4\text{-}46$$

With this information we can find the inverse transform of Eq. 4.4-45 by inverting each factor individually. The result is

$$y = A\left[1 - e^{-t/\tau} - \frac{t}{\tau}e^{-t/\tau}\right] \qquad 4.4\text{-}47$$

Thus if a function has repeated, or multiple, roots, it is necessary to differentiate the appropriate form of the expansion, Eqs. 4.4-40 and 4.4-41, in order to evaluate the coefficients. In fact, if a root is repeated n times, n differentiations will be required.

The preceding results can be generalized, and they are usually referred to as the Heaviside expansion theorems.

Heaviside Expansion Theorems

We consider some transformed expression

$$\bar{y}(s) = \frac{p(s)}{q(s)} \qquad 4.4\text{-}48$$

where $p(s)$ and $q(s)$ are polynomials in s and the order of $q(s)$ is higher than that of $p(s)$, and attempt to find the inverse transform.

1. For any unrepeated factor $(s - a)$ of $q(s)$, we can write the transformed expression as

$$\bar{y}(s) = \frac{p(s)}{q(s)} = \frac{p(s)}{Q(s)(s - a)} \qquad 4.4\text{-}49$$

where $Q(s)$ is the product of all factors of $q(s)$ except $(s - a)$. Then the contribution to the inverse transform, $y(t)$, caused by the factor $(s - a)$ is

$$y_a(t) = \frac{p(a)}{Q(a)} e^{at} \qquad 4.4\text{-}50$$

An alternate expression, which is identical to the preceding result, is

$$\frac{p(a)}{(dq/ds)|_a} e^{at} \qquad 4.4\text{-}51$$

2. For a repeated factor of $q(s)$, say $(s - a)^n$, we can write the transformed expression in the form

$$\bar{y}(s) = \frac{p(s)}{q(s)} = \frac{\phi(s)}{(s - a)^n} \qquad 4.4\text{-}52$$

where $\phi(s)$ is the ratio of $p(s)$ to all factors of $q(s)$ except $(s - a)^n$. Then the contribution of this repeated factor to the inverse transform $y(t)$ is

$$\begin{aligned}
y_a(t) &= \left[\frac{(d^{n-1}\phi/ds^{n-1})|_a}{(n-1)!} + \frac{(d^{n-2}\phi/ds^{n-2})|_a}{(n-2)!}\frac{t}{1!} + \cdots \right. \\
&\quad \left. + \frac{(d\phi/ds)|_a}{1!}\frac{t^{n-2}}{(n-2)!} + \phi(a)\frac{t^{n-1}}{(n-1)!} \right] e^{at} \\
&= e^{at} \sum_{i=1}^{n} \left[\frac{d^{n-i}\phi/ds^{n-i}}{(n-i)!} \right]_a \frac{t^{(i-1)}}{(i-1)!} \qquad 4.4\text{-}53
\end{aligned}$$

3. For a quadratic factor of $q(s)$, say $[(s + a)^2 + b^2]$, we can write the transformed expression in the form

$$\bar{y}(s) = \frac{p(s)}{q(s)} = \frac{\psi(s)}{(s + a)^2 + b^2} \qquad 4.4\text{-}54$$

where $\psi(s)$ is the ratio of $p(s)$ to all factors in $q(s)$ except $[(s + a)^2 + b^2]$.

Then the contribution of this quadratic factor to the inverse transform $y(t)$ is

$$y_{-a \pm jb}(t) = \frac{e^{-at}}{b}(\psi_i \cos bt + \psi_r \sin bt) \qquad \text{4.4-55}$$

where ψ_r and ψ_i are the real and imaginary parts of $\psi(s)$ when $s = -a + jb$.

To illustrate the application of these Heaviside theorems we consider our previous example

$$\tilde{y} = \frac{A}{s}\frac{1}{(\tau s + 1)^2} \qquad \text{4.4-37}$$

For this case we have one root at $s = 0$ and a pair of repeated roots at $s = 1/\tau$. Considering the unrepeated factor first, we write

$$\tilde{y} = \frac{p(s)}{Q(s)(s-a)} = \left[\frac{1}{(\tau s + 1)^2}\right]\frac{1}{s}$$

so that $p(s) = 1$, $Q(s) = 1/(\tau s + 1)^2$, and the contribution of this factor to the solution is

$$y_{(0)} = \frac{1}{(\tau \cdot 0 + 1)^2}e^{0 \cdot t} = 1$$

For the repeated root we have

$$\tilde{y} = \frac{\phi}{(s-a)^n} = \frac{A/\tau^2 s}{(s + 1/\tau)^2}$$

Hence $\phi = A/\tau^2 s$. The root is repeated twice, so we need two terms from the expansion in Eq. 4.4-53

$$y_a = \left[\frac{(d\phi/ds)|_a}{1!} + \phi\Big|_a\frac{t}{1!}\right]e^{at}$$

Since $d\phi/ds = -A/\tau^2 s^2$, and $(d\phi/ds)|_{-1/\tau} = -A$, then the contribution of the repeated factor will be

$$y_{(-1/\tau)} = \left[-A - A\frac{t}{\tau}\right]e^{-t/\tau}$$

Combining the two expressions, we have

$$y(t) = A\left[1 - e^{-t/\tau} - \frac{t}{\tau}e^{-t/\tau}\right]$$

which is identical to our previous result, Eq. 4.4-47.

Limit Theorems

As mentioned several times in previous discussions, a step input to a process will force the system to go to some new steady state operating condition. Although this final operating level can be determined by inverting the Laplace transform for the system output and examining the solution (see

Eq. 4.3-15), it would be advantageous to be able to calculate this value directly from the transformed expression. It is possible to develop a relationship of this kind by considering certain limits of the definition of the Laplace transform of a derivative

$$\int_0^\infty \frac{df}{dt} e^{-st}\, dt = s\tilde{f}(s) - f(0) \qquad 4.2\text{-}17$$

If we take the limit of this equation as s approaches zero, we obtain

$$\lim_{s\to 0} \int_0^\infty \frac{df}{dt} e^{-st}\, dt = \lim_{s\to 0} [s\tilde{f}(s)] - f(0)$$

For most functions of engineering interest, it is permissible to interchange the order of the integration and limiting operations so that the equation becomes

$$\int_0^\infty \frac{df}{dt}\, dt = \lim_{s\to 0} [s\tilde{f}(s)] - f(0)$$

After evaluating the integral, we find that

$$\lim_{t\to\infty} [f(t)] - f(0) = \lim_{s\to 0} [s\tilde{f}(s)] - f(0)$$

Hence

$$\lim_{t\to\infty} [f(t)] = \lim_{s\to 0} [s\tilde{f}(s)] \qquad 4.4\text{-}56$$

This result is known as the *final value theorem*, and it is applicable for most functions of interest to engineers provided that the roots of the denominator of $s\tilde{f}(s)$ have negative real parts. If this last restriction is not met, the system is unstable and the output becomes unbounded as t approaches infinity.

It is interesting to note that if we again consider the definition of the Laplace transform of a derivative, but consider the limit of this expression as s approaches infinity rather than zero,

$$\lim_{s\to\infty} \int_0^\infty \frac{df}{dt} e^{-st}\, dt = \lim_{s\to\infty} [s\tilde{f}(s)] - f(0)$$

for most functions the term on the left-hand side of the equation approaches zero and we obtain

$$\lim_{s\to\infty} [s\tilde{f}(s)] = \lim_{t\to 0} [f(t)] = f(0) \qquad 4.4\text{-}57$$

This result is known as the *initial value theorem* and is sometimes useful in the analysis of engineering problems.

Example 4.4-4 Final value of step response

The expression for the Laplace transform of the step response of a first-order system is given by Eq. 4.3-14.

$$\tilde{y} = \frac{A}{s}\left(\frac{1}{\tau s + 1}\right) \qquad 4.3\text{-}14$$

Hence, according to the final value theorem,

$$\lim_{t \to \infty} [y(t)] = \lim_{s \to 0} [s\tilde{y}(s)] = \lim_{s \to 0} \left[\frac{A}{\tau s + 1} \right] = A \qquad \text{4.3-58}$$

which agrees with the result obtained from Eq. 4.3-15.

SECTION 4.5   RESPONSE OF SECOND-ORDER SYSTEMS

In our previous discussion of the diagonalization of matrix equations, we found that it was always possible to put the equations into the form

$$\frac{dy_i}{dt} = \lambda_i y_i + f_i(t), \qquad i = 1, 2, \ldots, n \qquad \text{4.5-1}$$

However, if λ_i was one root of a pair of complex conjugate roots, say $\lambda_1 = \alpha + j\omega$ and $\lambda_2 = \alpha - j\omega$, Eq. 4.5-1 implies that y_i must be a complex variable. For this case, we avoided the additional complexity of handling complex state variables by making a second transformation

$$y_1 = z_1 + jz_2 \qquad y_2 = z_1 - jz_2 \qquad \text{3.3-94}$$

to obtain a new canonical form

$$\frac{dz_1}{dt} = \alpha z_1 - \omega z_2 + \frac{f_1 + f_2}{2} = \alpha z_1 - \omega z_2 + g_1$$

$$\frac{dz_2}{dt} = \omega z_1 + \alpha z_2 + \frac{f_1 + f_2}{2j} = \omega z_1 + \alpha z_2 + g_2 \qquad \text{3.3-95}$$

where g_1 and g_2 represent linear combinations of the original forcing functions.

Although it would be possible to find solutions for this pair of coupled, first-order differential equations, we will combine them into a single second-order equation. Differentiating the second expression in the set gives

$$\frac{d^2 z_2}{dt^2} = \omega \frac{dz_1}{dt} + \alpha \frac{dz_2}{dt} + \frac{dg_2}{dt} \qquad \text{4.5-2}$$

Then, eliminating dz_1/dt by using the first equation gives

$$\frac{d^2 z_2}{dt^2} = \omega(\alpha z_1 - \omega z_2) + \alpha \frac{dz_2}{dt} + \omega g_1 + \frac{dg_2}{dt} \qquad \text{4.5-3}$$

Finally, solving the second equation in the set for z_1 and using the result to eliminate this variable from the equation above gives

$$\frac{d^2 z_2}{dt^2} = \alpha \left(\frac{dz_2}{dt} - \alpha z_2 \right) - \omega^2 z_2 + \alpha \frac{dz_2}{dt} + \omega g_1 + \frac{dg_2}{dt} - \alpha g_2 \qquad \text{4.5-4}$$

or

$$\frac{d^2 z_2}{dt^2} - 2\alpha \frac{dz_2}{dt} + (\alpha^2 + \omega^2) z_2 = \frac{dg_2}{dt} + \omega g_1 - \alpha g_2 \qquad \text{4.5-5}$$

A similar approach can be used to show that

$$\frac{d^2z_1}{dt^2} - 2\alpha\frac{dz_1}{dt} + (\alpha^2 + \omega^2)z_1 = \frac{dg_1}{dt} - \alpha g_1 - \omega g_2 \qquad 4.5\text{-}6$$

Of course, the application of differential operators leads to the same results.

Rather than study the behavior of the foregoing equations, we will consider the general equation

$$\tau^2\frac{d^2y}{dt^2} + 2\gamma\tau\frac{dy}{dt} + y = f(t) \qquad 4.5\text{-}7$$

Clearly, this relationship can be reduced to either Eq. 4.5-5 or Eq. 4.5-6 by letting

$$\tau = \frac{1}{\sqrt{\alpha^2 + \omega^2}} \qquad \gamma = -\frac{\alpha}{\sqrt{\alpha^2 + \omega^2}} \qquad 4.5\text{-}8$$

and by letting $f(t)$ be equal to the right-hand side of Eq. 4.5-5 or 4.5-6 divided by $(\alpha^2 + \omega^2)$. This expression is usually taken as the standard form for second-order equations in the literature, for it is applicable to cases of two real roots and complex conjugate roots, whereas the expressions given by Eq. 4.5-5 or 4.5-6 are limited to complex roots. Although our study will be restricted to simple forcing functions $f(t)$, the response caused by more complicated inputs can be determined by slight modifications of the analysis.

Taking the Laplace transform of Eq. 4.5-7 gives the result

$$(\tau^2 s^2 + 2\gamma\tau s + 1)\tilde{y} = \tilde{f} \qquad 4.5\text{-}9$$

so that the *transfer function* becomes

$$\frac{\tilde{y}}{\tilde{f}} = H(s) = \frac{1}{\tau^2 s^2 + 2\gamma\tau s + 1} \qquad 4.5\text{-}10$$

After finding the roots of the quadratic term in the denominator

$$\lambda_1 = -\frac{\gamma}{\tau} + \frac{1}{\tau}\sqrt{\gamma^2 - 1} \qquad \lambda_2 = -\frac{\gamma}{\tau} - \frac{1}{\tau}\sqrt{\gamma^2 - 1} \qquad 4.5\text{-}11$$

we can write the transfer function as

$$\frac{\tilde{y}}{\tilde{f}} = H(s) = \frac{1}{\tau^2\{s + [(\gamma - \sqrt{\gamma^2 - 1})/\tau]\}\{s + [(\gamma + \sqrt{\gamma^2 - 1})/\tau]\}} \qquad 4.5\text{-}12$$

The nature of the characteristic roots depends on the value of γ; that is, if $0 < \gamma < 1$ we obtain complex conjugate roots, if $\gamma = 1$ we obtain two equal roots, and if $\gamma > 1$ we obtain real roots. Thus we expect to get a different kind of response for different values of γ. This parameter is called the *damping coefficient*, and the response when $0 < \gamma < 1$ is called *underdamped*, the response when $\gamma = 1$ is called *critically damped*, and the response when $\gamma > 1$ is called *overdamped*.

Impulse Response

For an impulse input, we know from Eq. 4.2-15 that

$$\tilde{f} = A \qquad\qquad 4.5\text{-}13$$

The transform of the output becomes

$$\tilde{y} = \frac{A}{\tau^2\{s + [(\gamma - \sqrt{\gamma^2 - 1})/\tau]\}\{s + [(\gamma + \sqrt{\gamma^2 - 1})/\tau]\}} \qquad 4.5\text{-}14$$

We can find the inverse transform by using the Heaviside theorems, Eqs. 4.4-48 through 4.4-55, where the appropriate form of the theorem depends on the value of γ. The results are

Underdamped response $0 < \gamma < 1$

$$y(t) = \frac{A}{\tau} \frac{1}{\sqrt{1 - \gamma^2}} e^{-\gamma t/\tau} \sin \frac{\sqrt{1 - \gamma^2}\,t}{\tau} \qquad 4.5\text{-}15$$

Critically damped response $\gamma = 1$

$$y(t) = \frac{A}{\tau^2} t e^{-t/\tau} \qquad\qquad 4.5\text{-}16$$

Overdamped response $\gamma > 1$

$$y(t) = \left[\frac{A}{\tau\sqrt{\gamma^2 - 1}}\right]\left[\exp\left(-\frac{\gamma t}{\tau}\right)\right]\left[\sinh\left(\sqrt{\gamma^2 - 1}\,\frac{t}{\tau}\right)\right] \qquad 4.5\text{-}17$$

or

$$y(t) = \left[\frac{A}{2\tau\sqrt{\gamma^2 - 1}}\right]\left\{\exp\left[-\left(\gamma - \sqrt{\gamma^2 - 1}\right)\frac{t}{\tau}\right]\right.$$
$$\left. - \exp\left[-\left(\gamma + \sqrt{\gamma^2 - 1}\right)\frac{t}{\tau}\right]\right\} \qquad 4.5\text{-}18$$

Providing that γ and τ are positive quantities, the system output will always return to the original steady state value after an impulse upset. A sketch of some response curves is given in Figure 4.5-1. The graph shows that the underdamped system response approaches the original steady state operating condition in an oscillatory manner, taking on values both above and below the original output. This kind of behavior was not observed for the impulse response of a first-order system, and it is due to the presence of the pair of complex conjugate roots. The response of a critically damped system shows the fastest return to steady state operation with no oscillation, and the overdamped system returns more slowly. It is interesting to note that the solution for the overdamped plant, Eq. 4.5-18, involves the sum of two exponential terms, $e^{\lambda_1 t}$ and $e^{\lambda_2 t}$, and therefore can be interpreted as a linear combination of the response of two first-order systems (see Eqs. 4.3-10 and

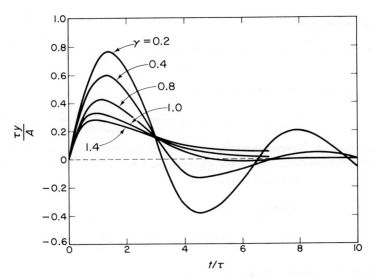

Figure 4.5-1. Impulse response of second-order systems.

4.3-11). The validity of this relationship can be established by comparing the solution obtained using the canonical transformation with the result above.

Step Response

For a step input, Eq. 4.2-3 gives that

$$\tilde{f} = \frac{A}{s} \qquad\qquad 4.5\text{-}19$$

and the transform of the output becomes

$$\tilde{y} = \frac{A}{\tau^2 s\{s + [(\gamma - \sqrt{\gamma^2 - 1})/\tau]\}\{s + [(\gamma + \sqrt{\gamma^2 - 1})/\tau]\}} \qquad 4.5\text{-}20$$

Using Heaviside's theorems once again, we find that the outputs for the three cases are

Underdamped response $0 < \gamma < 1$

$$y = A - \left[\frac{A}{\sqrt{1 - \gamma^2}}\right]\left[\exp\left(-\frac{\gamma t}{\tau}\right)\right]\sin\left(\frac{\sqrt{1 - \gamma^2}\,t}{\tau} + \phi\right) \qquad 4.5\text{-}21$$

where

$$\phi = \tan^{-1}\left(\frac{\sqrt{1 - \gamma^2}}{\gamma}\right) \qquad\qquad 4.5\text{-}22$$

Critically damped response

$$y = A - A\left(1 + \frac{t}{\tau}\right)e^{-t/\tau} \qquad\qquad 4.5\text{-}23$$

Overdamped response

$$y = A - Ae^{-\gamma t/\tau}\left[\cosh\left(\frac{\sqrt{\gamma^2 - 1}\,t}{\tau}\right) + \frac{\gamma}{\sqrt{\gamma^2 - 1}}\sinh\left(\frac{\sqrt{\gamma^2 - 1}\,t}{\tau}\right)\right] \qquad 4.5\text{-}24$$

or

$$y = A - \frac{A}{2\sqrt{\gamma^2 - 1}}\left\{(\gamma + \sqrt{\gamma^2 - 1})\exp\left[-\frac{(\gamma - \sqrt{\gamma^2 - 1})t}{\tau}\right]\right.$$
$$\left. - (\gamma - \sqrt{\gamma^2 - 1})\exp\left[-\frac{(\gamma + \sqrt{\gamma^2 - 1})t}{\tau}\right]\right\} \qquad 4.5\text{-}24$$

A study of these equations indicates that, providing that γ and τ are positive, the system approaches a new steady state operating level A as time approaches infinity. Some typical solutions are plotted in Figure 4.5-2. The results for the underdamped case show that the output initially overshoots the final steady state value and then approaches it in an oscillatory manner. This kind of behavior (i.e., the possibility of overshoot) does not appear in first-order systems and cannot be predicted from a steady state analysis. Since it is apparent that serious problems can be encountered if certain variables, such as reactor temperature or pressure, are allowed to exceed their design values greatly, during a start-up procedure for example, we find another

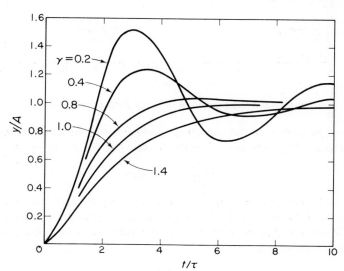

Figure 4.5-2. Step response of second-order systems.

reason for at least obtaining an estimate of the dynamic characteristics of a new plant.

The response of the critically damped system shows the fastest approach to the new steady state with no oscillations, whereas the output of an over-damped plant has a more sluggish approach to the ultimate value. In some situations we would like to change the steady state operating level as quickly as possible. Although the critically damped response is faster than the over-damped, the underdamped response is the fastest. Normally we do not mind if the system oscillates somewhat, but we cannot tolerate an excessive amount of overshoot. In order to be able to discuss this kind of a compromise in quantitative terms, it is common practice to introduce several new terms. The quantities used in the definitions are shown in Figure 4.5-3.

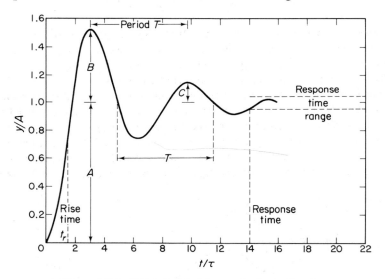

Figure 4.5-3. Step response of an underdamped system.

Natural frequency. If the damping coefficient γ is set equal to zero, the general equation for the second-order system, Eq. 4.5-7, becomes

$$\tau^2 \frac{d^2 y}{dt^2} + y = f(t) \qquad\qquad 4.5\text{-}25$$

For this case the complex conjugate roots have no real part, and the system generates oscillations that do not decay as time approaches infinity. From Eq. 4.5-21 we find that the frequency of these undamped oscillations, called the *natural frequency*, is

$$\omega_n = \frac{1}{\tau} \qquad\qquad 4.5\text{-}26$$

The ratio of the actual frequency for a damped system to the natural frequency is

$$\frac{\omega}{\omega_n} = \sqrt{1 - \gamma^2} \qquad\qquad 4.5\text{-}27$$

For low values of the damping coefficient $\gamma < 0.3$, there will be less than a 10 percent difference between these two values. (see Figure 4.5-4.)

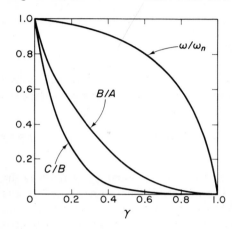

Figure 4.5-4. The effect of damping coefficient on response parameters.

Period of oscillation. Since the frequency of the oscillation, given by Eq. 4.5-21, is

$$\omega = \frac{\sqrt{1 - \gamma^2}}{\tau} \text{ radians per time} \qquad\qquad 4.5\text{-}28$$

and since

$$\omega = 2\pi f \qquad\qquad 4.5\text{-}29$$

where f is in cycles per time, the *period of the oscillation, T,* or the time per cycle is

$$T = \frac{1}{f} = \frac{2\pi\tau}{\sqrt{1 - \gamma^2}} \qquad\qquad 4.5\text{-}30$$

The period can be determined from experimental data by measuring the time between successive maxima or minima on the response curve or the time between alternate crossings of the final steady state operating level.

Overshoot. The *overshoot* is defined as the ratio of the amount that the maximum value of the first peak exceeds the final steady state value to the final steady state value—that is, the ratio of B to A on Figure 4.5-3. An analytical expression for this quantity can be determined directly from the solution for the step response, Eq. 4.5-21. The result is

$$\text{Overshoot} = \frac{B}{A} = \exp\left(-\frac{\pi\gamma}{\sqrt{1 - \gamma^2}}\right) \qquad\qquad 4.5\text{-}31$$

and this equation is plotted in Figure 4.5-4.

Decay ratio. The *decay ratio* is defined as the ratio of the amount that the maximum value of the second peak exceeds the final steady state value to the amount that the first peak exceeds this value—that is, the ratio of C to B on Figure 4.5-3. Again, the solution for the transient response can be used to derive an analytical expression for the decay ratio

$$\text{Decay ratio} = \frac{C}{B} = \exp\left(-\frac{2\pi\gamma}{\sqrt{1-\gamma^2}}\right) \qquad 4.5\text{-}32$$

This result is plotted in Figure 4.5-4.

Rise time. The *rise time* is defined as the time required for the system output to first attain the final steady state value. From the transient response solution, we find that the rise time t_r must be

$$t_r = \frac{\tau}{\sqrt{1-\gamma^2}}\tan^{-1}\left(\frac{\gamma}{\sqrt{1-\gamma^2}}\right) \qquad 4.5\text{-}33$$

Response time. The *response time* of the system usually is defined as the time required for the system output to always remain within $\pm 5\%$ of the final steady state value. In other words, it is a measure of the time required for the plant to finally achieve the new steady state value, and it is somewhat arbitrary, for the approach is asymptotic. This same difficulty was encountered in the case of first-order systems where the response time was given as three to five time constants.

A more complete discussion of the physical significance of these terms will be deferred until we consider the techniques for designing control systems.

Frequency Response

For sinusoidal inputs, $f(t) = A\sin\omega_0 t$ and

$$\tilde{f} = \frac{A\omega_0}{s^2 + \omega_0^2} \qquad 4.3\text{-}16$$

where ω_0 is the forcing frequency. The equation for the transform of the output signal becomes

$$\tilde{y} = \left(\frac{1}{\tau^2 s^2 + 2\tau\gamma s + 1}\right)\left(\frac{A\omega_0}{s^2 + \omega_0^2}\right) \qquad 4.5\text{-}34$$

Although it is possible to use the Heaviside theorems to find the inverse transform, we are most interested in the pseudo-steady state frequency response, and therefore we can follow the approach described in Example 4.4-3 and simply substitute $j\omega_0$ for s in the transfer function. Thus

$$H(j\omega_0) = \frac{1}{\tau^2(j\omega_0)^2 + 2\tau\gamma(j\omega_0) + 1} = \frac{1}{1 - \tau^2\omega_0^2 + 2\tau\gamma\omega_0 j} \qquad 4.5\text{-}35$$

After rationalizing this complex number, we obtain

$$H(j\omega_0) = \frac{1 - \tau^2\omega_0^2 - 2\tau\gamma\omega_0 j}{(1 - \tau^2\omega_0^2)^2 + (2\tau\gamma\omega_0)^2} \qquad 4.5\text{-}36$$

Hence, according to Eq. 4.4-34, the system gain is

$$|H(j\omega_0)| = \sqrt{\frac{(1 - \tau^2\omega_0^2)^2 + (2\tau\gamma\omega_0)^2}{[(1 - \tau^2\omega_0^2)^2 + (2\tau\gamma\omega_0)^2]^2}}$$

$$= \frac{1}{\sqrt{(1 - \tau^2\omega_0^2)^2 + (2\tau\gamma\omega_0)^2}} \qquad 4.5\text{-}37$$

and the phase angle is

$$\phi = \tan^{-1}\left(-\frac{2\tau\gamma\omega_0}{1 - \tau^2\omega_0^2}\right) \qquad 4.5\text{-}38$$

Also, from Eq. 4.4-36, the solution for the pseudo-steady state response is

$$y(t) = \frac{A}{\sqrt{(1 - \tau^2\omega_0^2)^2 + (2\tau\gamma\omega_0)^2}} \sin(\omega_0 t + \phi) \qquad 4.5\text{-}39$$

It is a simple matter to determine the qualitative nature of these results. Considering the phase angle first, we find that as ω_0 approaches zero, the phase angle will approach zero; when $\tau\omega_0 = 1$, the phase angle must be equal to $-90°$ for all values of the damping coefficient; and when ω_0 becomes very large and approaches infinity, the argument of the arc tangent is always positive and approaches zero so that ϕ approaches $-180°$. Thus the maximum phase lag for a second-order system is $-180°$, which is just twice the value for a first-order system.

The gain curve can be described in a similar fashion. At very low values of ω_0—that is, as ω_0 approaches zero—the system gain approaches unity. Also, as ω_0 approaches infinity, the gain curve becomes asymptotic to the line $|H(j\omega_0)| = 1/\tau^2\omega_0^2$. Since a Bode plot is a graph of the logarithm of $|H(j\omega_0)|$, or $|G|$, plotted against the logarithm of $\tau\omega_0$, the line will have a slope of -2 and will pass through the point where $|G| = 1$ when $\tau\omega_0 = 1$. The shape of the curve between these two asymptotes depends on the value of γ, as expected. We can get some additional insight into the behavior of the curves, however, if we differentiate the expression for the system gain with respect to $\tau\omega_0$ in order to determine the location of possible maxima of the curves. This leads to the result

$$(\tau\omega_0)_{max} = \sqrt{1 - 2\gamma^2} \qquad 4.5\text{-}40$$

so that a maximum will occur when γ is less than 0.707. The value of the maximum gain can then be found by substituting this expression back into Eq. 4.5-37

$$|G_{max}| = \frac{1}{2\gamma\sqrt{1 - \gamma^2}} \qquad 4.5\text{-}41$$

The Bode diagram is given in Figure 4.5-5

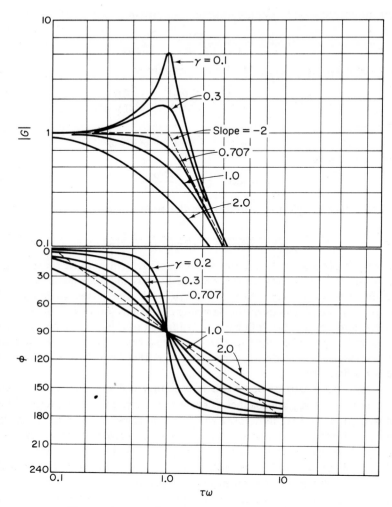

Figure 4.5-5. Bode diagram for a second-order system.

Our analysis of first-order systems showed that the amplitude of the output sine wave was always smaller than that of the input signal; that is, first-order systems always tend to damp out fluctuations. However, if the damping factor, γ, is less than 0.707 for a second-order system, the system gain is greater than unity and any fluctuations entering the system will be amplified. This phenomenon is called resonance and is commonly observed in physical systems. For example, the vibration of a car at certain speeds, the vibration of a bridge at certain wind conditions, the ability of a child to "pump" a swing, the ability of a tenor (in the movies) to shatter a mirror by singing a certain note, and so on, all may be attributed to resonance. The maximum gain occurs at what is called the resonant frequency, ω_r, and from Eq. 4.5-40 we find that

$$\omega_r = \frac{1}{\tau}\sqrt{1 - 2\gamma^2}, \qquad 0 < \gamma < 0.707 \qquad\qquad \text{4.5-42}$$

It is clear from Eq. 4.5-41 that the maximum gain becomes unbounded as γ approaches zero, so that a system failure will be obtained at these conditions. The rate of growth of the amplitude of this case can be determined by solving the equation

$$\frac{d^2y}{dt^2} + \omega^2 x = A \sin \omega t \qquad\qquad \text{4.5-43}$$

The complementary solution is just $y_c = C_1 \sin \omega t + C_2 \cos \omega t$, and the particular integral must be chosen as $y_p = C_3 t \sin \omega t + C_4 t \cos \omega t$, which becomes unbounded as t approaches infinity.

The solutions for the step and impulse response for the case where $\gamma > 1$ (i.e., two real roots) could be written in terms of a linear combination of the response of two first-order systems. If we write the transfer function in the form

$$H(s) = \frac{1}{\tau^2\{s + [(\gamma - \sqrt{\gamma^2 - 1})/\tau]\}\{s + [(\gamma + \sqrt{\gamma^2 - 1})/\tau]\}}$$

$$= \frac{1}{\tau^2(s - \lambda_1)(s - \lambda_2)} \qquad\qquad \text{4.5-44}$$

for this case, and let

$$-\lambda_1 = \frac{1}{\tau_1}, \quad -\lambda_2 = \frac{1}{\tau_2}, \quad \tau_1\tau_2 = \frac{1}{\tau^2} \qquad\qquad \text{4.5-45}$$

which is analogous to the first-order case (see Eq. 4.3-11), the transfer function becomes

$$H(s) = \frac{1}{(\tau_1 s + 1)(\tau_2 s + 1)} \qquad\qquad \text{4.5-46}$$

Now, substituting $j\omega_0$ for s, we obtain

$$H(j\omega_0) = \frac{1}{(1 + j\omega_0\tau_1)(1 + j\omega_0\tau_2)} = \frac{1 - \omega_0^2\tau_1\tau_2 - j\omega_0(\tau_1 + \tau_2)}{(1 + \tau_1^2\omega_0^2)(1 + \tau_2^2\omega_0^2)} \qquad \text{4.5-47}$$

The phase angle for the frequency response is

$$\tan \phi = -\frac{\omega_0(\tau_1 + \tau_2)}{1 - \omega_0^2\tau_1\tau_2} = \frac{(-\omega_0\tau_1) + (-\omega_0\tau_2)}{1 - (-\omega_0\tau_1)(-\omega_0\tau_2)} \qquad \text{4.5-48}$$

Letting

$$\tan \phi_1 = (-\omega_0\tau_1) \qquad \tan \phi_2 = (-\omega_0\tau_2) \qquad\qquad \text{4.5-49}$$

and using the trigonometric identity

$$\tan (\phi_1 + \phi_2) = \frac{\tan \phi_1 + \tan \phi_2}{1 - \tan \phi_1 \tan \phi_2} \qquad\qquad \text{4.5-50}$$

we see that the total phase angle ϕ is just the sum of the phase angles of two first-order systems.

From Eq. 4.5-47 we find that the system gain is

$$|H(j\omega_0)| = \sqrt{\frac{(1 - \omega_0^2\tau_1\tau_2)^2 + \omega_0^2(\tau_1 + \tau_2)^2}{(1 + \tau_1^2\omega_0^2)^2(1 + \tau_2^2\omega_0^2)^2}} \qquad 4.5\text{-}51$$

After some manipulation, this result can be written

$$|H(j\omega_0)| = \frac{1}{\sqrt{(1 + \tau_1^2\omega_0^2)(1 + \tau_2^2\omega_0^2)}} \qquad 4.5\text{-}52$$

Also,

$$\log|H(j\omega_0)| = \log\left(\frac{1}{\sqrt{1 + \tau_1^2\omega_0^2}}\right) + \log\left(\frac{1}{\sqrt{1 + \tau_2^2\omega_0^2}}\right) \qquad 4.5\text{-}53$$

Thus if we have a second-order system with two real roots, we can find the frequency response by graphically adding the gains and phase angles of two first-order systems on a Bode plot (see Figure 4.5-6).

If we actually have two first-order systems in series, the system equations

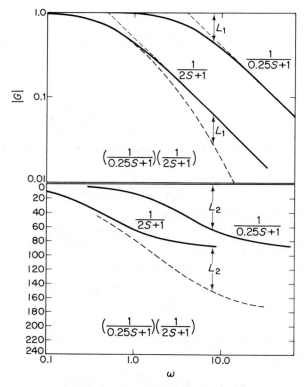

Figure 4.5-6. Frequency response of an overdamped system.

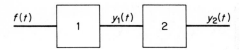

Figure 4.5-7. First-order systems in series.

can be written as

$$\tau_1 \frac{dy_1}{dt} + y_1 = f(t) \qquad\qquad 4.5\text{-}54$$

$$\tau_2 \frac{dy_2}{dt} + y_2 = y_1 \qquad\qquad 4.5\text{-}55$$

Taking the Laplace transform of the equations gives

$$(\tau_1 s + 1)\tilde{y}_1 = \tilde{f} \qquad\qquad 4.5\text{-}56$$

$$(\tau_2 s + 1)\tilde{y}_2 = \tilde{y}_1 \qquad\qquad 4.5\text{-}57$$

and after eliminating $\tilde{y}_1$, the transfer function becomes

$$\frac{\tilde{y}}{\tilde{f}} = \frac{1}{(\tau_1 s + 1)(\tau_2 s + 1)} \qquad\qquad 4.5\text{-}58$$

which is identical to Eq. 4.5-56. This kind of a system is called *noninteracting* because whatever process is occurring in the second system has no affect on the first. Alternatively, a system described by the pair of equations

$$\frac{dx_1}{dt} = a_{11}x_1 + a_{12}x_2 + f_1 \qquad\qquad 4.5\text{-}59$$

$$\frac{dx_2}{dt} = a_{21}x_1 + a_{22}x_2 + f_2 \qquad\qquad 4.5\text{-}60$$

which can be put into the standard form

$$\tau^2 \frac{d^2y}{dt^2} + 2\gamma\tau \frac{dy}{dt} + y = f(t) \qquad\qquad 4.5\text{-}7$$

using the procedure described in Eqs. 4.5-2 through 4.5-7, is called an *interacting* system because both equations depend on both dependent variables. However, as was shown above, this plant also can be written in terms of an equivalent system of two first-order processes in series.

Example 4.5-1 Step response of a manometer

An astute sophomore once asked, Why is it so easy for manometers to overflow?

Solution

Let h represent the height above the rest position, A equal the cross-sectional area of the manometer tube, R equal tube radius, p equal the liquid density, P equal gas pressure, L equal total length of liquid in the system, and μ stand for liquid viscosity. Assuming that the gas density is

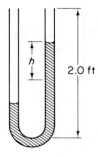

Figure 4.5-8. Manometer.

small compared to the liquid density and that the flow in the manometer is laminar, a momentum balance gives the result[1]

$$\frac{\rho AL}{g_c}\frac{d^2h}{dt^2} = PA - \rho A2h\frac{g}{g_c} - \frac{8\mu L}{R^2 g_c}\frac{dh}{dt}A \qquad 4.5\text{-}61$$

where Poiseuille's law, Eq. 4.3-67, has been used to relate the frictional pressure drop to the flow velocity in the last term. After some rearranging, we obtain

$$\frac{L}{2g}\frac{d^2h}{dt^2} + \frac{4L\mu}{\rho R^2 g}\frac{dh}{dt} + h = \frac{Pg_c}{2\rho g} \qquad 4.5\text{-}62$$

Now if we let

$$\tau^2 = \frac{L}{2g} \quad \text{or} \quad \tau = \sqrt{\frac{L}{2g}} \qquad 4.5\text{-}63$$

$$2\gamma\tau = \frac{4L\mu}{\rho R^2 g} \quad \text{or} \quad \gamma = \frac{2L\mu}{\rho R^2 g}\sqrt{\frac{2g}{L}} \qquad 4.5\text{-}64$$

and

$$f(t) = \frac{Pg_c}{2\rho g} \qquad 4.5\text{-}65$$

our equation becomes

$$\tau^2\frac{d^2h}{dt^2} + 2\gamma\tau\frac{dh}{dt} + h = f(t) \qquad 4.5\text{-}66$$

If we consider a 2.0-ft-long manometer made from $\frac{1}{4}$-in. I.D. glass tubing half-filled with water for a case where $\rho = 1.0$ gram/cc and $\mu = 1.0$ cP, then

$$\tau = \sqrt{\frac{2}{2(32.2)}} = 0.1756 \text{ sec}^{-1}$$

and

$$\gamma = \frac{2\mu L}{\rho R^2 g\tau} = \frac{2(0.01)(2)(2.54)(12)}{1.0(0.25/2)^2(2.54)^2 980(0.1756)} = 0.07$$

[1] A more accurate derivation indicates that the first term should be multiplied by $\frac{4}{3}$; see p. 229 of the reference given in footnote 3 on p. 53.

According to Eq. 4.5-31, the overshoot for a step change in pressure will be

$$\text{Overshoot} = \exp\left(-\frac{\pi\gamma}{\sqrt{1-\gamma^2}}\right) = \exp\left(-\frac{\pi(0.07)}{\sqrt{1-0.07^2}}\right) = 0.80$$

Thus for a system where we suddenly increase the pressure on one side of the manometer by 0.50 psi, which would make the difference in water level 14 in. at steady state conditions, the manometer will overflow during the initial transient period; that is,

$$\text{Total level change} = 14 + 0.8(14) = 25.2 \text{ in.}$$

which is greater than the height of the manometer.

From the preceding example we find that the damping coefficient of a manometer is very small, so that the manometer response will be very oscillatory and high overshoots will be encountered. Therefore it is common practice to introduce additional damping into the system by inserting a constriction at the bottom of the manometer tube.

Example 4.5-2   Start-up of a nonisothermal, continuous-stirred-tank reactor

A procedure for determining the optimum steady state design of a nonisothermal reactor was discussed in detail in Example 2.1-2. If we consider a case where

$C_A = 0.072$ (\$)(sq cm)$^{-1}$(hr)$^{-1}$ $(-\Delta H) = 5505$ (cal)(g mole)$^{-1}$

$C_f = 0.0149$ (\$)(g mole)$^{-1}$ $k_0 = 5 \times 10^{-4}e^{81.5}$ (sec)$^{-1}$

$C_H = 0.01$ (\$)(1000 g)$^{-1}$ $G = 1440$ (g mole)(hr)$^{-1}$

$C_V = 0.0735$ (\$)(hr)$^{-1}$(liter)$^{-1}$ $T_f = 300$ (°K)

$C_p\rho = 1.0$ (cal)(cc)$^{-1}$(°K)$^{-1}$ $T_H = 373$ (°K)

$C_{pH} = 1.0$ (cal)(g)$^{-1}$(°K)$^{-1}$ $U = 1.0$ (cal)(sq cm)$^{-1}$(sec)$^{-1}$(°K)$^{-1}$

$E = 59,850$ (cal)(g mole)$^{-1}$ and $A_f = 0.01$ (g mole)(cc)$^{-1}$,

then the optimum design values can be shown to be

$T = 367$ (°K) $V = 400,000$ liters

$T_0 = 369$ (°K) $Q_H = 1150$ (cal)(sec)$^{-1}$

$k = 5 \times 10^{-4}$ (sec)$^{-1}$ $q_H = 287.5$ (g)(sec)$^{-1}$

$A = 0.002$ (g mole)(cc)$^{-1}$ $A_H = 287.5$ sq cm

$q = 50.0$ (cc)(sec)$^{-1}$

Once the optimum design has been established, it is necessary to devise a start-up procedure for the reactor. Perhaps the most straightfoward approach would be to fill the reactor with the feed material, turn on the agitator, adjust the inlet and effluent flow rates to the desired values, and then to turn on the steam. Estimate the response of the reactor for this start-up procedure.

Solution

The dynamic equations for the reactor were developed in Example 3.1-3.

$$V\frac{dA}{dt} = q(A_f - A) - kVA \tag{3.1-23}$$

$$VC_p\rho\frac{dT}{dt} = qC_p\rho(T_f - T) + (-\Delta H)kVA + \frac{UA_H Kq_H}{1 + Kq_H}(T_H - T) \tag{3.1-24}$$

where

$$K = \frac{2C_{pH}\rho_H}{UA_H} \qquad k = k_0 \exp\left(-\frac{E}{RT}\right) \tag{3.1-25}$$

In order to obtain a first estimate of the system response, we will assume that no reaction takes place in the vessel during the filling time. Thus the initial conditions will be taken as

$$A = A_f, \quad T_H = T_f, \quad q_H = 0, \quad T = T_f \tag{4.5-67}$$

with all other variables set at their optimum steady state values. At time equal to zero, we suddenly change q_H and T_H to their optimum settings. Although it is possible to linearize the dynamic equations around this initial set of steady state conditions and solve for the response to the step changes in q_H and T_H, we will follow the alternate approach and linearize around the final optimum steady state values and compute the response for a given set of initial conditions. The results of both procedures would be identical if we solved the nonlinear dynamic equations. However, the linearization procedure will give an incorrect value for the system gain, as we demonstrated in Example 4.3-1, so that the final steady state conditions determined from the step response will not be the same as the optimum steady state. In addition, this alternate approach will provide an illustration of the manner in which initial conditions must be carried through the analysis.

The equations can be linearized by following the procedure given in Example 3.3-1. Letting

$$x_1 = \frac{A - A_s}{A_{fs}}, \quad x_2 = \frac{C_p\rho(T - T_s)}{(-\Delta H)A_{fs}}, \quad x_{1f} = \frac{A_f - A_{fs}}{A_{fs}}$$

$$x_{2f} = \frac{C_p\rho(T_f - T_{fs})}{(-\Delta H)C_p\rho}, \quad x_{2H} = \frac{C_p\rho(T_H - T_{Hs})}{(-\Delta H)A_{fs}} \tag{4.5-68}$$

$$u_1 = \frac{q - q_s}{q_s}, \quad u_2 = \frac{q_H - q_{Hs}}{q_{Hs}}, \quad \tau = \frac{q_s t}{V}$$

and expanding the right-hand sides of Eqs. 3.1-23 and 3.1-24 in a Taylor series up to the first-order terms, we obtain

$$\frac{dx_1}{d\tau} = a_{11}x_1 + a_{12}x_2 + b_{11}u_1 + c_{11}x_{1f} \qquad \text{3.3-12}$$

$$\frac{dx_2}{d\tau} = a_{21}x_1 + a_{22}x_2 + b_{21}u_1 + b_{22}u_2 + c_{22}x_{2f} + c_{23}x_{2H} \qquad \text{3.3-13}$$

where

$$a_{11} = -\left(1 + \frac{k_sV}{q_s}\right), \quad a_{12} = -\frac{E(-\Delta H)k_sVA_s}{q_sC_p\rho RT_s^2}, \quad a_{21} = \frac{k_sV}{q_s}$$

$$a_{22} = -\left[1 - \frac{E(-\Delta H)k_sVA_s}{q_sC_p\rho RT_s^2} + \frac{UA_HKq_{Hs}}{q_sC_p\rho(1 + Kq_{Hs})}\right]$$

$$b_{11} = \frac{A_{fs} - A_s}{A_{fs}}, \quad b_{21} = \frac{C_p\rho(T_{fs} - T_s)}{(-\Delta H)A_{fs}} \qquad \text{4.5-69}$$

$$b_{22} = \frac{UA_HKq_{Hs}C_p\rho(T_{Hs} - T_s)}{q_sC_p\rho(-\Delta H)A_{fs}(1 + Kq_{Hs})^2}$$

$$c_{11} = 1, \quad c_{22} = 1, \quad c_{23} = \frac{UA_HKq_{Hs}}{q_sC_p\rho(1 + Kq_{Hs})}$$

For the problem of interest, all the system inputs are at their steady state values, so that $u_1 = u_2 = x_{1f} = x_{2f} = x_{2H} = 0$, and we know the initial values of x_1 and x_2. (With the alternate formulation, we set the initial conditions to zero and consider step changes in u_2 and x_{2H}.) Thus the approximate equations can be written as

$$\frac{V}{q}\frac{dx_1}{dt} = a_{11}x_1 + a_{12}x_2 \qquad x_1(0) = x_{1_0} \qquad \text{4.5-70}$$

$$\frac{V}{q}\frac{dx_2}{dt} = a_{21}x_1 + a_{22}x_2 \qquad x_2(0) = x_{2_0} \qquad \text{4.5-71}$$

where again we have used real time as the independent variable rather than dimensionless time. At this point we could make the canonical transformation described in Example 3.3-4. However, in order to demonstrate that the same result can be obtained in a variety of ways, we will take the Laplace transform of the preceding equations. This leads to the result

$$\frac{V}{q}(s\tilde{x}_1 - x_{1_0}) = a_{11}\tilde{x}_1 + a_{22}\tilde{x}_2 \qquad \text{4.5-72}$$

$$\frac{V}{q}(s\tilde{x}_2 - x_{2_0}) = a_{21}\tilde{x}_1 + a_{22}\tilde{x}_2 \qquad \text{4.5-73}$$

or

$$\left(\frac{V}{q}s - a_{11}\right)\tilde{x}_1 - a_{12}\tilde{x}_2 = \frac{V}{q}x_{1_0} \qquad \text{4.5-74}$$

$$-a_{21}\tilde{x}_1 + \left(\frac{V}{q}s - a_{22}\right)\tilde{x}_2 = \frac{V}{q}x_{2_0} \qquad \text{4.5-75}$$

Using Cramer's rule to solve the equations, we obtain

$$
\begin{aligned}
\tilde{x}_1 &= \frac{(V^2/q^2)sx_{1_0} + (V/q)(a_{12} - a_{22}x_{1_0})}{(V^2/q^2)s^2 - (V/q)(a_{11} + a_{22})s + a_{11}a_{22} - a_{12}a_{21}} \\
&= \frac{[(V^2/q^2)sx_{1_0} + (V/q)(a_{12}x_{2_0} - a_{22}x_{1_0})][1/(a_{11}a_{22} - a_{12}a_{21})]}{[(V^2/q^2)/(a_{11}a_{22} - a_{12}a_{21})]s^2} \\
&\qquad - \{[(a_{11} + a_{22})(V/q)]/(a_{11}a_{22} - a_{12}a_{21})\}s + 1
\end{aligned}
\tag{4.5-76}
$$

$$
\begin{aligned}
\tilde{x}_2 &= \frac{(V^2/q^2)sx_{2_0} + (V/q)(a_{21}x_{1_0} - a_{11}x_{2_0})}{(V^2/q^2)s^2 - (V/q)(a_{11} + a_{22})s + a_{11}a_{22} - a_{12}a_{21}} \\
&= \frac{[(V^2/q^2)sx_{2_0} + (V/q)(a_{21}x_{1_0} - a_{11}x_{2_0})][1/(a_{11}a_{22} - a_{12}a_{21})]}{[(V^2/q^2)/(a_{11}a_{22} - a_{12}a_{21})]s^2} \\
&\qquad - \{[(a_{11} + a_{22})(V/q)]/(a_{11}a_{22} - a_{12}a_{21})\}s + 1
\end{aligned}
\tag{4.5-77}
$$

Obviously both expressions have the same general form

$$
\tilde{y} = \frac{\tilde{f}}{\tau^2 s^2 + 2\gamma\tau s + 1}
\tag{4.5-78}
$$

We can find the inverse transforms once the nature of the characteristic roots has been specified. From Eq. 4.5-11, we know that

$$
\lambda_1 = -\frac{\gamma}{\tau} + \frac{1}{\tau}\sqrt{\gamma^2 - 1} \qquad \lambda_2 = -\frac{\gamma}{\tau} - \frac{1}{\tau}\sqrt{\gamma^2 - 1}
\tag{4.5-11}
$$

For the problem under investigation we find that, from Eqs. 4.5-69,

$$
a_{11} = -\left[1 + \frac{400{,}000}{2000(50)}\right] = -5
$$

$$
a_{12} = -\frac{59{,}280(5505)(400{,}000)(0.002)(5 \times 10^{-4})}{(50)(1)(1.987)(367)^2} = -9.77
$$

$$
a_{21} = -\frac{5 \times 10^{-4}(400{,}000)}{50} = 4
$$

$$
\begin{aligned}
a_{22} &= -\left\{1 - \frac{59{,}280(5505)(5 \times 10^{-4})(400{,}000)0.002}{50(1)(1.987)(367)^2}\right. \\
&\qquad \left. + \frac{1.0(287.5)[2(287.5)/287.5]}{50(1)[1 + (2/287.5)287.5]}\right\} \\
&= -(1 - 9.77 + 3.83) = 4.94
\end{aligned}
$$

and

$$
\tau = \sqrt{\frac{(V/q)^2}{a_{11}a_{22} - a_{12}a_{21}}} = \sqrt{\frac{(400{,}000/50)^2}{-5(4.94) - (-9.77)4}} = 2.11 \times 10^3
$$

$$
\gamma = -\frac{1}{2\tau}\left[\frac{(a_{11} + a_{22})(V/q)}{a_{11}a_{22} - a_{12}a_{21}}\right] = \frac{-1}{4.22 \times 10^3}\left[\frac{(-5 + 4.94)(400{,}000/50)}{14.38}\right]
$$

$$
= 7.911 \times 10^{-3}
$$

Also

$$
\lambda = -\frac{1}{\tau}[\gamma \pm \sqrt{\gamma^2 - 1}]
$$

$$= \frac{-1}{2.11 \times 10^3}[7.911 \times 10^{-3} \pm \sqrt{(7.911 \times 10^{-3})^2 - 1}]$$

$$= \frac{-1}{2.11 \times 10^{-3}}(7.911 \times 10^{-3} \pm j)$$

Thus the roots are complex conjugates.

The transforms of both outputs have the form

$$\tilde{y} = \frac{c_1 s + c_2}{\tau^2 s^2 + 2\gamma\tau s + 1} = \frac{(c_1 s + c_2)/\tau^2}{[s + (\gamma/\tau)]^2 + [(1 - \gamma^2)/\tau^2]} \qquad 4.5\text{-}79$$

Using the Heaviside theorem for complex roots, we obtain the inverse transform

$$y(t) = e^{-\gamma t/\tau}\left[\frac{c_1}{\tau^2}\cos\left(\frac{\sqrt{1 - \gamma^2}t}{\tau}\right) + \frac{c_2\tau - c_1\gamma}{\tau^2\sqrt{1 - \gamma^2}}\sin\left(\frac{\sqrt{1 - \gamma^2}t}{\tau}\right)\right] \qquad 4.5\text{-}80$$

Since

$$x_{1_0} = \frac{A_f - A_s}{A_s} = \frac{0.010 - 0.002}{0.010} = 0.80$$

$$x_{2_0} = \frac{C_p\rho(T_f - T_s)}{(-\Delta H)A_{fs}} = \frac{1.0(300 - 367)}{5505(0.01)} = -1.218$$

the constants for the composition response are (see Eq. 4.5-76)

$$\frac{c_1}{\tau^2} = \frac{(V/q)^2 x_{1_0}}{(a_{11}a_{22} - a_{12}a_{21})\tau^2} = x_{1_0} = 0.80$$

$$\begin{aligned}
\frac{c_2\tau - c_1\gamma}{\tau^2\sqrt{1 - \gamma^2}} &= \frac{(a_{12}x_{2_0} - a_{22}x_{1_0})(V/q)\tau}{\tau^2\sqrt{1 - \gamma^2}(a_{11}a_{22} - a_{12}a_{21})} - \frac{x_{1_0}\gamma}{\sqrt{1 - \gamma^2}} \\
&= \frac{[(-9.77)(-1.218) - (4.94)(0.80)](400{,}000/50)}{(2.11 \times 10^3)\sqrt{1 - (7.911 \times 10^{-3})^2}(14.38)} \\
&\quad - \frac{0.80(7.911 \times 10^{-3})}{\sqrt{1 - (7.911 \times 10^{-3})^2}} = 2.10
\end{aligned}$$

The corresponding values for the temperature response can be obtained from

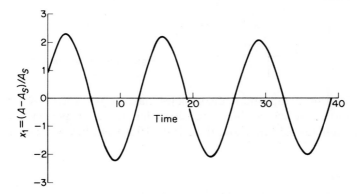

Figure 4.5-9. Start-up of a nonisothermal, continuous-stirred-tank reactor.

Eq. 4.5-77. The response curve for composition is plotted in Figure 4.5-9, and it is apparent from the graph that the reactor composition becomes negative for certain time intervals. This fact implies that the fluctuations are so large that our linear analysis is not valid. However, the analysis does serve the purpose of showing that we must look for a better start-up procedure because the approach to steady state conditions is so slow.

Example 4.5-3 Nonisothermal, continuous-stirred-tank reactor with an oscillating feed composition

Suppose that we consider the optimum stirred-tank reactor system described in the example above, but now assume that the reactant stream is prepared in two isothermal, batch catalytic reactors upstream of the CSTR, which operate sequentially. If the catalyst activity of the two reactors is different, the feed composition will vary somewhat. Providing that the reactors are properly designed, the average feed composition entering the CSTR will be $A_{fs} = 0.01$ and we will be able to supply this material at the desired rate. However, the actual feed composition will be a square wave. Find the

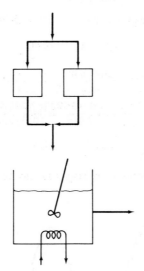

Figure 4.5-10. Stirred-tank reactor system.

effect that this composition disturbance has on the pseudo-steady state reactor for a case where the composition initially is 10 percent above the average value for 5.5 hours and then drops 10 percent below the average for 5.5 hours.

Solution

Although it is possible to determine the Laplace transform of the input signal by using the methods described in Section 4.2, we will approximate the input by a sum of sine waves; that is, we will calculate the Fourier expansion of the square wave and determine the system frequency response for these sinusoidal signals. This procedure should make it apparent that an estimate of the system behavior can be obtained for arbitrary periodic disturbances even in cases where it is not practical to determine their Laplace transforms.

If we subtract the average value from the feed composition disturbance function and divide the result by A_{fs}, we will obtain the function $x_{1f}(t)$. For simplicity we will select time equal to zero at a point where the feed

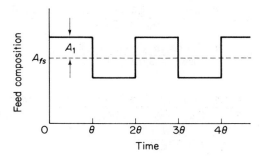

Figure 4.5-11. Feed-composition fluctuations.

composition is increasing, so that the square wave will be an odd function and the Fourier expansion will include only odd terms. Then we write the forcing function as an infinite series of sinusoidal functions

$$x_{1f}(t) = \sum_{n=1}^{\infty} b_n \sin \frac{n\pi t}{\theta} \qquad 4.5\text{-}81$$

The coefficients b_n are determined by multiplying both sides of the equation by $\sin (m\pi t/\theta)$, integrating over one period, and recognizing that the orthogonality property of trigonometric functions will make all but one term equal to zero. Thus we obtain

$$b_n = \frac{2}{\theta} \int_0^{\theta} x_{1f}(t) \sin \frac{m\pi t}{\theta} \, dt \qquad 4.5\text{-}82$$

or

$$b_n = \frac{2(A_1 - A_{fs})/A_{fs}}{\theta} \left[-\frac{\theta}{n\pi} \cos \frac{n\pi t}{\theta} \Big|_0^{\theta} \right]$$

$$= \frac{2(A_1 - A_{fs})}{A_{fs}} \left(\frac{1 - \cos n\pi}{n\pi} \right) = \begin{cases} 0, & \text{if } n \text{ is even} \\ \frac{4}{n\pi} \left(\frac{A_1 - A_{fs}}{A_{fs}} \right), & \text{if } n \text{ is odd} \end{cases} \qquad 4.5\text{-}83$$

and we can write an approximate expression for the forcing function as

$$x_{1f}(t) = \frac{4}{\pi} \left(\frac{A_1 - A_{fs}}{A_{fs}} \right) \left[\sin \frac{\pi t}{\theta} + \frac{1}{3} \sin \frac{3\pi t}{\theta} + \frac{1}{5} \sin \frac{5\pi t}{\theta} + \cdots \right] \qquad 4.5\text{-}84$$

For a case where all the system inputs except feed composition remain constant at their steady state values, the approximate equations describing the system are

$$\frac{V}{q}\frac{dx_1}{dt} = a_{11}x_1 + a_{12}x_2 + c_{11}x_{1f} \qquad \text{3.3-12}$$

$$\frac{V}{q}\frac{dx_2}{dt} = a_{21}x_1 + a_{22}x_2 \qquad \text{3.3-13}$$

Taking the Laplace transforms of these equations and solving for the transforms of the outputs gives

$$\tilde{x}_1 = \frac{[(V/q)s - a_{22}][\tilde{x}_{1f}/(a_{11}a_{22} - a_{12}a_{21})]}{\tau^2 s^2 + 2\gamma\tau s + 1} \qquad \text{4.5-85}$$

$$\tilde{x}_2 = \frac{a_{21}\tilde{x}_{1f}/(a_{11}a_{22} - a_{12}a_{21})}{\tau^2 s^2 + 2\gamma\tau s + 1} \qquad \text{4.5-86}$$

where

$$\tilde{x}_{1f} = \frac{4}{\pi}\left(\frac{A_1 - A_{fs}}{A_{fs}}\right)\left\{\frac{\pi/\theta}{s^2 + (\pi/\theta)^2} + \frac{1}{3}\left[\frac{3\pi/\theta}{s^2 + (3\pi/\theta)^2}\right]\right.$$
$$\left. + \frac{1}{5}\left[\frac{5\pi/\theta}{s^2 + (5\pi/\theta)^2}\right] + \cdots\right\}$$

Since the transforms of both outputs may be written as

$$\tilde{y} = \frac{(c_1 s + c_2)\tilde{x}_{1f}}{\tau^2 s^2 + 2\gamma\tau s + 1} \qquad \text{4.5-87}$$

and since it is possible to divide the transform of the input into three or more separate terms, the steady state frequency response can be determined by substituting $j\omega_n$ for s. Thus

$$\tilde{y} = \frac{c_1 j\omega_n + c_2}{\tau^2(-\omega_n)^2 + 2\gamma\tau(j\omega_n) + 1}$$
$$= \frac{[c_2(1 - \tau^2\omega_n^2) + c_1 2\gamma\tau\omega_n^2] + j[c_1\omega_n(1 - \tau^2\omega_n^2) - c_2 2\gamma\tau\omega_n]}{(1 - \tau^2\omega_n^2)^2 + (2\gamma\tau\omega_n)^2} \qquad \text{4.5-88}$$

General expressions for the system gain and phase shift can now be established, using our previous methods. However, we note that $c_1 = 0$ in the equation for the transform of the effluent temperature, Eq. 4.5-86, and therefore we can use the solution we developed previously, Eq. 4.5-39, for this variable. The solution for the composition response can be determined by substituting the known solution for $x_2(t)$ into Eq. 3.3-13 and solving for $x_1(t)$. When the forcing function is $f(t) = A \sin \omega_0 t$, the solution is

$$y(t) = \frac{A}{\sqrt{(1 - \tau^2\omega_0^2)^2 + (2\tau\gamma\omega_0)^2}} \sin(\omega_0 t + \phi) \qquad \text{4.5-39}$$

$$\phi = \tan^{-1}\left(\frac{-2\tau\gamma\omega_0}{1 - \tau^2\omega_0^2}\right) \qquad \text{4.5-38}$$

and this solution will be valid for each input frequency. Hence the complete solution is

$$x_2(t) = \frac{4a_{21}(A_1 - A_{fs})/A_{fs}}{\pi(a_{11}a_{22} - a_{12}a_{21})} \left\{ \frac{1}{\sqrt{[1 - (\tau^2\pi^2/\theta^2)]^2 + (2\tau\gamma\pi/\theta)^2}} \sin\left[(\pi t/\theta) + \phi_1\right] \right.$$

$$+ \frac{1/3}{\sqrt{[1 - (\tau^2 3^2\pi^2/\theta^2)]^2 + (6\tau\gamma\pi/\theta)^2}} \sin\left[(3\pi t/\theta) + \phi_2\right]$$

$$\left. + \frac{1/5}{\sqrt{[1 - (\tau^2 5^2\pi^2/\theta^2)]^2 + (10\tau\gamma\pi/\theta)^2}} \sin\left[(5\pi t/\theta) + \phi_3\right] \right\}$$

where

$$\phi_1 = \tan^{-1}\left[\frac{-2\tau\gamma\pi/\theta}{1 - (\tau^2\pi^2/\theta^2)}\right], \quad \phi_2 = \tan^{-1}\left[\frac{-6\tau\gamma\pi/\theta}{1 - (9\tau^2\pi^2/\theta^2)}\right]$$

$$\phi_3 = \tan^{-1}\left[\frac{-10\tau\gamma\pi/\theta}{1 - (25\tau^2\pi^2/\theta^2)}\right]$$

For the problem in question

$$a_{21} = 4, \quad a_{11}a_{22} - a_{12}a_{21} = 14.38, \quad \tau = 2.11 \times 10^3, \quad \gamma = 7.911 \times 10^{-3}$$

$$\theta = 55(3600) = 1.98 \times 10^4 \qquad \frac{A_1 - A_{fs}}{A_{fs}} = 0.10$$

so that the solution is

$$x_2(t) = 0.0399 \sin (1.58 \times 10^{-4}t - 0.342)$$
$$+ 0.659 \sin (4.76 \times 10^{-4}t - 117.5)$$
$$+ 0.00393 \sin (7.94 \times 10^{-4}t - 179.2)$$

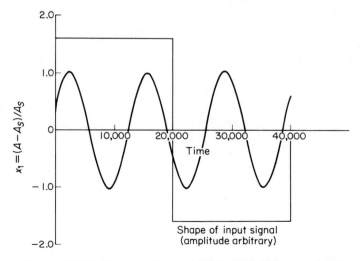

Figure 4.5-12. Response of a nonisothermal, continuous-stirred-tank reactor to a square wave in feed composition.

The corresponding result for composition is obtained by solving Eq. 3.3-13 for x_1. A plot of this solution is given in Figure 4.5-12.

At first glance, see Figure 4.5-12, the results obtained for the response to a square wave seem somewhat surprising. The fundamental component of the input signal—that is, the term having the same frequency as the apparent frequency of the square wave—has the highest amplitude, and therefore we might expect it to provide the major contribution to the output. However, the frequency of the next harmonic term is close to the resonant frequency and the system gain is very high because of the low-damping coefficient. Thus the amplification of this component more than offsets the lower amplitude of its input, and it becomes the predominant term in the solution. As a result, the frequency of the output appears to be higher than the input frequency; that is, the period of the output signal appears to be 13,200 rather than 39,600. It can be shown that forcing functions that are not periodic contain components at all frequencies rather than only discrete multiples of a fundamental frequency; consequently it is necessary to be cautious if a system can exhibit resonance.

Example 4.5-4 Measurement error

In order to verify our dynamic models, we must perform some experiments. However, the measuring equipment used to obtain the data will also have dynamic characteristics; therefore the observations obtained in the test runs might not give a direct test of the model. An estimate of the error introduced by the dynamics of the measuring system can be determined by solving the dynamic equations for the measuring instruments along with the system equations.

As a simple example of this procedure, we suppose that a test is being made to establish that a perfect mixing model is adequate to describe the dynamics of a stirred heater. If we make a step change in the feed temperature to the tank and measure the tank temperature with a thermometer, the equations describing the system are

$$VC_p\rho\frac{dT}{dt} = qC_p\rho(T_f - T) + UA(T_H - T) \qquad \text{4.5-89}$$

$$V_m C_{pm}\rho_m\frac{dT_m}{dt} = hA_t(T - T_m) \qquad \text{4.5-90}$$

Letting

$$\tau_1 = \frac{VC_p\rho}{qC_p\rho + UA}, \quad \tau_2 = \frac{V_m C_{pm}\rho_m}{hA_t}$$

$$K_1 = \frac{qC_p\rho}{qC_p\rho + UA}, \quad K_2 = \frac{UA}{qC_p\rho + UA} \qquad \text{4.5-91}$$

the system equations can be written

$$\tau_1\frac{dT}{dt} + T = K_1 T_f + K_2 T_H \qquad \text{4.5-92}$$

$$\tau_2 \frac{dT_m}{dt} + T_m = T \qquad\qquad 4.5\text{-}93$$

Now, letting

$$x_1 = T - T_s, \quad x_2 = T_m - T_{ms}$$
$$u = K_1(T_f - T_{fs}), \quad T_H - T_{Hs} = 0 \qquad\qquad 4.5\text{-}94$$

the equations become

$$\tau_1 \frac{dx_1}{dt} + x_1 = u \qquad\qquad 4.5\text{-}95$$

$$\tau_2 \frac{dx_2}{dt} + x_2 = x_1 \qquad\qquad 4.5\text{-}96$$

Taking the Laplace transforms, we obtain

$$\tilde{x}_1 = \frac{\tilde{u}}{(\tau_1 s + 1)} \qquad\qquad 4.5\text{-}97$$

$$\tilde{x}_2 = \frac{\tilde{x}_1}{(\tau_2 s + 1)} = \frac{\tilde{u}}{(\tau_1 s + 1)(\tau_2 s + 1)} \qquad\qquad 4.5\text{-}98$$

For a step change of amplitude A, $\tilde{u} = A/s$, and using Heaviside's theorem the system output is

$$x_2 = A - \left(\frac{A}{\tau_1 - \tau_2}\right)(\tau_1 e^{-t/\tau_1} - \tau_2 e^{-t/\tau_2}) \qquad\qquad 4.5\text{-}99$$

If τ_2 is very small compared with τ_1, this equation is approximately equal to

$$x_2(t) = x_1(t) = A(1 - e^{-t/\tau_1}) \qquad\qquad 4.5\text{-}100$$

and the measuring system (the thermometer) does not introduce any error.

Examples 4.5-5 The effect of recycle

If we examine the start-up of a series of two isothermal, continuous-stirred-tank reactors with a recycle stream, the equations describing the system are

$$V\frac{dx_1}{dt} = qx_f + Qx_2 - (q + Q)x_1 - kVx_1 \qquad\qquad 4.5\text{-}101$$

$$V\frac{dx_2}{dt} = (q + Q)x_1 - (q + Q)x_2 - kVx_2 \qquad\qquad 4.5\text{-}102$$

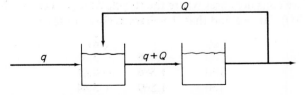

Figure 4.5-13. Isothermal reactors with a recycle stream.

Considering a case where initially solvent is running through the system at steady state conditions and at time equal to zero we switch to a feed stream having a composition x_f, the Laplace transforms of the system equations can be put into the form

$$\left(s + \frac{q+Q}{V} + k\right)\tilde{x}_1 = \frac{q}{V}\tilde{x}_f + \frac{Q}{V}\tilde{x}_2 \qquad \text{4.5-103}$$

$$\left(s + \frac{q+Q}{V} + k\right)\tilde{x}_2 = \frac{q+Q}{V}\tilde{x}_1 \qquad \text{4.5-104}$$

After some manipulation, the transform of the effluent composition can be shown to be

$$\tilde{x}_2 = \frac{(q/V)[(q+Q)/V]\tilde{x}_f}{\{s + [(q+Q)/V] + k\}^2 - (Q/V)[(q+Q)/V]} \qquad \text{4.5-105}$$

The roots of the denominator are

$$\lambda = -\left(\frac{q+Q}{V} + k\right) \pm \frac{Q}{V}\left(1 + \frac{q}{Q}\right)^{1/2} \qquad \text{4.5-106}$$

and providing that the recycle flow is not equal to zero, the output response for a step input, $\tilde{x}_2 = x_f/s$, is

$$x_2(t) = \frac{q}{V}\left(\frac{q+Q}{V}\right)\frac{x_f}{\lambda_1\lambda_2}\left[1 - \frac{1}{\lambda_1 - \lambda_2}(\lambda_1 e^{\lambda_2 t} - \lambda_2 e^{\lambda_1 t})\right] \qquad \text{4.5-107}$$

For the case where the recycle flow is equal to zero, the transform of the output is

$$\tilde{x}_2 = \frac{(q/V)^2\tilde{x}_f}{[s + (q/V) + k]^2} \qquad \text{4.5-108}$$

and the step response of the system is

$$x_2(t) = \frac{(q/V)^2 x_f}{[(q/V) + k]^2}\left\{1 - \left[1 + \left(\frac{q}{V} + k\right)t\right]\exp\left[-\left(\frac{q}{V} + k\right)t\right]\right\} \qquad \text{4.5-109}$$

If we consider a particular system where $q = 100$ ft^3/hr, $V = 100$ ft^3, $k = 1.0$ hr^{-1}, $x_f = 1.0$ mole/ft^3, the response of the series configuration is

$$x_2(t) = \frac{(100/100)^2(1)}{[(100/100) + 1]^2}\left\{1 - \left[1 + \left(\frac{100}{100} + 1\right)t\right]\exp\left[-\left(\frac{100}{100} + 1\right)t\right]\right\}$$

$$= \frac{1}{4}[1 - (1 + 2t)e^{-2t}]$$

This result is plotted in Figure 4.5-14.

Now if we consider cases where the recycle flow Q takes on the values 50, 100, and 150 ft^3/hr, we find that the characteristic roots are

Q	λ_1	λ_2
50	−1.636	−3.36
100	−1.586	−4.41
150	−1.560	−5.44

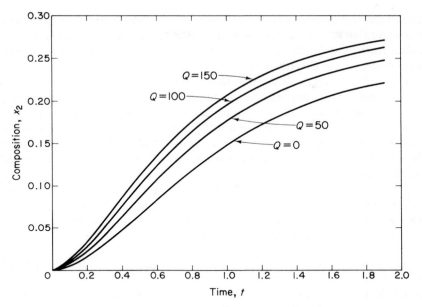

Figure 4.5-14. Response of two reactors in series.

Hence the equations for the system response for the three cases are

$$Q = 50 \qquad x_2 = 0.273\left[1 - \frac{1}{1.728}(3.364e^{-1.636t} - 1.636e^{-3.364t})\right]$$

$$Q = 100 \qquad x_2 = 0.286\left[1 - \frac{1}{2.828}(4.414e^{-1.586t} - 1.586e^{-4.414t})\right]$$

$$Q = 150 \qquad x_2 = 0.294\left[1 - \frac{1}{3.98}(5.44e^{-1.56t} - 1.56e^{-5.44t})\right]$$

These results are also plotted in Figure 4.5-14. It is apparent from the graph that the conversion obtained from the reactor system decreases as the recycle flow increases, as is expected from elementary reactor-design considerations. Also, the graph shows that the system response is more rapid as the recycle flow increases, for the apparent flow rate through the reactors increases. However, it would be more realistic to compare systems that gave the same overall conversion.

Estimating Time Constants from Experimental Data

In previous discussions we assumed that it was always possible to derive a theoretical model for the process. However, in certain situations this approach is impractical, and it is better to develop an empirical model of the system dynamics. Since the step response of a plant generally is the simplest experiment that provides dynamic information, it would be helpful to have a procedure that can be used to estimate the system parameters (the time constants) from experimental data. Several methods for this purpose have been developed and are described below.

Our earlier results indicate that the overshoot or decay ratio of an underdamped system only depends on the damping coefficient. Thus it is a simple matter to determine this system parameter from experimental data (see Figure 4.5-3 and Eqs. 4.5-31 and 4.5-32). Once the damping coefficient has been determined, the time constant τ can be obtained by measuring either the period of the oscillations or the rise time and using Eq. 4.5-30 or 4.5-33 to calculate τ. Very few chemical processes are underdamped, but closed-loop systems containing a controller normally exhibit this kind of behavior.

The most commonly occurring second-order systems in chemical engineering are overdamped. Consequently, we will consider four methods that can be used to find the system parameters. First, however, we let

$$\tau_1 = \frac{\tau}{\gamma - \sqrt{\gamma^2 - 1}} = \frac{-1}{\lambda_1} \qquad \tau_2 = \frac{\tau}{\gamma + \sqrt{\gamma^2 - 1}} = \frac{-1}{\lambda_2} \qquad 4.5\text{-}110$$

so that the step response of the overdamped system, Eq. 4.5-24, can be written as

$$y = A - \left(\frac{A}{\tau_1 - \tau_2}\right)(\tau_1 e^{-t/\tau_1} - \tau_2 e^{-t/\tau_2}) \qquad 4.5\text{-}111$$

Then we desire to calculate τ_1 and τ_2 from a set of measured values of y at various times t.

Slope-intercept Method. If the logarithm of the percentage incomplete response, $(1 - y/A)$, is plotted against time, the curve will approach $[\tau_1/(\tau_1 - \tau_2)]e^{-t/\tau_1}$ as t approaches infinity—that is, since the magnitude of λ_1 is always smaller than λ_2 (see Eq. 4.5-11), then τ_1 is always greater than τ_2 (see Eq. 4.5-110), and the quantity

$$1 - \frac{y}{A} = \left[\frac{\tau_1}{\tau_1 - \tau_2}e^{-t/\tau_1} - \frac{\tau_2}{\tau_1 - \tau_2}e^{-t/\tau_2}\right] \qquad 4.5\text{-}112$$

will approach the first term on the right-hand side of the equation as t approaches infinity. Thus if we extrapolate this curve back to the ordinate and find the time required for the curve to drop to 36.8 percent of the extrapolated value, this time will be equal to τ_1 (see Figure 4.5-15). By subtracting the experimental values from the extrapolated curve, we will obtain a plot of the logarithm of $[\tau_2/(\tau_1 - \tau_2)]e^{-t/\tau_2}$ versus t. Again, the time required for the ordinate to decrease to 36.8 percent of its original value will be equal to τ_2.

This method works quite well if the two time constants are very different from one another. In some cases, depending on the accuracy of the data, it is possible to use this approach to estimate the three time constants of a third-order system. The great advantage of this approach is that all the data

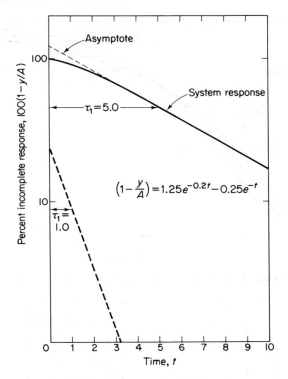

Figure 4.5-15. Estimation of system time constants.

are used to estimate the time constants and the graph indicates the quality of the fit of the second-order model to the data.

Method of Harriott. Harriott[2] plotted the fractional response of the system y/A against $t/(\tau_1 + \tau_2)$ for various ratios of τ_2/τ_1 and found that all the curves intersected approximately at $y/A = 0.73$ when $t/(\tau_1 + \tau_2) = 1.3$ (see Figure 4.5-16). Thus by measuring the time required for the system to reach 73 percent of its final value, $t_{0.73}$, the sum of the system time constants can be calculated

$$\tau_1 + \tau_2 = \frac{t_{0.73}}{1.3} \qquad\qquad 4.5\text{-}113$$

Harriott then plotted the fractional response at $t/(\tau_1 + \tau_2) = 0.5$ against the ratio of the time constants, since the curves in Figure 4.5-16 show the greatest deviation at this point (see Figure 4.5-17). Hence the value of y/A when $t = 0.5(\tau_1 + \tau_2)$ can be determined from the experimental data and the value of $\tau_1/(\tau_1 + \tau_2)$ can be read from Figure 4.5-17. If the fractional response

[2] P. Harriott, *Process Control*, p. 47, McGraw-Hill, N.Y., 1964.

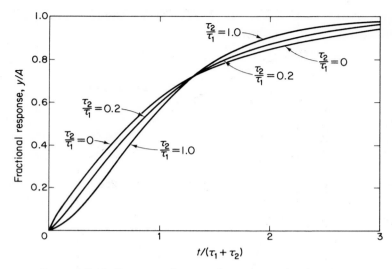

Figure 4.5-16. Response of an overdamped second-order system.

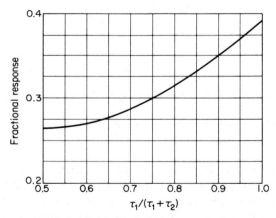

Figure 4.5-17. Fractional response at $t/(\tau_1 + \tau_2) = 0.5$.

is less than 0.26 at this point, the system must be at least third-order and the method is not applicable.

Method of Oldenbourg and Sartorius. Oldenbourg and Sartorius[3] developed a method based on the value of the maximum slope and the inflection point of the fractional response curve. After the fractional response, y/A, has been plotted against time, the slope at the inflection point is extrapolated to the limits of the graph and the values τ_A and τ_B determined (see

[3] R. C. Oldenbourg and H. Sartorius, *The Dynamics of Automatic Controls*, p. 77 ASME, N.Y., 1948.

Figure 4.5-18). These values are related to the system time constants by the equations

$$\tau_B = \tau_1 + \tau_2 = \left(1 + \frac{\tau_2}{\tau_1}\right)\tau_1 \qquad\qquad 4.5\text{-}114$$

$$\tau_A = \tau_1 \left(\frac{\tau_1}{\tau_2}\right)^{(\tau_2/\tau_1)/[1-(\tau_2/\tau_1)]} \qquad\qquad 4.5\text{-}115$$

Since these equations cannot be solved explicitly for τ_1 and τ_2, Oldenbourg and Sartorius prepared a special graph that can be used to find the solution with a minimum amount of effort (see Figure 4.5-19). The ratio τ_B/τ_A is used to construct a straight line having a slope of minus unity on this second graph, and the intersection of this straight line with the curve gives the values of τ_1/τ_A and τ_2/τ_A so that τ_1 and τ_2 can be calculated. If the straight line is tangent to the curve, the two time constants are equal.

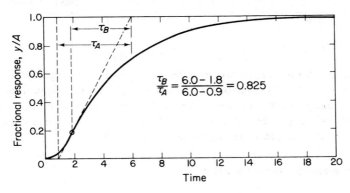

Figure 4.5-18. Response of an overdamped system.

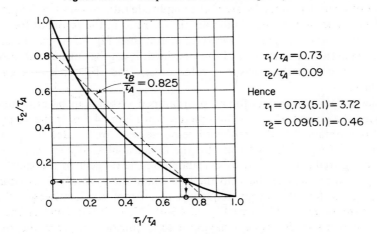

Figure 4.5-19. Auxiliary diagram.

Application of the Method of Moments

We will discuss the method of moments in considerable detail in the next chapter, but it is appropriate to note here that we can use this approach to estimate the system time constants. For example, if we plot the fractional incomplete response versus time and measure the area under the curve, sometimes called the zeroth moment, μ_0, we see from Eq. 4.5-112 that

$$\mu_0 = \int_0^\infty \left(1 - \frac{y}{A}\right) dt = \int_0^\infty \left[\frac{\tau_1}{\tau_1 - \tau_2}e^{-t/\tau_1} - \frac{\tau_2}{\tau_1 - \tau_2}e^{-t/\tau_2}\right] dt$$

$$= \tau_1 + \tau_2 \qquad\qquad 4.5\text{-}114$$

Also, the mean value of this curve, which is proportional to the first moment, is

$$\mu_1 = \frac{\int_0^\infty t[1 - (y/A)]\, dt}{\mu_0} = \frac{\int_0^\infty t\{[\tau_1/(\tau_1 - \tau_2)]e^{-t/\tau_1} - [\tau_2/(\tau_1 - \tau_2)]e^{-t/\tau_2}\}\, dt}{\mu_0}$$

so that

$$\mu_0\mu_1 = \tau_1^2 + \tau_1\tau_2 + \tau_2^2 \qquad\qquad 4.5\text{-}115$$

Therefore we can use the measured values of μ_0 and μ_1 to calculate τ_1 and τ_2. Similarly, we can use higher-order moments to check our calculations, or we can apply the procedure to first-order models, with or without a dead time. This approach might be superior to the others if noise in the output signal makes it difficult to read the dead time from the data.

Use of Frequency Response Data

The system parameters can also be estimated from frequency response data. For an underdamped system, the amplitude ratio at very low frequencies can be divided into the maximum amplitude ratio at the resonant frequency to give the maximum gain of the system, and then Eq. 4.5-41 can be solved for γ. Once this parameter has been obtained, it can be substituted into Eq. 4.5-42, along with the measured value of ω_r, and the value of τ can be calculated.

The system parameters for an overdamped system can be estimated by applying the inverse of the method pictured in Figure 4.5-6. If the amplitude ratio data obtained in the frequency response tests are divided by the value at very low frequencies, we can plot the dimensionless system gain against the forcing frequency. Next, by extrapolating the tangent to the gain curve at very high frequencies back to the line where the system gain is equal to unity, we can find the corner frequency, ω_{c_2}. From our knowledge of first-order systems, $\omega_{c_2} = 1/\tau_2$ (see Figure 4.3-4). Now, if we subtract the gain curve for a first-order system having a time constant τ_2 from the overall gain curve, we will be left with a curve for another first-order system. By again sketching in the asymptote and extrapolating this line, we can find a second value for a corner frequency, ω_{c_1}, and calculate a second time constant, $\tau_1 = 1/\omega_{c_1}$.

Of course, it is possible to devise other methods for estimating the system parameters from experimental data. Since the system equations are known, they can be manipulated in various ways and put into forms where the system parameters have simple relationships to experimentally observable quantities. Although frequency response experiments normally are more accurate than step response measurements, they are more time consuming, for pseudo-steady state operation must be obtained at each forcing frequency. Moreover, a number of experiments must be made over a wide range of frequencies in order to obtain a good description of the gain and phase curves. Thus it is often a good idea to measure the system step response first and then use the estimates of the system parameters calculated from these data to plan a set of critical frequency response experiments.

SECTION 4.6   RESPONSE OF STAGED SYSTEMS

Although it is always possible, at least conceptually, to use the canonical transformation to uncouple a large number of simultaneous equations and write them in terms of single first- and second-order systems, there are cases where it is both simpler and faster to make a direct attack on the problem. A direct approach is particularly advantageous when the coefficient matrix is bi- or tri-diagonal, since for these cases the eigenvalues, or characteristic roots, can be obtained analytically. Rather than describe the matrix manipulations leading to the characteristic roots and the analytical solutions for the system response, we will show that the theory of finite differences and difference-differential equations can be used to find the analytical expressions for the dynamic response. This approach has received relatively little attention in the control literature, but it provides a very powerful tool for treating the staged systems commonly encountered in chemical plants.

Series of First-Order Systems

The simplest problem we can consider is one where we have a number of first-order systems arranged in series. Letting the time constant of the nth stage be τ_n, denoting the dependent variable leaving a stage by the stage number, and including the possibility of being able to control each stage, a balance on the nth stage gives the equation

$$\tau_n \frac{dy_n}{dt} + y_n = \alpha_n y_{n-1} + \beta_n u \qquad \text{4.6-1}$$

and this same equation will apply to every stage.

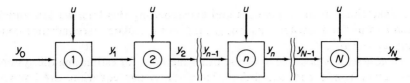

Figure 4.6-1. Series of first-order systems.

If we assume that the system is originally operating at some steady state condition and that the dependent variable represents deviations from steady state, the Laplace transform of Eq. 4.6-1 gives the result

$$(\tau_n s + 1)\tilde{y}_n = \alpha_n \tilde{y}_{n-1} + \beta_n \tilde{u} \qquad 4.6\text{-}2$$

Thus the complete set of equations can be written as

$$(\tau_1 s + 1)\tilde{y}_1 = \alpha_1 \tilde{y}_0 + \beta_1 \tilde{u}$$

$$(\tau_2 s + 1)\tilde{y}_2 = \alpha_2 \tilde{y}_1 + \beta_2 \tilde{u}$$

$$\cdot$$
$$\cdot$$
$$\cdot \qquad\qquad 4.6\text{-}3$$

$$(\tau_n s + 1)\tilde{y}_n = \alpha_n \tilde{y}_{n-1} + \beta_n \tilde{u}$$

$$\cdot$$
$$\cdot$$
$$\cdot$$

Now, we can solve the first equation in the set for $\tilde{y}_1$ in terms of $\tilde{y}_0$ and $\tilde{u}$ and then use this result to eliminate $\tilde{y}_1$ from the second equation. This procedure can be repeated until we obtain an explicit expression for $\tilde{y}_n$ in terms of $\tilde{y}_0$ and $\tilde{u}$, the system inputs. Hence

$$\tilde{y}_n = \frac{(\alpha_n \alpha_{n-1} \cdots \alpha_2 \alpha_1)\tilde{y}_0}{(\tau_n s + 1)(\tau_{n-1} s + 1) \cdots (\tau_1 s + 1)}$$

$$+ \left[\frac{\beta_n(\tau_{n-1} s + 1)(\tau_{n-2} s + 1) \cdots (\tau_1 s + 1)}{(\tau_n s + 1)(\tau_{n-1} s + 1) \cdots (\tau_1 s + 1)} \right.$$

$$+ \alpha_n \frac{\beta_{n-1}(\tau_{n-2} s + 1)(\tau_{n-3} s + 1) \cdots (\tau_1 s + 1)}{(\tau_n s + 1)(\tau_{n-1} s + 1) \cdots (\tau_1 s + 1)} + \cdots$$

$$\left. + \frac{\beta_1 \alpha_1 \alpha_2 \cdots \alpha_{n-1}}{(\tau_n s + 1)(\tau_{n-1} s + 1) \cdots (\tau_1 s + 1)} \right]\tilde{u} \qquad 4.6\text{-}4$$

or, for a case where all of the time constants and system gains are equal, the transfer function for inlet disturbances is

$$\frac{\tilde{y}_n}{\tilde{y}_0} = \frac{\alpha^n}{(\tau s + 1)^n} \qquad 4.6\text{-}5$$

and the transfer function for control variable changes is

$$\frac{\tilde{y}_n}{\tilde{u}} = \frac{\beta \alpha^{n-1} \sum_{r=0}^{n-1} \alpha^{-r}(\tau s + 1)^r}{(\tau s + 1)^n} \qquad 4.6\text{-}6$$

Since the response to feed composition disturbances is the simplest expression, we will consider first the system response for various forcing functions $\tilde{y}_0$.

Impulse Response

For an impulse input, $\tilde{y}_0 = A$ and

$$\tilde{y}_n = \frac{A\alpha^n}{(\tau s + 1)^n} = \frac{A(\alpha/\tau)^n}{(s + 1/\tau)^n} \qquad 4.6\text{-}7$$

Using the Heaviside expansion theorem for a repeated factor, Eq. 4.4-53, we find that the inverse transform is

$$y(t) = \frac{A\alpha^n}{\tau^n(n-1)!}t^{n-1}e^{-t/\tau} \qquad 4.6\text{-}8$$

This result is plotted in Figure 4.6-2 for a case where the sum of the time

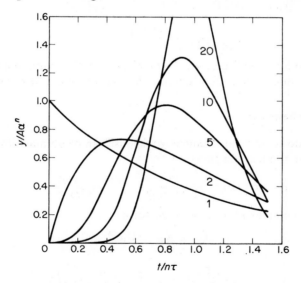

Figure 4.6-2. Impulse response of n first-ordr systems in series.

constants is taken as a constant value $\tau_T = n\tau$. The graph shows that the impulse response becomes sharper as n becomes large.

Step Response

For a step input, $\tilde{y}_0 = A/s$ and the transform of the output becomes

$$\tilde{y}_n = \frac{A\alpha^n}{s(\tau s + 1)^n} = \frac{A\alpha^n}{\tau^n s(s + 1/\tau)^n} \qquad 4.6\text{-}9$$

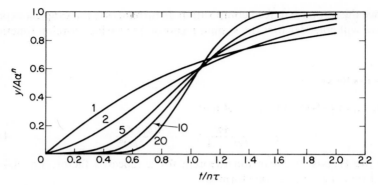

Figure 4.6-3. Step response of n first-order systems in series.

Again, using Heaviside's theorem, we find that

$$y(t) = A\alpha^n\left\{1 - e^{-t/\tau}\left[1 + \frac{t}{\tau} + \frac{1}{2!}\left(\frac{t}{\tau}\right)^2 + \cdots\right.\right.$$
$$\left.\left. + \frac{1}{(n-1)!}\left(\frac{t}{\tau}\right)^{n-1}\right]\right\} \qquad\text{4.6-10}$$

This result is plotted in Figure 4.6-3. It is clear from the graph that the system output looks more like a step function delayed in time as n becomes large.

Frequency Response

The pseudo-steady state frequency response can be obtained by substituting $j\omega$ for s in the transfer function

$$H(j\omega) = \frac{\alpha^n}{(1 + j\tau\omega)^n} = \frac{\alpha^n(1 - j\tau\omega)^n}{(1 + \tau^2\omega^2)^n} \qquad\text{4.6-11}$$

If we write a complex number in polar form

$$C_1 + jC_2 = C_3 e^{j\phi} \qquad\text{4.6-12}$$

where

$$C_3 = \sqrt{C_1^2 + C_2^2} \qquad \phi = \tan^{-1}\left(\frac{C_2}{C_1}\right) \qquad\text{4.6-13}$$

so that

$$(1 - j\tau\omega)^n = (\sqrt{1 + \tau^2\omega^2}\, e^{j\phi})^n \qquad \phi = \tan^{-1}(-\tau\omega) \qquad\text{4.6-14}$$

then Eq. 4.6-11 can be written as

$$H(j\omega) = \frac{\alpha^n(\sqrt{1 + \tau^2\omega^2})^n e^{jn\phi}}{(1 + \tau^2\omega^2)^n} = \frac{\alpha^n e^{jn\phi}}{\sqrt{1 + \tau^2\omega^2}^{\,n}} \qquad\text{4.6-15}$$

Hence the system gain is

$$H(j\omega) = \left(\frac{\alpha}{\sqrt{1 + \tau^2\omega^2}}\right)^n = \alpha^n |G|^n \qquad\text{4.6-16}$$

the phase angle is

$$\theta = n \tan^{-1}(-\tau\omega) \qquad\qquad 4.6\text{-}17$$

and the system output is (see Eq. 4.4-36)

$$y(t) = \left(\frac{\alpha}{\sqrt{1+\tau^2\omega^2}}\right)^n \sin(\omega t + \theta) \qquad\qquad 4.6\text{-}18$$

A Bode plot based on Eqs. 4.6-16 and 4.6-17 is given in Figure 4.6-4. The equations show that the phase angle varies from $0°$ to $-n(90°)$ as the forcing frequency varies from 0 to infinity—that is, the maximum phase angle for n first-order systems in series is just n times the maximum phase angle for a first-order system. Also, we see that the gain curve is bounded by two asymptotic lines. One asymptote is given by the line $|G| = 1$ in the region from $\omega\tau = 0$ to $\omega\tau = 1$ and the other passes through the point $|G| = 1$ when $\omega\tau = 1$ and has a slope of $-n$—that is, just n times the slope of a first-order system.

The system response for various changes in the control variable $\tilde{u}$ can be obtained similarly. The solutions for the three forcing functions treated above

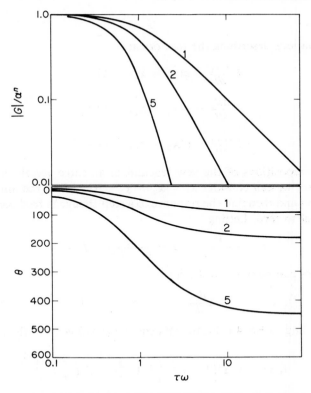

Figure 4.6-4. Bode plot for n first-order systems in series.

are presented as Exercise 51 at the end of this chapter. It should be noted at this point, however, that the transfer function for the two kinds of inputs are quite different; therefore we cannot expect to obtain exactly the same kind of behavior.

Example 4.6-1 Start-up of a series of isothermal, continuous-stirred tank reactros

Suppose that we have a battery of three continuous-stirred-tank reactors in series, which initially is operating at steady state conditions and is making some particular product. However, because of a changing market we want to use the reactors to make some new product. If we introduce the new feed material at the same flow rate, assume that the reactions have no effect on one another, and consider a case where the new reaction is first-order and isothermal, we would like to estimate the time required to obtain a salable product. As a specific example, find the time required to reach the new steady state for a case where $V_1 = V_2 = V_3 = 3000$ liters, $q = 1652$ liters/hr, $x_f = 0.1$ g mole/liter, and $k = 0.20$ hr^{-1}.

Solution

The equations describing the reactors are

$$V\frac{dx_1}{dt} = q(x_f - x_1) - kVx_1$$

$$V\frac{dx_2}{dt} = q(x_1 - x_2) - kVx_2$$

$$V\frac{dx_3}{dt} = q(x_2 - x_3) - kVx_3$$

Since the compositions of the new reactant in all three vessels initially are equal to zero, we can consider $x_1 = x_2 = x_3 = x_f = 0$ as the initial steady state solution and then find the response to a step change in feed composition at time equal to zero. Letting

$$\tau = \frac{V}{q + kV}, \quad \alpha = \frac{q}{q + kV}, \quad \beta = 0$$

the equations can be put into the form

$$\tau\frac{dy_n}{dt} + y_n = \alpha y_{n-1}$$

Thus, according to Eq. 4.6-10, the effluent composition from the last vessel, $n = 3$, is

$$x_3(t) = x_f \alpha^3 \left\{ 1 - e^{-t/\tau}\left[1 + \frac{t}{\tau} + \frac{1}{2!}\left(\frac{t}{\tau}\right)^2 \right] \right\}$$

As t approaches infinity x_3 approaches $x_f \alpha^3$, and the approach will be asymptotic. Now, if we assume that we can sell the product when $x_3(t)$ has reached 90 percent of its final value, we find from Figure 4.6-3 that

$$\frac{t}{n\tau} = 1.80$$

Since

$$\tau = \frac{V}{q + kV} = \frac{3000}{1652 + 0.2(3000)} = 1.333$$

the start-up time is

$$t = 3\tau(1.80) = 4(1.80) = 7.20 \text{ hr}$$

Finite Difference Equations

Although it was a simple task to determine the system transfer function for the process described above, our main purpose in this section is to introduce some of the ideas of the calculus of finite differences so that we will be able to handle more complicated equations describing staged systems. We first notice that if the stirred-tank reactors in the preceding example were placed very close together and if a great many more reactors were added to the battery, the sequence of stirred-tank reactors would somewhat resemble an isothermal, tubular reactor. The reactant composition would vary with the distance from the system inlet; but instead of changing in a continuous fashion, it drops in discrete amounts as the material passes from stage to stage. Similarly, the distance from the system inlet really should not be considered as a continuous variable, for the conditions everywhere within a single stage are the same. Thus we have a system where the independent variable (distance from the system inlet) is restricted to certain discrete values (the stage numbers) and the dependent variable (reactant composition) only has meaning at the discrete values of the independent variable. Clearly, the variation of gas and liquid compositions with height in a plate absorption column, or, in fact, the changes occurring in any staged process, exhibit this kind of behavior.

We say that any relationship between an independent variable that assumes discrete values only and a dependent variable that exists only at these discrete values of the independent variable is a *finite-difference equation*. Finite-difference equations are classified in the same way as ordinary differential equations. For example, the equation

$$a_0 y_0 + a_1 y_1 + a_2 y_2 + \cdots + a_n y_n = \phi(n) \qquad\qquad 4.6\text{-}19$$

is a linear, finite-difference equation, for the dependent variable enters the equation in a linear fashion. If $a_0, a_1, \ldots, a_n$ are functions of n, we say that the equation has variable coefficients, whereas if these coefficients are constants, the equation has constant coefficients. If $\phi(n)$ is a function of n or a constant, the equation is nonhomogeneous, whereas if $\phi(n) = 0$, the equation is homogeneous.

Finite-difference equations are solved in much the same way as ordinary differential equations. We first set $\phi(n) = 0$ to obtain the reduced equation and then seek the solution, called the complementary solution, of this simplified equation. The complete solution will be the sum of the complementary and some particular solution. In an analogous manner to ordinary differential equations, we might assume that we can find a solution of a homogeneous, linear, finite-difference equation having constant coefficients that has the form

$$y = e^{mn} \qquad\qquad 4.6\text{-}20$$

However, normally it is more convenient to let

$$z = e^m \qquad\qquad 4.6\text{-}21$$

and assume that the solution is

$$y = z^n \qquad\qquad 4.6\text{-}22$$

Substituting this expression into the reduced form of Eq. 4.6-19 gives

$$a_0 z^0 + a_1 z + a_2 z^2 + \cdots + a_n z^n = 0 \qquad\qquad 4.6\text{-}23$$

This equation will be satisfied for those values of z that are the roots of this nth order polynomial, called the characteristic equation. If all the roots are real and distinct, the complementary solution is simply

$$y_n = c_1 z_1^n + c_2 z_2^n + \cdots + c_n z_n^n \qquad\qquad 4.6\text{-}24$$

where $c_1, c_2, \ldots, c_n$ are arbitrary constants, which must be determined from the boundary conditons for the problem, and $z_1, z_2, \ldots, z_n$ are the roots of the characteristic equation. Hence the solution contains n linearly independent functions, $z_i^n = e^{m_i n}$, which is identical to the results obtained in the theory of ordinary differential equations.

If the characteristic equation has repeated real roots, it is necessary to modify the analysis somewhat in order to obtain n linear independent functions for the complementary solution. It can be shown that the portion of the solution corresponding to a root z_i repeated $(r + 1)$ times is

$$y = z_i^n (c_0 + c_1 n + c_2 n^2 + \cdots + c_r n^r) \qquad\qquad 4.6\text{-}25$$

Also, the portion of the solution corresponding to a pair of complex conjugate roots, say $z_i = r_1 + jr_2$ or $z_i = r(\cos\theta + j\sin\theta)$, can be put into the form

$$y = r^n (c_1 \cos\theta n + c_2 \sin\theta n) \qquad\qquad 4.6\text{-}26$$

Both results are essentially the same as the solutions of the corresponding cases in ordinary differential equations.

Once the complementary solution has been established, it is still necessary to find a particular integral if the original finite-difference equation is nonhomogeneous. As might be expected, the method of undetermined coefficients normally can be used for this purpose. The forms chosen for the particular solution are identical to those used in ordinary differential equations and the approach is the same. Also, the method of variation of parameters can be extended to the analysis of finite-difference equations.

Even though the steady state operation of a staged process can often be represented by an equation similar to Eq. 4.6-19, the dynamic operation of the system must contain a time-derivative term for the accumulation in the system is a continuous function of time. An equation like

$$\frac{dy_n}{dt} + a_0 y_0 + a_1 y_1 + a_2 y_2 + \cdots + a_n y_n = \phi_n \qquad \text{4.6-27}$$

is called a *difference–differential* equation, since it contains both an independent variable that only takes on discrete values, n, and one that is continuous, time. If we take the Laplace transform of this equation, we obtain

$$a_0 \tilde{y}_0 + a_1 \tilde{y}_1 + a_2 \tilde{y}_2 + \cdots + (a_n + s)\tilde{y}_n = \tilde{\phi}_n \qquad \text{4.6-28}$$

which is just an nth-order difference equation containing the Laplace parameter s in one of the coefficients. By again assuming a solution having the form of Eq. 4.6-22, we obtain the characteristic equation

$$a_0 + a_1 z + \cdots + (a_n + s)z^n = 0 \qquad \text{4.6-29}$$

Providing that we can solve for the roots of this equation in terms of the parameter s, we can establish the complementary solution for the transform of the system output. Once the particular solution is added to this result and the arbitrary constants have been fixed to match the specified boundary conditions, we attempt to invert the transform and thereby determine the time dependence of the output.

Of course, the foregoing outline of finite-difference equations is far from complete. However, it does provide a sufficient background to be able to obtain at least a first estimate of the dynamic response of many staged systems. Some examples of the technique are given below and in the exercises at the end of this chapter. Although these illustrate the power of the method, an interested reader should refer to one of the standard texts on finite-difference calculus in order to obtain a more rigorous background in the theory.[1]

Example 4.6-2 Start-up of a series of isothermal, continuous-stirred-tank reactors

Find the solution of Example 4.6-1, using finite-difference methods.

Solution

Since the general equation describing the nth tank in the series is

$$\tau \frac{dy_n}{dt} + y_n - \alpha y_{n-1} = 0$$

---

[1] See, for example, Ch. 9 of the reference in footnote 5 on p. 86, Ch. 4 of the reference given in footnote 6 on p. 86, or Ch. 9 of H. S. Mickley, T. K. Sherwood, and C. E. Reed, *Applied Mathematics in Chemical Engineering*, McGraw-Hill, N. Y., 1957.

we have a linear difference-differential equation with constant coefficients. Taking the Laplace transform of this equation, we obtain

$$(\tau s + 1)\tilde{y}_n - \alpha\tilde{y}_{n-1} = 0$$

Assuming a solution like

$$\tilde{y}_n = z^n$$

and substituting this solution into the difference equation, we find that

$$(\tau s + 1)z^n - \alpha z^{n-1} = 0$$

or

$$z^{n-1}[(\tau s + 1)z - \alpha] = 0$$

This equation will be satisfied if we let

$$z = \frac{\alpha}{\tau s + 1}$$

so that the solution must be

$$\tilde{y}_n = C_1\left(\frac{\alpha}{\tau s + 1}\right)^n$$

The arbitrary constant is determined by recognizing that at the inlet to the system, $n = 0$, the feed composition y_0 undergoes a step input, $\tilde{y}_0 = A/s$. Thus

$$\tilde{y}_0 = \frac{A}{s} = C_1\left(\frac{\alpha}{\tau s + 1}\right)^0$$

or

$$C_1 = \frac{A}{s}$$

Therefore the transform of the effluent composition of the nth tank is

$$\tilde{y}_n = \frac{A}{s}\left(\frac{\alpha}{\tau s + 1}\right)^n$$

which is identical to our previous result and can be inverted in the same way to give $y_n(t)$.

Example 4.6-3 Start-up of a plate, gas absorption unit

A dynamic model for a plate, gas absorption unit was developed in Example 3.1-2.

$$H\frac{dx_n}{dt} = Lx_{n+1} - (L + mV)x_n + mVx_{n-1}$$

where x_n is the composition of transferable component in the liquid phase, L is the liquid flow rate, V is the vapor flow rate, m is the slope of the vapor-liquid equilibrium line (the distribution coefficient), and H is the liquid holdup on a plate. If we consider a situation where gas and liquid are initially passing in countercurrent flow through the column at the desired steady state flow rate, $L = $ (lb moles)/(hr) and $V = $ (lb moles)/(hr), and at time equal to zero we introduce a gas mixture of known composition into the bottom of the plate column, we want to be able to estimate the time for the system to reach steady state conditions.

Solution

Let

$$\alpha = \frac{L}{mV} \qquad \tau = \frac{H}{mV}$$

so that the difference-differential equation describing the system can be written

$$\tau \frac{dx_n}{dt} = \alpha x_{n+1} - (\alpha + 1)x_n + x_{n-1}$$

For a case where the initial value of x_n on every plate is equal to zero, the Laplace transform of the preceding equation gives the result

$$\alpha \tilde{x}_{n+1} - (\tau s + \alpha + 1)\tilde{x}_n + \tilde{x}_{n-1} = 0$$

which is a linear difference equation having constant coefficients. Assuming a solution

$$\tilde{x}_n = z^n$$

and substituting this solution into the difference equation, we obtain

$$\alpha z^{n+1} - (\tau s + \alpha + 1)z^n + z^{n-1} = 0$$

or

$$z^{n-1}[\alpha z^2 - (\tau s + \alpha + 1)z + 1] = 0$$

Let us call the two roots of this quadratic equation z_1 and z_2. The roots cannot be evaluated at this point, for the Laplace parameter s is an unknown quantity. However, we can write the solution in the form

$$\tilde{x}_n = c_1 z_1^n + c_2 z_2^n$$

and we can easily show that

$$z_1 + z_2 = \frac{\tau s + \alpha + 1}{\alpha} \qquad z_1 z_2 = \frac{1}{\alpha}$$

The arbitrary constants in the solution, c_1 and c_2, must be evaluated so that the boundary conditions on the coumn are satisfied. According to the nomenclature used in Example 3.1-2, the liquid enters the column at plate N and the vapor enters at plate 1. Since the entering liquid composition is always zero and the inlet vapor composition suddenly increases from zero to some value y_f, we must have, at $t = 0$,

$$x_{N+1} = 0, \quad y_0 = y_f = mx_0 + b, \quad x_0 = \frac{y_f - b}{m} = A$$

The Laplace transforms of these quantities are, at $t = 0$,

$$\tilde{x}_{N+1} = 0 \qquad \tilde{x}_0 = \frac{A}{s}$$

Substituting these relationships into our general solution gives the two equations

$$0 = c_1 z_1^{N+1} + c_2 z_2^{N+1} \qquad \frac{A}{s} = c_1 + c_2$$

Thus we find that

$$c_2 = \frac{z_1^{N+1}(A/s)}{z_1^{N+1} - z_2^{N+1}} \qquad c_1 = -\frac{z_2^{N+1}(A/s)}{z_1^{N+1} - z_2^{N+1}}$$

and the complete solution for the transform of the liquid composition must be

$$\tilde{x}_n = -\frac{A}{s}\left(\frac{z_2^{N+1}z_1^n - z_1^{N+1}z_2^n}{z_1^{N+1} - z_2^{N+1}}\right)$$

Now we must find some way of inverting this Laplace transform, and during this inversion we need to remember that both z_1 and z_2 are functions of the Laplace parameter s. We can use Heaviside's expansion theorem for the inversion, once we have determined the roots of the denominator. By inspection we find that the demoninator is equal to zero when $s = 0$ or when $z_1^{N+1} = z_2^{N+1}$. The second result requires that $|z_1| = |z_2|$; but since z_1 and z_2 are the roots of a quadratic equation, they are either equal to each other or complex conjugates. Thus if we let

$$z_1 = re^{+j\theta} \qquad z_2 = re^{-j\theta}$$

where r is the magnitude of the complex number and θ its phase angle, the equation for the complex conjugate roots becomes

$$z_1^{N+1} - z_2^{N+1} = 0 = r^{N+1}\{\exp[(N+1)j\theta] - \exp[-(N+1)j\theta]\}$$
$$= 2jr^{N+1}\sin(N+1)\theta$$

Obviously this last equation can be satisfied whenever

$$(N+1)\theta = k\pi$$

where k is an integer. Now if we substitute $z_1 = re^{j\theta}$ and $z_2 = re^{-j\theta}$ into the original expressions for z_1 and z_2, we find that

$$z_1 z_2 = \frac{1}{\alpha} = r^2$$

$$z_1 + z_2 = \frac{\tau s + \alpha + 1}{\alpha} = r(e^{j\theta} + e^{-j\theta}) = 2r\cos\theta$$

Thus from the first expression we see that the magnitude of the complex conjugate roots must be

$$r = \frac{1}{\sqrt{\alpha}}$$

Substituting this result and $\theta = k\pi/(N+1)$ into the second relationship, we find that the only values of the Laplace parameter that will lead to a valid solution—that is, the roots of the denominator—are

$$s_k = -\frac{\alpha + 1 - 2\sqrt{\alpha}\,\cos\,[k\pi/(N+1)]}{\tau}$$

This expression gives a distinct set of values of s_k for $k = 0, 1, \ldots, (N+1)$. However, it repeats these if integral values of k greater than $(N+1)$ are substituted. Also, when $k = 0$ or $k = (N+1)$, the equations for z_1 and z_2,

$$z_1 = re^{+j\theta} \qquad z_2 = re^{-j\theta}$$

both predict that $z_1 = z_2$, so that these two sets of roots are not independent. The case where $z_1 = z_2$ appears to introduce some additional difficulty into the analysis since the expression

$$\tilde{x}_n = -\frac{A}{s}\left(\frac{z_2^{N+1}z_1^n - z_1^{N+1}z_2^n}{z_1^{N+1} - z_2^{N+1}}\right)$$

becomes indeterminant. However, L'Hôpital's rule can be used to show that $\tilde{x}_n$ remains finite as z_1 approaches z_2.

Thus it can be shown that the transform $\tilde{x}_n$ has $(N+1)$ **simple poles**

$$s = 0 \text{ and } s_k = -\frac{\alpha + 1 - 2\sqrt{\alpha}\,\cos\,[k\pi/(N+1)]}{\tau},$$

$$\text{for } k = 1, 2, \ldots, N.$$

If we apply the Heaviside theorem for the root $s = 0$, then

$$z_1 z_2 = \frac{1}{\alpha} \qquad z_1 + z_2 = \frac{\alpha + 1}{\alpha}$$

or

$$z_1 = 1 \qquad z_2 = \frac{1}{\alpha}$$

so that the contribution of this term to the solution is

$$x(t) = A\left[\frac{(1/\alpha)^{N+1} - (1/\alpha)^n}{(1/\alpha)^{N+1} - 1}\right] = A\left(\frac{\alpha^{N+1-n} - 1}{\alpha^{N+1} - 1}\right)$$

Similarly, the contribution corresponding to one of the roots s_k can be determined by first substituting the expressions $z_1 = re^{j\theta}$ and $z_2 = re^{-j\theta}$ into the equation for $\tilde{x}_n$

$$\tilde{x}_n = \frac{Ar^n}{s_k}\frac{\sin\,(N+1-n)\theta}{\sin\,(N+1)\theta}$$

Although we cannot write the denominator in terms of separate factors, we still can use Eq. 4.4-51 to find the contribution of each factor. Thus

$$x(t) = \frac{Ar^n}{s_k}\frac{\sin\,(N+1-n)\theta}{[(N+1)\cos\,(N+1)\theta](d\theta/ds)}e^{+s_k t}$$

where

$$\frac{d\theta}{ds} = -\frac{\tau}{2\alpha r \sin \theta}$$

since

$$2r \cos \theta = \frac{\tau s + \alpha + 1}{\alpha}$$

Substituting this expression and our previous results

$$(N + 1)\theta = k\pi, \quad r^2 = \frac{1}{\alpha}, \quad s_k = -\frac{\alpha + 1 - 2\sqrt{\alpha} \cos [k\pi/(N + 1)]}{\tau}$$

we find that the contribution to $x_n(t)$ of the root s_k must be

$$x(t) = \frac{A(-1)^k 2\alpha^{[-(n-1)/2]} \sin [k\pi/(N + 1)] \sin \{(N + 1 - n)[k\pi/(N + 1)]\}}{(N + 1)\{\alpha + 1 - 2\sqrt{\alpha} \cos [k\pi/(N + 1)]\}}$$

$$\times \exp \left\{ \left[-\alpha + 1 - 2\sqrt{\alpha} \cos \left(\frac{k\pi}{N + 1} \right) \right] \frac{t}{\tau} \right\}$$

Then if we sum the solutions over all k and add our previous result for $s = 0$, we obtain the equation for the effluent composition $x_n(t)$

$$x_n(t) = A\left(\frac{\alpha^{N+1-n} - 1}{\alpha^{N+1} - 1} \right)$$

$$+ A \sum_{k=1}^{N} \frac{(-1)^k 2\alpha^{[-(n-1)/2]} \sin [k\pi/(N+1)] \sin \{[k\pi(N+1-n)]/(N+1)\}}{(N + 1)\{\alpha + 1 - 2\sqrt{\alpha} \cos [k\pi/(N + 1)]\}}$$

$$\times \exp \left\{ -\left[\alpha + 1 - 2\sqrt{\alpha} \cos \left(\frac{k\pi}{N + 1} \right) \right] \frac{t}{\tau} \right\}$$

Once the liquid compostions have been determined, the corresponding vapor compositions can be calculated, using the equilibrium relationship, $y_n = mx_n$.

For a particular problem[2,3] where $m = 0.72$, $L = 40.8$, $V = 66.7$, $H = 75$, $N = 6$, $x_7 = 0$, $y_0 = 0.3$, we find that

$$\alpha = \frac{L}{mV} = \frac{40.8}{0.72(66.7)} = 0.85 \qquad \tau = \frac{H}{mV} = \frac{75}{0.72(66.7)} = 1.561$$

Some other values of interest are

k	θ	$2\sqrt{\alpha} \cos \theta$	λ_i	$\sin^2 \theta$	$\dfrac{(-1)^k \alpha^{5/2} \sin^2 \theta}{7(\alpha + 1 - 2\sqrt{\alpha} \cos \theta)}$
1	0.449	1.661	−0.121	0.188	−0.428
2	0.898	1.150	−0.449	0.611	0.374
3	1.346	0.410	−0.922	0.950	−0.283
4	1.795	−0.410	−1.448	0.950	0.180
5	2.244	−1.150	−1.922	0.611	−0.087
6	2.693	−1.661	−2.249	0.188	0.023

[2] L. Lapidus and N. R. Amundson, *Ind. Eng. Chem.*, **42**, 1071 (1950).
[3] A. Acrivos and N. R. Amundson, *Ind. Eng. Chem.*, **47**, 1533 (1955).

so that the solution for the response is

$$y_6 = 0.72x_6 = 0.72(0.417)\left(\frac{0.85 - 1}{0.85^7 - 1}\right)$$

$$+ 0.72(0.417)[-0.428e^{-0.121t} + 0.374e^{-0.449t} - 0.283e^{-0.922t}$$

$$+ 0.180e^{-1.448t} - 0.087e^{-1.922t} + 0.023e^{-2.249t}]$$

This result is plotted in Figure 4.6-5.

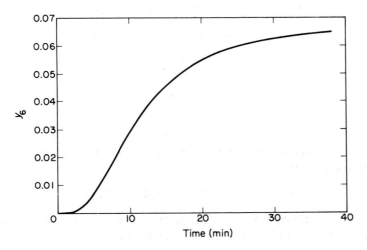

Figure 4.6-5. Step response of gas absorber.

SECTION 4.7   THE CONVOLUTION INTEGRAL AND WEIGHTING FUNCTIONS

Thus far we have emphasized the use of Laplace transforms to obtain estimates of the dynamic response of various systems. One might expect, however,
that it should be possible to obtain the same results through an analysis in
the time domain. This, in fact, is the case, and the appropriate analysis requires an evaluation of the convolution integral between the system-weighting
function and the process input. In order to gain some insight into this
approach, a good understanding of the convolution integral must be developed.

Convolution Integral

One of the most important theorems concerning Laplace transforms
states that the product of two Laplace transforms of arbitrary functions of

time must be the same as the Laplace transform of their convolution integral

$$\mathscr{L}[f(t)]\mathscr{L}[g(t)] = \mathscr{L}\left[\int_0^t f(t-\lambda)g(\lambda)\,d\lambda\right]$$

$$= \mathscr{L}\left[\int_0^t f(\lambda)g(t-\lambda)\,d\lambda\right] \qquad 4.7\text{-}1$$

To show that this theorem is valid, we first write the definition of the product of two transforms using different "dummy variables" of integration

$$\bar{f}(s)\bar{g}(s) = \left[\int_0^\infty e^{-sv}f(v)\,dv\right]\left[\int_0^\infty e^{-su}g(u)\,du\right]$$

Since the second integral is independent of v, we can write

$$\bar{f}(s)\bar{g}(s) = \int_0^\infty \int_0^\infty e^{-s(v+u)}f(v)g(u)\,dv\,du$$

$$= \int_0^\infty g(u)\left[\int_0^\infty e^{-s(v+u)}f(v)\,dv\right]du \qquad 4.7\text{-}2$$

If we let

$$t = v + u \qquad dt = dv \qquad 4.7\text{-}3$$

in the inner integral, where we can consider u to be a constant for the integration, we obtain

$$\bar{f}(s)\bar{g}(s) = \int_0^\infty g(u)\left[\int_u^\infty e^{-st}f(t-u)\,dt\right]du \qquad 4.7\text{-}4$$

Now if we interchange the order of integration and make the appropriate changes in the limits of integration as indicated in Figure 4.7-1, we find that

$$\bar{f}(s)\bar{g}(s) = \int_0^\infty \left[\int_0^t e^{-st}f(t-u)g(u)\,du\right]dt$$

$$= \int_0^\infty e^{-st}\left[\int_0^t f(t-u)g(u)\,du\right]dt \qquad 4.7\text{-}5$$

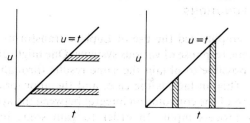

Figure 4.7-1. Integration limits.

Since this last result is simply the Laplace transform of the term in brackets, we can write

$$\bar{f}(s)\bar{g}(s) = \mathscr{L}\left[\int_0^t f(t-u)g(u)\,du\right] \qquad 4.7\text{-}6$$

which is the result we were trying to prove. Of course, the other form of Eq.

4.7-1 can be obtained in an identical manner by interchanging the functions $f(t)$ and $g(t)$.

Example 4.7-1 Response of a first-order system

The equation describing the response of a first-order system can be written as

$$\frac{dx}{dt} + ax = f(t) \qquad x(0) = 0$$

where $f(t)$ is the system input. Develop an expression for the time dependence of the system output for an arbitrary forcing function.

Solution

Taking the Laplace transform of the equation and solving for the transform of the output gives

$$\tilde{x}(s) = \frac{\tilde{f}(s)}{s + a}$$

or

$$\tilde{x}(s) = \mathscr{L}[e^{-at}]\mathscr{L}[f(t)]$$

Thus, to find the inverse transform of $\tilde{x}(s)$, we must invert the product of two Laplace transforms. However, from Eq. 4.7-1 we see that this is merely the convolution integral of the two time functions

$$x(t) = \int_0^t f(t - \lambda)e^{-a\lambda} \, d\lambda$$

which is a general expression relating the output to an arbitrary input $f(t)$.

Weighting Functions

Obviously the results of the simple example just described can be generalized to more complicated systems. Since the transfer function for any process is defined as the ratio of the Laplace transform of the system output to the input, we know that we can always write

$$\frac{\tilde{y}}{\tilde{f}} = H(s) \qquad\qquad 4.7\text{-}7$$

or

$$\tilde{y}(s) = H(s)\tilde{f}(s) \qquad\qquad 4.7\text{-}8$$

For a case where the input is a unit impulse, $\tilde{f}(s) = 1$, the output response will be some function of time $h(t)$, which is determined by inverting the transfer function

$$y(t) = \mathscr{L}^{-1}[H(s)] = h(t) \qquad\qquad 4.7\text{-}9$$

Also, for the arbitrary forcing function $f(t)$, we must have

$$y(t) = \mathscr{L}^{-1}[H(s)\tilde{f}(s)]$$

$$= \int_0^t f(t - \lambda)h(\lambda)\,d\lambda \qquad\qquad 4.7\text{-}10$$

In other words, response of a system to an arbitrary input can be determined by evaluating the convolution integral of the input and the impulse response of the system. The impulse response $h(t)$ is often called the weighting function of the system.

In order to gain a clearer picture of this procedure, let us suppose that we measure the impulse response of some process and obtain the weighting function sketched in Figure 4.7-2. For simplicity we will consider a discrete approximation of the weighting function, and examine what happens as we pass to the continuous case. Also, we consider that some arbitrary forcing function is made up of a series of impulses where the magnitude of each impulse is equal to the area under one of the small rectangles in Figure 4.7-3. Again, for simplicity, we will restrict our attention to the response to a step input that enters the system at time equal zero.

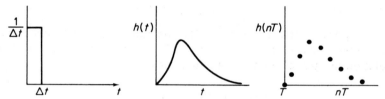

Figure 4.7-2. Weighting function.

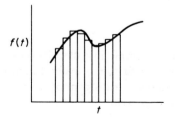

Figure 4.7-3. Forcing function.

Obviously, at time zero the output will be zero. Then the step is introduced, but according to the weighting function we do not see any immediate result. After one time interval has elapsed, the output will be equal to the sum of the effects caused by a new impulse entering the process, which has a zero contribution, and the value of the weighting function after one time interval, caused by the original impulse. Proceeding one more time interval, we find that we must add the responses caused by a new impulse, the value of the weighting function after one interval caused by the second impulse,

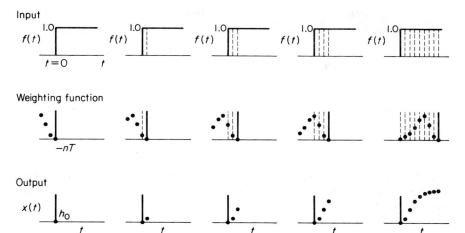

Figure 4.7-4. Convolution process.

and the value of the weighting function after two intervals caused by the original impulse.

It is clear from Figure 4.7-4 that we can carry out this calculation by plotting the weighting function backward in time (it is folded back), multiplying the values of the weighting function by the corresponding input values, and summing the results to get one output value. Then the position of the weighting curve is shifted and the procedure is repeated to get the next output value. Thus the approximate value of the output at some time $t = t_1$ may be written as

$$x(t_1) = \sum_{n=0}^{k} h_n f(t_1 - nT) = h_0 f(t_1) + h_1 f(t_1 - T)$$
$$+ h_2 f(t_1 - 2T) + \cdots + h_k f(t_1 - kT) \qquad 4.7\text{-}11$$

If we let the size of the discrete intervals approach zero, the preceding summation will approach the integral

$$x(t_1) = \int_0^{t_1} h(\lambda) f(t_1 - \lambda)\, d\lambda \qquad 4.7\text{-}12$$

which is the convolution integral.

Example 4.7-2 Residence time distribution of a reactor

It is a common practice to measure the residence time distribution of reactors in order to gain some insight into the amount of fluid mixing that takes place inside the reactor. The normal procedure is to introduce some finite amount of inert tracer material within a very short time interval and then measure the tracer composition as a function of time in the effluent from the reactor. From our previous discussions it is clear that the residence time

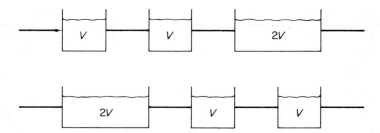

Figure 4.7-5. Series of stirred-tank reactors.

distribution is just the impulse response of the system. Suppose, then, that we want to show that the residence time distribution of the two reactor configurations sketched in Figure 4.7-5 are the same.

Solution

If we fill the reactors with an inert material having the same properties as the reacting mixture, the equations describing the first configuration are

$$V\frac{dx_1}{dt} = q(x_f - x_1)$$

$$V\frac{dx_2}{dt} = q(x_1 - x_2)$$

$$2V\frac{dx_3}{dt} = q(x_2 - x_3)$$

while the equations describing the second configuration are

$$2V\frac{dx_1}{dt} = q(x_f - x_1)$$

$$V\frac{dx_2}{dt} = q(x_1 - x_2)$$

$$V\frac{dx_3}{dt} = q(x_2 - x_3)$$

Letting $V/q = \tau$, taking the Laplace transforms, and finding the system transfer function either by eliminating the intermediate variables or by solving the difference equations, we obtain

$$\frac{\tilde{x}_3}{\tilde{x}_f} = \frac{1}{(\tau s + 1)^2(2\tau s + 1)}$$

for both cases. Hence the residence time distribution for both reactor configurations must be identical. For a unit impulse input, $\tilde{x}_f = 1$, so that

$$\tilde{x}_3 = \frac{1}{(\tau s + 1)^2(2\tau s + 1)}$$

and using Heaviside's theorem, we find that the residence time distribution is

$$x_3(t) = \frac{2}{\tau} e^{-t/2\tau} - \frac{1}{\tau}\left(2 + \frac{t}{\tau}\right)e^{-t/\tau}$$

If complete segregation in a reactor having this residence time distribution is assumed, the reactor conversion can be determined by multiplying the solution of the batch reactor equation by the residence time distribution and integrating the product from zero to infinity.

Solution of Matrix Differential Equations

In previous discussions we noted that we could always approximate the model of a dynamic process by the set of linear matrix differential equations

$$\frac{d\mathbf{x}}{dt} = \mathbf{Ax} + \mathbf{Bu} + \mathbf{Cv} \qquad 4.7\text{-}13$$

Also, for systems having distinct and real characteristic roots, it was possible to make a canonical transformation and completely uncouple the equations to obtain a set of first-order equations in terms of real variables

$$\frac{d\mathbf{y}}{dt} = \mathbf{\Lambda y} + \mathbf{P}^{-1}\mathbf{Bu} + \mathbf{P}^{-1}\mathbf{Cv} \qquad 4.7\text{-}14$$

or

$$\frac{dy_i}{dt} = \lambda_i y_i + f_i \qquad i = 1, 2, \ldots, n \qquad 4.7\text{-}15$$

where f_i is the ith row of the column vector $(\mathbf{P}^{-1}\mathbf{Bu} + \mathbf{P}^{-1}\mathbf{Cv})$. The solution of this simple first-order equation is

$$y_i = c_i e^{\lambda_i t} + e^{\lambda_i t} \int f_i(\tau) e^{-\lambda_i \tau} \, d\tau \qquad 4.7\text{-}16$$

where τ is a dummy variable of integration and c_i is an integration constant. Since the integral is independent of t, we can also write the solution as

$$y_i = c_i e^{\lambda_i t} + \int f_i(\tau) \exp\left[+ \lambda_i(t - \tau)\right] d\tau \qquad 4.7\text{-}17$$

and it is apparent that the integral term is the convolution integral.

It can also be shown that the solution of Eq. 4.7-13 is

$$\mathbf{x}(t) = [\exp(\mathbf{A}t)]\mathbf{x}_0 + \int_0^t [\exp \mathbf{A}(t - \tau)][\mathbf{Bu}(\tau) + \mathbf{Cv}(\tau)] \, d\tau \qquad 4.7\text{-}18$$

so that the matrix quantity $[\exp \mathbf{A}(t - \tau)]$ must be the matrix of weighting functions for the process. For some simple plants, it is possible to determine explicitly the elements of this matrix by using Sylvester's theorem[1] to simplify

---

[1] See, for example, footnotes 4, 5, and 6, p. 86, or footnote 3, p. 184.

the matrix exponential. However, the results obtained by using the canonical transformation normally require less manipulation.

SECTION 4.8   SUMMARY

In this chapter we developed analytical solutions for the response of first- and second-order systems to impulse, step, and sinusoidal inputs. The theory of Laplace transforms can be used to simplify the analysis, and therefore some of the important theorems of transform operations were discussed in detail. In addition, an introduction to the theory of finite-difference calculus was presented, for it provides a powerful approach for studying the dynamics of staged operations. Furthermore, the relationship between the transfer function and weighting function methods of representing process dynamics was described.

The results for the impulse and step response of a first-order system showed that the system output approached some steady state condition (the original value for an impulse and a new steady state for a step) in an asymptotic manner; that is, theoretically it takes an infinite length of time for the system to arrive at steady state. Thus it was necessary to define a characteristic time for the process, the time constant, and to assume that a practical engineering estimate of the time required to reach steady state would be in the range of three to five time constants. In the analysis of the frequency response of a first-order system, we found that the output fluctuations always have the same frequency as the forcing function, that the ratio of the amplitude of the output to the input signal was always less than unity (the ratio approaches zero as the input frequency increases), and that the output signal always lagged behind the input.

The response of a second-order system was shown to be dependent on the nature of the characteristic roots. If both roots are negative and real, the system acts like two first-order systems placed in series. For this case, the impulse response is smoother than the corresponding result for a first-order system and the step response exhibits an inflection point. Also, the frequency response shows that the amplitude ratio is less than unity and that the output signal lags behind the input, where the maximum lag is twice as high as that for a first-order system. Again, the amplitude of the output fluctuations was found to approach zero as the frequency of the forcing function increased.

For the case where the characteristic roots were complex conjugates, and, in particular, for systems having a damping coefficient of less than 1.0, a completely new kind of behavior was observed. In response to an impulse input, the system output would increase but then would drop below the original value and return to steady state in an oscillatory manner. Similarly, the response to a step change in the input would initially overshoot the final value and then oscillate toward this new steady state. The frequency response analysis showed that the amplitude of the output signal could be significantly larger than the input amplitude if the frequency of the forcing function was close to the resonant frequency of the system. The fact that complex conjugate roots lead to a different system behavior is to be expected, for a special canonical transformation is required for this case.

Although it is always possible to obtain a first estimate of the dynamic response of a plant by using the canonical transformation to write the complete set of dynamic equations in terms of first- and second-order systems, there are situations where other methods can be used to obtain a more direct solution to the problem. In particular, if the coefficient matrix is bi- or tri-diagonal, the calculus of finite differences provides a simpler approach. The results for a series of first-order systems show that the impulse and step response seem to approach an impulse and a step output as the number of units in series increases.

Similarly, it is often possible to find analytical solutions for the dynamic response of some distributed parameter processes; and whenever this can be accomplished, it is always a faster approach than differencing the partial differential equation and then using matrix or finite-difference techniques to find an approximate solution. The appropriate analytical techniques, together with some examples, are discussed in the next chapter.

QUESTIONS FOR DISCUSSION

What is the relationship between the development and experimental verification of a dynamic model and the theory of the statistical design of experiments? What is your concept of a "critical" experiment? If experimental evidence does not agree with the predictions of a dynamic model, how do you go about selecting a new set of assumptions for another model?

EXERCISES

1. (A) Normally it is a fairly simple task to open (close) a valve in such a way that one of the input variables increases (decreases) as a linear function of time. Therefore this *ramp input* is often used to study the dynamic characteristics of a plant. Determine the response of a first-order linear system to a ramp input.

2. (A‡) Design an experiment that will make it possible to estimate the time constant of the isothermal reactor described in Example 4.3-1 from the ramp response of the system. Discuss as many ways as you can think of for treating the data.

3. (A‡) The response to a ramp input can also be used to establish the dynamic behavior of second-order linear processes. Find the ramp response for the three kinds of second-order plants—that is, underdamped, critically damped, and over-damped. Describe how you would treat the experimental measurements to establish the two parameters in the system equation.

4. (A‡) In addition to being a style leader on the East Coast, with the nudie look and the Indian head band, Lonesome Polecat[1] became one of the outstand-ing intellectuals at Harvard (although it is debatable whether he progressed or the other students regressed). In fact, there was not a chemical engineer on campus half as bright as he. Moreover, he discovered that his purpose in life was to bring joy to others in a way that only he, and his good friend Hairless Joe, could do. Together they would mass produce Kickapoo Joy Juice so that intellectuals everywhere could "blow their minds" in a way that made "Hash" seem like Ju Ju Bes.

The first step in this project would be to develop a continuous process. There would be some problems doing so, such as how to feed the rattlesnakes and saber-tooth tigers through pipe lines and into the reacting vat. Nevertheless, with a suffi-ciently large grant from the National Science Foundation, he would find a way. What is more, he would even determine the optimum start-up procedure in order to maximize the production and increase the number of souls he could elevate to a higher state of unconciousness.

Although it would be a simple task for him to apply the calculus of variations to establish the optimal start-up policy, he thought that Hairless Joe might follow a case-study approach with greater ease. According to his flow sheet, in the first process unit he wants to add the mixture of crushed bat wings and yak horns to the seepage from Cesspool Sally's Chicken Fat and Flour Works. The vat contains a steam coil consisting of 10 turns of 1-in. O.D. tubing 5 ft in diameter, so that the feed mixture can be heated to 94°C. Steam is available at 110°C from the Skunk Works, and his copy of the Intentionally Confusing Tables (ICT) indicates that a reasonable value for the overall heat-transfer coefficient between the coil and the perfectly mixed contents of the vat is 120 (Btu)/(hr)(ft²)(°F). The inlet temperature of solid and fluid is 20°C, the specific heat of the feed mixture can be taken as unity, the desired steady state flow rate of mixture is 1000 lb/hr, the maximum feed rate is three times the desired value, and the tank capacity is 5000 lb.

Lonesome Polecat realizes that one possible start-up procedure for this unit would be to fill the tank with the feed mixture in the proper proportions, start the agitator, commence feeding the solid and liquids at the desired feed rate while simultaneously withdrawing slurry at this same rate, and then turn on the steam.

An alternate start-up procedure would be to turn on the steam when the tank was empty, fill the tank, and then adjust the feed and effluent flow rates to the desired values.

At this point Lonesome Polecat passed out. Can you calculate the required start-up time for the two procedures? Also, suggest at least one other start-up approach, which hopefully will have a shorter start-up time, and compare the results.

---

[1] The characters are from Li'l Abner by Al Capp, who taught me the fun of learning philosophy through humor.

5. (A) Of course, the calculated values obtained in the foregoing problem would not be the exact values that we would measure with a thermometer, for the thermometer dynamics will make the observed temperature different from the true value. Consider a thermometer having a $\frac{1}{2}$-in. long and $\frac{1}{8}$-in. diameter bulb filled with mercury, which is immersed in the vat. The heat-transfer coefficient between the fluid and the bulb is 600 (Btu)/(hr)(ft²)(°F). Plot the observed temperature versus time for the first start-up procedure.

6. (A‡) In Example 4.3-1 we calculated the dynamic response of an isothermal, continuous-stirred-tank reactor for a case where the feed composition first increased by 10 percent for 1 hour and then decreased by 10 percent for 1 hour, before returning to its original value. Plot the response of the unit for cases where the disturbances last 1 minute and 1 day. Also, plot the response for 50 percent changes lasting 1 hour.

7. (A) If you wanted to use a frequency response technique to verify experimentally the time constant of the isothermal reactor described in Example 4.3-1, what frequency range would you select for the input signal?

8. (A*) One method for measuring the pressure within a unit is to connect a lead filled with fluid from the measuring point to a bellows. The bellows is designed so that expansion can take place only in one direction, and a spring is installed to oppose this expansion. If we place a pointer on the end of the bellows, we can relate the observed displacement to the pressure changes at the measuring point (see Figure 4.8-1).

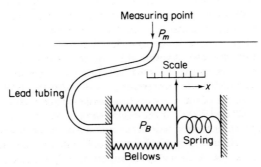

Figure 4.8-1. Pressure measuring device.

As a first approximation, it is possible to develop a very simple, dynamic model for this measuring device by neglecting the inertia of the bellows and spring. For this case, the accumulation of mass within the bellows must be equal to the mass entering from the lead, which we assume to be by laminar flow because the lead tubing normally is very small. Thus

$$\rho \pi R_B^2 \frac{dx}{dt} = \rho \frac{\pi R^4 (P_m - P_B)}{8 \mu L}$$

where ρ is the fluid density in the bellows and lead tubing, R_B is the radius of the bellows, x is the displacement of the end of the bellows, R is the radius of the pressure lead, P_m is the pressure at the measuring point, P_B is the pressure within the bellows, μ is the viscosity of the fluid in the lead tubing, and L is the length of lead tubing used. Similarly, for the case with no inertia, a force balance indicates that the changing-pressure force on the end of the bellows must be balanced by the spring

$$P_B \pi R_B^2 - K(x - x_0) = 0$$

where K is a spring constant, x_0 is an initial displacement, and both of these quantities must be determined by calibrating the device. We can eliminate P_B between the two equations, and in this way obtain a first-order model relating the observed displacement to the pressure at the measuring point.

Consider a device of this type for a case where $R = 0.1$ in., $R_B = 0.5$ in., the bellows and tubing are filled with water, a 1-in. expansion of the bellows represents an increase in pressure of 15 psi at steady state conditions, and the scale position is adjustable. What would be the maximum allowable distance from the measuring point to the control room if we hoped to observe sinusoidal pressure oscillations in the range of 1 Hz (cycle per second)? How does the length vary with the frequency of the oscillations?

9. (A*) In the preceding problem we neglected the inertia of the moving parts. Are our previous results changed significantly if we assume that the system has an effective mass of 5 grams located at the point where the spring and bellows are connected?

10. (A*) The current interest in air-pollution problems indicates that it is essential to gain a better understanding of the combustion process in an internal combustion engine. We expect that the kinetics of the reactions taking place will be exponential functions of temperature, so that it is important to be able to measure the exact values of temperature in the cylinder. Suppose that we attempt to use a thermocouple for this purpose. Assume that the thermocouple bead is spherical 0.02 in. in diameter, its heat capacity is 0.1 (cal)/(g)(°C), its density is 8.5 (g)/(cm³), and the heat-transfer coefficient is approximately 800 Btu/(hr)(ft²)(°F). The actual temperature in the cylinder of the reciprocating, single-stroke cycle engine is believed to oscillate between 500 and 700°C. If the engine runs at 100 rpm, which means that 100 times per minute an explosion will occur and a maximum temperature will be observed, estimate whether or not the thermocouple will provide satisfactory data.

11. (A*) During the discussion with your supervisor of your calculations from the preceding problem, he *suggests* that you install the thermocouple inside a well in order to prevent high-temperature corrosion. Thus you decide to estimate the reponse of the thermocouple when it is inside a well 0.5 in. long, made of 0.25-in. O. D. tubing, which has a wall thickness of 0.66 in. The heat capacity of the metal well is 0.12 (cal)/(g)(°C), its density is 9.0 (g)(cm³), the heat-transfer coefficient between the well and the hot gases is 750 (Btu)/(hr)(ft²)(°F), and the heat-transfer coefficient between the well and the thermocouple is 3.0 (Btu)/(hr)(ft³)(°F).

12. (B*) As a matter of curiosity, you decide to study the effect of the shape of the input signal on the response of the bare thermocouple described in Exercise 10. In addition to your previous result, it seems reasonable to consider a case where during each cycle the explosion would instantaneously raise the temperature from 500 to 700°C, and then the temperature would decay in a linear manner to the former value. Plot the solution for this case and compare it to the solution you obtained before.

13. (A) *The Handbook for Problem Plagerizers* reveals that an interesting frequency response problem was published by Johnson.[2] He describes a plant test on a gas-fired oil heater, where the response of the hot-oil effluent temperature to sinusoidal variations in the gas-flow control value was measured. When 8 to 10 psig fluctuations were introduced in the air pressure to the gas-valve actuator, for a case where the oil entered the heater at 200 °F at a rate of 120 gpm, the values of the effluent oil temperature were recorded, and they are listed in Table 4.8-1.

Prepare a Bode plot for the system, and from the plot develop an approximate dynamic model that can be used to describe the system.

TABLE 4.8-1
MEASURED VALUES OF EFFLUENT OIL TEMPERATURE

Period (min)	Steady State	10	2.5	1.0	0.5
Maximum oil temperature (°F)	440	436	430	425	423
Minimum oil temperature (°F)	404	408	415	419	421
Lag, percent of period	0	5	22	51	92

14. (A) Johnson also presents a problem where a dynamic experiment is used to obtain basic information about a process. An attempt is being made to measure the heat-transfer coefficient between a thermistor and a gas stream. It is observed that the thermistor begins to respond immediately when the gas temperature is suddenly changed from 6.6 to 14.9°C, that it takes 7.1 sec for the thermistor to approach 90 percent of the way to its new steady state value when the gas velocity is 20 ft/sec., and that it takes only 3.0 sec. for a 90 percent approach when the gas velocity is 40 ft/sec. The relationship between heat-transfer coefficient and gas velocity is expected to have the form $h = Cv^n$, and you are to use the data given above to evaluate the parameters in this expression.

15. (A‡) Summer jobs are one of the most fascinating parts of a chemical engineer's training. In fact, it is my belief that summer jobs and homework problems provide the real sustenance of the whole educational experience. There are times, however, when summer jobs do not work out too well, as is illustrated in a problem of Levenspiel.[3]

Levenspiel describes a case where two nice boys, both chemical engineering students, were part of a summer training program. They were assigned to work on a "practical" problem; namely, scraping the paint off an operating reactor and then repainting the unit. Unfortunately, however, they had taken a course in kinetics and were always talking about incomplete mixing in backmix reactors, residence time distributions, and other such nonsense. It was almost as if the existing reactor, which was made from a salvaged gasoline storage tank 8 ft in diameter and 18 ft long with 800 ft³ of usable volume, and the outboard motor used as an agitator were not properly designed by that plant engineer who died 20 years ago.

[2] E. F. Johnson, *Automatic Process Control*, p. 79, McGraw-Hill, N.Y., 1967.
[3] See p. 307 of the reference in footnote 2 on p. 12.

Nevertheless, the two boys said that they were willing to work overtime if they would be allowed to measure the residence time distribution. After some high-level management discussions, and after the boys said that they would pay for the lost product and the equipment needed for the tests, they finally received permission to proceed with their crazy scheme. Everyone watched them cart gallon bottles of India ink into the plant; and every once in a while, in between their paint-chipping activities, they would dump a gallon of ink into the reactor, along with the 20 cfm of feed, and then take readings on the outlet pipe with a flashlight and a photographer's exposure meter. It took several attempts to get the kind of data they wanted; finally, however, one night they left the plant all smiles and said they knew of a way the company could improve its production rate. They had an appointment to explain their scheme to the boss the next day, but, unfortunately, early that morning when they were scraping paint again the reactor sprang a leak and both boys dissolved in the corrosive spray. Only the data listed below remained.

Concentration* of Ink at Reactor Outlet	with Jumps up to 500	90	55	35	20	10	7	3	1
Time (min)	0–20	10	20	30	40	50	60	80	100

* One gallon of ink uniformly distributed throughout the reactor gives a concentration reading of 100.

Can you figure out what to do with these numbers? These two boys deserve a a fitting memorial, so think of as many ways to treat the data as you can and write a nice report.

16. (A) Apply the four methods discussed in Section 4.5 for estimating time constants from experimental data to the step-response curve for a gas absorber shown in Figure 4.6-5. Plot your second-order models and compare them to the original curve.

17. (A) Staffin and Chu[4] studied the dynamic response of a single-stage extraction unit. After some manipulation, they report one set of data as

$c(t)$	0.9	0.85	0.75	0.68	0.54	0.44	0.33	0.20	0.14
t	5	7.5	10	15	22	30	45	60	75

$c(t)$	0.12	0.07	0.05	0.04
t	90	105	120	150

Use at least three procedures to estimate the time constant of the unit.

18. (B) One of the most interesting features of process dynamics problems is that in some cases the problem statement can be quite startling. An example of this type is presented by Agnew,[5] who tells us that one day it started snowing heavily, but at a constant rate. A snowplow starts out at noon, proceeds 2 miles the first hour, and clears another mile in the second hour. Agnew then poses the question, What time did it start snowing? The problem is simple to solve, if we just assume that a snowplow can remove snow at a constant rate, but this book would seem incomplete without some reference to an "Agnew snow job."

[4] H. K. Staffin and J. C. Chu, *AIChE Journal*, **10**, 98 (1964).
[5] R. P. Agnew, *Differential Equations*, McGraw-Hill, N.Y. 1942.

19. (B*) A favorite "trick" question often asked on Ph.D. oral examinations is, Does a cup of coffee cool faster if you add the cream initially, or wait until the end of the cooling period? Common practice dictates that the cream should be added at the outset, but a more careful consideration reveals that the opposite policy may be better. In other words, if we recognize that the rate of heat loss, both by convection and radiation, from the coffee will be greatest when the temperature driving force is largest, we see that the effect of adding the cream at the start of the cooling process is to decrease the cooling rate.

(a) We can consider a question to be a "trick" question whenever it contains an implicit assumption that is misleading. Examine the statement of the problem carefully and see if you agree with all of its premises. Then, attempt to reformulate the problem, so that the commonly accepted procedures will be correct.

(b) One of the best techniques for solving a "trick" problem is to pose a different "trick" question with the opposite answer. For example, we could say that our original analysis was incomplete because we did not consider the coffee cup. With this new system—that is, the coffee and cup at some initial temperature—we would expect that the initial rate of heat loss from the coffee, primarily by radiation, would exceed that from the cup wall. Thus the cup would transfer heat both to the air and to the coffee. We can increase the rate of heat loss from the cup wall to the coffee by adding the cream at the start of the process; therefore this is the most efficient procedure.

What implicit assumptions in this problem formulation may be invalid?

(c) In order to resolve the conflicting arguments presented above, we should be able to predict the rate of cooling quantitatively. One engineering approach to the problem would be to develop a solution by the method of successive approximations. We could commence our analysis by considering the cooling of only the coffee. For this case, and with a number of simplifying assumptions, we would expect to obtain a first-order, nonlinear differential equation. Also, we know that the equation will become linear if we introduce the radiation heat-transfer coefficient. Thus by comparing these two solutions, we can gain some insight into the importance of the nonlinear term. Next, we can consider a simplified model for the the cooling of the cup and the coffee. The solution of this problem will enable us to understand the importance of the wall capacity. If the problem warranted additional work, we could modify our simplifying assumptions and take other effects into account.

Derive the dynamic models mentioned above. Carefully list the assumptions you use in each case. Select a set of *realistic* values for the system parameters, use whatever process correlations are necessary, and solve the equations for the simplified cases—that is, ordinary differential equations. What conclusions do you draw from your results? (Note that $T^4 - T_a^4 = [T^2 + T_a^2][T^2 - T_a^2]$.)

20. (A‡) Determine the response of the manometer described in Example 4.5-1 if the water is replaced by mercury. Repeat the calculation, using carbon tetrachloride. What conclusions can you draw about the speed of approach to steady state conditions and the failure due to overshoot for the various manometer fluids?

Next, consider cases where the water-filled manometer has lengths of 1.0 and 3.0 ft. Finally, repeat the original analysis with $\frac{1}{8}$-in. I.D. and $\frac{1}{2}$-in. O.D. tubes. Which variables have the greatest effect on the dynamic performance?

21. (B*) Estimate the start-up time for the reactor system described in Exercise 3.5-8 if the only material available initially is at the feed composition $A_f = 1.0$ lb mole/cu ft^3. Use the matrix approach for uncoupling the equations to obtain the solution. Also, determine the start-up time if we initially fill the reactors with the product material. Plot the composition in each vessel versus time for the two start-up techniques.

22. (B*) Estimate the response of the reactor system described in Exercise 3.5-8 to a disturbance in feed composition, where the feed composition increases by 10 percent for one hour, drops 10 percent below its initial value for one hour, and then returns to its original value. Plot your results and compare them to Example 4.3-1 and Exercise 4.8-6.

23. (B*) In Exercise 3.5-8 we found that two continuous-stirred-tank reactors in series could provide a more profitable system than a single reactor. We arbitrarily chose the reactors to have equal volumes, although, in practice, we would attempt to establish the optimum sizes.

Suppose we now consider two other cases: (a) where the first reactor has a volume of 750 ft^3 and the second has a volume of 250 ft^3 and (b) where the first reactor is 250 ft^3 and the second is 750 ft^3. Calculate the steady state operating conditions and profits, linearize the dynamic equations around the steady states, and estimate the response to the disturbance described in the previous exercise. Plot the response curves and compare them to the curves obtained before.

24. (B) In Example 4.5-1 we showed that the dynamic model for a manometer could be written as

$$\tau^2 \frac{d^2h}{dt^2} + 2\gamma\tau\frac{dh}{dt} + h = f(t) \qquad\qquad 4.5\text{-}66$$

If we define two new dependent variables as

$$x_1 = h \qquad x_2 = \frac{dx_1}{dt} = \frac{dh}{dt}$$

we see we can write the single second-order equation as two first-order equations— that is,

$$\frac{dx_1}{dt} = x_2 \qquad \tau^2\frac{dx_2}{dt} + 2\gamma\tau x_2 + x_1 = f(t)$$

It takes only a slight rearrangement to put these equations into matrix form.

Using the parameters given in Example 4.5-1, find the transformation matrix that can be used to put the manometer equations into canonical form. Develop the complete solutions of the canonical equations and plot your results. Then transform these curves back into the original set of variables.

25. (B*) Coughanowr and Koppel[6] discuss the simulation of a laboratory-scale gas absorber containing two plates on an analog computer. The dynamic equations used to describe the unit are given in Exercise 3.5-9. At steady state conditions, an air-SO_2 mixture containing 2 mole percent SO_2 entered the bottom of the column at a rate of 0.051 lb mole/min, while pure water was introduced into the top of the column at a rate of 0.90 lb mole/min. The column was operated at 25°C and 1.0

[6] D. R. Coughanowr and L. B. Koppel, *Process Systems Analysis and Control*, pp. 338, 469, McGraw-Hill, N.Y., 1965.

atm so that the equilibrium relationship was approximated by the expression $y = 27x$ — 0.00324, where y and x are mole fractions. The values of H and τ were experimentally measured to be $H = 0.11$ lb mole and $\tau = 4$ sec. Coughanowr and Koppel used these values when they numerically soved the complete set of four equations for a case where the inlet composition of SO_2 in the air stream was doubled and for a case where the water flow rate was decreased to 0.60 lb mole/min. We are interested to see if we can use linear systems theory to estimate the response.

First, solve for the steady state compositions in the liquid phase on each plate. Next, linearize the two material balance equations in Exercise 3.5-9 around these steady state conditions. Then, considering only the two material balance equations, estimate the step response for the two cases. Finally, solve the set of four linear equations for the step response for each case. Plot your results and compare them to the graphs presented by Coughanowr and Koppel (p. 473).

26. (B*) Under what conditions can the dynamic characteristics of the liquid flow through the column described in the previous problem be neglected? Is it possible to make any general statements about tray hydraulics for similar plants? How does the response change if you use the modification suggested by Lees (see footnote 2, p. 106)?

27. (B*) Suppose the plate efficiencies (see Exercise 2, p. 104, for the definition of this quantity) of the trays in Exercise 25 above are 0.75 rather than unity. Estimate the response of the column for the two types of step inputs and compare your results with those obtained earlier.

28. (B) One of the real pleasures in having children is that you get to go to a Fun House again without looking foolish. The simple pleasure of watching them attempt to navigate turntables, find their way through mazes, walk across air pillows in the dark, and so on is a continuing joy.

The reason for including this preamble is that it was while watching my kids tumbling in a rotating barrel that I finally realized how to do a homework problem I had been assigned some 15 years previously when a student at Johns Hopkins. The problem was given by Professor H. E. Hoelscher[7] and is described below. I certainly hope that it does not take you 15 years to solve it. Hoelscher's problem is as follows:

A horizontal rotating drum is being used as a cooling unit for a hot granular solid. The drum is 4 ft long, 2 ft in diameter, and is well insulated both on the outer surface and the ends. It contains, initially, an amount of solid equal to one-third the volume of the drum, and it rotates at such a rate that one-third of the solid in the drum is continually falling through the air, while the remaining two-thirds is on the bottom. Air, at 80°F, enters one end of the drum at a rate of 20 cfh and leaves through a solid's filter at the other end. If the initial solid's temperature is 300 °F, calculate the time required to reduce the solid's temperature to 100°F.

You may assume that the air is perfectly mixed in the drum by the action of the falling solid; the solid's distribution is uniform throughout the length of the drum, so the same amount is held in the spray everywhere along the length; the solids are perfect spheres; all heat transfer from the solid to the air is by convection; the volume of the flights causing the solid's motion may be neglected; and the solid's

[7] H. E. Hoelscher. Class notes from John Hopkins University, 1955.

bed on the bottom of the drum may be assumed flat. The data needed for the problem are true density of solid $= 165$ lb/ft^3, bulk density of bed 110 lb/ft^3, average particle size $= 160$ microns, specific heat of solid $= 0.15$ (Btu)/(lb)($°$F); and heat-transfer coefficient between solid and air $= 10$ (Btu)/(hr)(ft^2)($°$F) for both the solid on the bottom and the solid in the air.

29. (B*) In our numerous discussions of the dynamics of continuous-stirred-tank reactors, we limited our attention to the concentration of the reactants and we neglected to discuss how to calculate the compositions of the products. For the simplest case of an irreversible, isothermal reaction, such as we studied in Example 4.3-1, we might attempt simply to write a material balance equation for both the reactant and the product

$$V\frac{dA}{dt} = q(A_f - A) - kVA^2$$

$$V\frac{dB}{dt} = q(B_f - B) + kVA^2$$

Considering a case where the feed stream does not contain any product, $B_f = 0$, the second equation becomes

$$V\frac{dB}{dt} = -qB + kVA^2$$

At steady state conditions we see that

$$B_s = \left(\frac{kV}{q}\right)A_s^2$$

and when we linearize the equations around the steady state operating point, after dividing by V, we obtain

$$\frac{d(A - A_s)}{dt} = -\left(\frac{q}{V} + 2kA_s\right)(A - A_s) + \left(\frac{q}{V}\right)(A_f - A_{fs})$$

$$\frac{d(B - B_s)}{dt} = (2kA_s)(A - A_s) - \left(\frac{q}{V}\right)(B - B_s)$$

Of course, we can make the dependent variables dimensionless by dividing both equations by some constant reference composition, such as A_{fs}, and we would follow this procedure if we were actually attempting to develop a solution rather than studying how to calculate the time dependence of the product concentration.

At first it seems as if it will be necessary to solve a set of second-order equations—that is, to find the solution for two first-order, coupled equations. After some additional inspection, however, we notice that the material balance for the reactant A is identical to our previous expression. Therefore it must have the same solution, and we can use this result to eliminate $(A - A_s)$ from the product material balance. In this way we obtain

$$\frac{d(B - B_s)}{dt} + \frac{q}{V}(B - B_s) = f(t)$$

It is always a simple matter to solve this equation for the concentration of B. Our only problem now is to try to guess if we always obtain a simple result of this type.

Before attempting to generalize our analysis, it is helpful to recognize that there is a simpler approach. If we merely add the two original material balance equations (or, in general, combine them in such a way that the reaction rate term is eliminated), we find that

$$V\frac{dA}{dt} + V\frac{dB}{dt} = qA_f + qB_f - qA - qB$$

Defining a new composition variable

$$C = A + B$$

and dividing by the volume, we have

$$\frac{dC}{dt} + \frac{q}{V}C = \frac{q}{V}(A_f + B_f)$$

Again, this is a simple equation to solve, so that we can always find the total composition $(A + B)$ as a function of time. Knowing this quantity, and determining A as a function of time, makes it a simple job to calculate the time dependence of B.

Consider a reaction

$$aA + bB \longrightarrow rR + sS$$

described by an nth-order rate equation, $r_A = -kA^n$, in an isothermal, continuous-stirred tank reactor, and develop expressions for the compositions of components B, R, and S in terms of the limiting reactant A and time.

30. (B*) See if you can extend the results of the problem described above to the set of complex reactions

$$A + B \longrightarrow C + D$$
$$A + C \longrightarrow R + S$$

Next, consider the case of reversible reactions.

31. (B) In Chapter 7 we will demonstrate how a feedback control system can be used to alter the dynamic characteristics of a process. Without worrying about the details of these changes, however, we can gain some initial insight into the kinds of behavior we might anticipate by studying another problem published by Johnson[2]. He describes a set of experiments where a step change was made in the set-point of a controller installed on a large fractionating column. For a case where the controller gain was low, $K_c = 3.1$, the response was not oscillatory and the observed output is given in Table 4.8-2.

After the new steady state operation was achieved, the controller gain was increased to $K_c = 5$ and the set-point was returned to its original value; that is, a reverse step was introduced. For this case, a damped oscillatory output was observed, with the first peak occurring at an instrument reading of 46.4 percent after 13.6 min and the second peak at 42.3 percent after 39.8 min.

Develop approximate dynamic models for the behavior of the fractionator and

TABLE 4.8-2

SYSTEM RESPONSE

Time (min)	0	4	8	12	16	20	120
Percent of instrument scale	41.2	39.0	36.4	34.6	33.5	32.8	31.1

controller for the two cases. Use as many methods as you can to evaluate the parameters in the models.

32. (B) Johnson[2] also discusses an interesting design problem based on dynamic considerations. He considers a perfectly mixed, molten salt bath operating in the vicinity of 500°C but fluctuating at ± 8.7°C with a period of 5.8 min. A spherical steel test vessel is to be placed in this bath. Assuming that the molten salt is noncorrosive and that the heat-transfer coefficient between the salt and the metal is 200 (Btu)/(hr)(ft²)(°F), how large would the test cell have to be so that its temperature would remain constant to within ± 0.05°C?

33. (B*) Our previous work has shown that virtually all process units will provide some damping of fluctuations in the input streams. Thus in some cases it will not be necessary to install surge tanks upstream of the unit. Consider the laboratory-scale, gas absorber described in Exercise 25 above and determine the frequency range of input oscillations that will be effectively damped by the unit. Solve the problem for both feed composition and liquid flow rate disturbances.

34. (B‡) In Example 4.6-3 we presented an analysis of the step response of a plate, gas absorber. The calculations were fairly lengthy, and therefore we expect that it would be advantageous to have a simple procedure for obtaining a rough estimate of the most important time constants. One approach we might try is to lump several of the stages together. In other words, we could pretend that there were only two trays, rather than six, and that the holdup per tray was $H_{\text{approx}} = 3(75) = 225$. Use these values to estimate the step response of the absorber, plot your solution, and compare the results to Figure 4.6-5. How do your eigenvalues compare with those for the six-plate system? Do you think this technique for lumping stages has any merit?

35. (B‡) The parameters for the optimum design of some nonisothermal, continuous-stirred-tank reactor problems are given in Table 4.8-3. Case I refers to a single irreversible reaction, similar to those considered previously. If we define a set of dimensionless variables as

$$x_1 = -\left(\frac{A - A_s}{A_s}\right), \quad x_2 = -\left(\frac{T - T_s}{T_s}\right)$$

$$u_1 = \left(\frac{q - q_s}{q_s}\right), \quad u_2 = \left(\frac{q_H - q_{Hs}}{q_{Hs}}\right), \quad \tau = \frac{q_s t}{V}$$

we can show that the linearized equations become

$$\begin{pmatrix} \dot{x}_1 \\ \dot{x}_2 \end{pmatrix} = \begin{pmatrix} -5.0 & -52.06 \\ 0.09195 & -0.4432 \end{pmatrix}\begin{pmatrix} x_1 \\ x_2 \end{pmatrix} + \begin{pmatrix} -4.0 & 0 \\ 0.1379 & -0.01248 \end{pmatrix}\begin{pmatrix} u_1 \\ u_2 \end{pmatrix}$$

Use these linearized equations to obtain a first estimate of an acceptable start-up procedure.

36. (B‡) Case II in Table 4.8-3 lists the parameters corresponding to the optimum steady state design for a single irreversible reaction in a nonisothermal, continuous-stirred-tank reactor. If we define the set of dimensionless variables given in the previous exercise, the linearized equations obtained are

$$\begin{pmatrix} \dot{x}_1 \\ \dot{x}_2 \end{pmatrix} = \begin{pmatrix} -3.333 & -37.58 \\ 0.112 & -0.5162 \end{pmatrix}\begin{pmatrix} x_1 \\ x_2 \end{pmatrix} + \begin{pmatrix} -2.333 & 0 \\ 0.20 & -0.03645 \end{pmatrix}\begin{pmatrix} u_1 \\ u_2 \end{pmatrix}$$

Use these equations to estimate the effect of sinusoidal fluctuations in the hot-fluid flow rate. Assume that the amplitude of the oscillations is 10 percent of the steady state value, and consider input frequencies $\omega = 0.75$, 1.5, and 7.5. Discuss your results.

37. (B) Case III in Table 4.8-3 lists the parameters corresponding to the optimum steady state design of a pair of parallel irreversible reactions, $A \rightarrow B$ and $A \rightarrow C$, where the first reaction giving the desired product is second order and the second

TABLE 4.8-3

REACTOR PARAMETERS

Parameter	Case I	Case II	Case III	Case IV	Units
System Inputs					
G_B	5000	5000	6000	4000	(g mole)/(hr)
T_f	300	300	300	300	(°K)
T_H or T_c	373	400	373	300	(°K)
U	200	300	300	500	(cal)/(hr)(cm^2)(°K)
A_f	0.005	0.005	0.01	0.01	(g mole)/(cm^3)
System Constants					
$C_{pf}\rho_f$	1.0	1.0	1.0	1.0	(cal)/(cm^3)(°K)
$C_{pH}\rho_H$	1.0	1.0	1.0	1.0	(cal)/(cm^3)(°K)
E_1	–	–	14,000	12,000	(cal)/(g mole)
E_2	9000	12,000	10,000	15,000	(cal)/(g mole)
$(-\Delta H_1)$	–	–	-5000	8000	(cal)/(g mole)
$(-\Delta H_2)$	8000	12,000	10,000	9000	(cal)/(g mole)
k_{10}	–	–	2.42×10^{11}	2.54×10^{10}	(cm^3)/(g mole)(hr)
k_{20}	1.517×10^6	2.285×10^7	2.005×10^6	2.35×10^{11}	(hr)$^{-1}$
Optimum Design					
G_c	–	–	5782.7	7.225×10^4	(g mole)/(hr)
A_H	5491	10,730	8557	6.075×10^4	(cm^2)
k_1	–	–	304.36	151.5	(cm^3)/(g mole)(hr)
k_2	3.362	2.316	0.8798	12.3	(hr)$^{-1}$
Q_H	2.0×10^7	4.714×10^7	4.792×10^7	-4.05×10^8	(cal)/(hr)
q_s	1.25×10^6	1.428×10^6	1.683×10^6	1.386×10^7	(cm^3)/(hr)
q_{Hs}	1.473×10^6	2.276×10^6	2.567×10^6	3.037×10^7	(cm^3)/(hr)
T	348	375	345	320	(°K)
T_0	359.4	379.3	354.3	313.3	(°K)
V	1.487×10^6	1.439×10^6	2.19×10^6	1.307×10^6	cm^3
A	0.001	0.0015	0.003	0.0045	(g mole)/(cm^3)
B	–	–	0.003564	2.885×10^{-4}	(g mole)/(cm^3)
Optimum Costs					
C_B	–	–	8×10^5	1×10^6	($)/(g mole)
C_C	–	–	6500	7000	($)/(g mole)
C_f	1.728×10^4	9521	2000	1000	($)/(g mole)
C_A	1800	2400	1200	1000	($)/(cm^2)(hr)
C_H	2.5	8.0	2	1	($)/(cm)^3
C_V	25	50	12	72	($)/(hr)(cm^3)
Profit	–	–	7.94×10^5	7.34×10^5	($)/(g mole)
Total cost	3.175×10^4	3.68×10^4	–	–	($)/(g mole)

reaction giving the byproduct is first-order, in a nonisothermal, continuous-stirred-tank reactor. If we linearize the system equations and introduce the dimensionless variables

$$x_1 = -\left(\frac{A - A_s}{A_s}\right), \quad x_2 = -\left(\frac{T - T_s}{T_s}\right), \quad x_3 = -\left(\frac{B - B_s}{B_s}\right)$$

$$u_1 = \left(\frac{q - q_s}{q_s}\right), \quad u_2 = \left(\frac{q_H - q_{Hs}}{q_{Hs}}\right), \quad \tau = \frac{q_s t}{V}$$

we obtain

$$\begin{pmatrix} \dot{x}_1 \\ \dot{x}_2 \\ \dot{x}_3 \end{pmatrix} = \begin{pmatrix} -4.521 & -41.11 & 0 \\ -3.86 \times 10^{-3} & -1.619 & 0 \\ 2.0 & 20.50 & -1.0 \end{pmatrix} \begin{pmatrix} x_1 \\ x_2 \\ x_3 \end{pmatrix} + \begin{pmatrix} -2.333 & 0 \\ 0.1304 & -0.0275 \\ 1.0 & 0 \end{pmatrix} \begin{pmatrix} u_1 \\ u_2 \end{pmatrix}$$

Determine the frequency range of oscillations in the feed rate and the heating fluid flow rate such that the optimum steady state composition of the desired component B will not change by more than 0.5 percent.

38. (B‡) Case IV in Table 4.8-3 corresponds to a reactor design problem similar to that described immediately above. The linearized equations for this case are

$$\begin{pmatrix} \dot{x}_1 \\ \dot{x}_2 \\ \dot{x}_3 \end{pmatrix} = \begin{pmatrix} -2.286 & -28.63 & 0 \\ 0.1609 & 1.1442 & 0 \\ 2.0 & 18.94 & -1 \end{pmatrix} \begin{pmatrix} x_1 \\ x_2 \\ x_3 \end{pmatrix} + \begin{pmatrix} -1.221 & 0 \\ 0.0625 & 0.03042 \\ 1.0 & 0 \end{pmatrix} \begin{pmatrix} u_1 \\ u_2 \end{pmatrix}$$

Establish an acceptable start-up procedure for the reactor based on this linear model. Evaluate the profit produced by the system for various start-up policies.

39. (C‡) Harris and Schechter[8] used an analog computer to determine the frequency response of a second-order reaction in a nonisothermal, continuous-stirred-tank reactor. They observed that the output signals from the computer were not symmetrical, which indicates that the system nonlinearities must be significant. Hence they averaged the maximum and minimum values of the output signal in order to estimate the system gain and to prepare an approximate Bode plot.

Using the values given in the paper, linearize the equations and plot the frequency response on a Bode diagram. Compare your results to the analog computer solutions presented in the paper. Is there any chance that the characteristics of the multipliers in their analog computer could be the cause of the distortion they observed in their output signals?

40. (B‡) As part of an experimental study of time optimal control, Javinsky and Kadlec[9] made dynamic measurements of composition and temperature for the saponification of ethyl acetate in a continuous-stirred-tank reactor surrounded by a cooling jacket. They compare their experimental data to analog computer solutions of the nonlinear equations and obtain an excellent agreement.

Using the parameters presented in their paper, linearize the equations around the steady state solutions corresponding to full cooling, no cooling, and steady state cooling. Solve these linear equations for various initial conditions and then plot the values of composition versus temperature at a particular time. Compare the

[8] J. T. Harris and R. S. Schechter, *Ind. Eng. Chem. Proc. Design & Develop.*, **2**, 245, (1963).

[9] M. A. Javinsky and R. H. Kadlec. Paper presented at the 63rd National AIChE Meeting, Preprint 9c, St. Louis, Missouri, February 1968.

three-phase plane plots you obtain in this way to their numerical, or experimental, solutions, and find the region where the linear analysis is valid.

41. (C*) Bilous and Amundson[10] studied a simple plant where a reactor was followed by a single-stage separation unit and the unreacted material in the rich phase was recylced. The dynamic equations describing the system were presented in Exercise 3.5-7, and a matrix formulation of the problem was asked for in Exercise 3.5-17. Using the parameters presented in the original paper, find the transformation matrix that can be used to diagonalize the equations. Next, solve this set of canonical equations for the transient response, and, finally, construct the solution for the original dependent variables. Plot the response for components A_1 and B_1, and compare your curves with the analog computer results given in the paper.

42. (C*) In Exercise 41 we looked for a solution of a set of six coupled differential equations. Obviously it would be desirable to find a way of simplifying the problem so that we did not have to handle so many equations at one time. A procedure that sometimes can be used for this purpose was described in Exercise 29 above. Is this technique helpful in simplifying Bilous and Amundson's problem?

Make certain you consider the possibility of an overall material balance for the whole plant, as well as for the individual units.

43. (C*) An interesting feature of the Bilous and Amundson problem is the presence of the recycle loop. Thus a sinusoidal disturbance in the feed composition entering the reactor will not only produce oscillations in the reactor and separator compositions but will also, in general, cause the concentration of the recycle stream entering the reactor to fluctuate. Suppose, however, that the damping of the inlet oscillations in the reactor and separator units is such that the amplitudes of the fluctuations in the recycle stream are negligible. For this case, we can solve for the dynamic behavior of the plant as though there were two units in series, rather than a coupled system, which obviously is much simpler. Using the parameters given in the original paper, determine the frequency range of input reactant compositions for which this simple uncoupling technique will be valid.

44. (C) A kinetic scheme for some of the condensation reactions taking place when formaldehyde (F) is added to sodium paraphenolsulfonate (M) in an alkaline-aqueous solution was described in Exercise 28, p. 110. Also, in that problem you were asked to derive a dynamic model for a continuous-stirred-tank reactor. Consider a case where the reactant concentrations of F and M are each 0.15 g mole/liter, the feed rate is 10.5 ml/min, and the volume of an isothermal, laboratory-scale reactor is 750 ml. Find the steady state compositions in the reactor, linearize the dynamic equations around this steady state, determine the transformation matrix that can be used to diagonalize the equations, solve the canonical equations for a start-up problem where the reactor initially contains only component M, and then transform your solution back in terms of the original variables.

45. (C‡) A crude dynamic model for a catalytic cracking unit has been discussed in Example 3.1-4 and in Section 3.4. Unfortunately, there seems to be a misprint in the list of system parameters given in the original paper,[11] so Schleckser[12] suggested using the values: $A_1 = 48.55$, $A_2 = 2501$, $C_{ps} = 0.15$, $C_{p1} = 75$, $C_{p2} = 6.0$,

---

[10] See footnote 1, p. 105.
[11] See footnote 8, p. 62.
[12] H. Schleckser. Personal communication.

$E_1 = 2.0 \times 10^4$, $E_2 = 2.66 \times 10^4$, $H_1 = 7.5 \times 10^3$, $H_2 = 5.0 \times 10^3$, $(+\Delta H_1) = +2.0 \times 10^4$, $(+\Delta H_2) = -2.25 \times 10^6$, $M_1 = 1.5 \times 10^5$, $M_2 = 1.0 \times 10^5$, $P_1 = 2837$, $P_2 = 2117$, $Q = -13,700$, $T_a = 500$, $T_0 = 1176$, $T_1 = 1360$, $T_2 = 1510$, $V_a = 1.200$, $V_0 = 0.300$, $V_1 = 0.315$, $V_2 = 1.310$, $W = 600$, $x_1 = 3.95 \times 10^{-5}$, $x_2 = 1.45 \times 10^{-5}$, $y_a = 0.21$, $y_0 = 0$, $y_1 = 0.5$, $y_2 = 0.01$, and $m = 16$, where the temperatures are in (degrees Rankine) and the units of the other quantities are given in Example 3.1-4. Is it possible to have any other steady state solutions for this set of parameters? Is the coefficient matrix of the linearized dynamic equations singular? What does it mean if this matrix is singular?

Find the transformation matrix that can be used to diagonalize the linearized equations, and use the canonical equations to develop a start-up procedure for the unit.

46. (C‡) Hougen and Walsh[13] have published an extensive discussion of pulse-testing procedures to establish the dynamic characteristics of a variety of process units. They point out that pulse testing is fast and inexpensive and that normally the data can be used to predict the frequency response within reasonable limits. The main difficulties with the method are selecting the proper pulse heights (to obtain the best compromise between overloading the system, or making its non-linear nature become apparent, and minimizing the noise in the output signal) and the proper pulse duration (to obtain a response fast enough so that all the plant dynamics are excited and yet not so fast that the dynamics of the measuring instruments are significant).

They provide some information on various pulse shapes, including a rectangular pulse, a half-sine wave, a triangular pulse, and a displaced cosine pulse. Suppose that we want to study the effect of the magnitude and duration of these pulses, as well as one described by the equation $v = m_1 t e^{-m_2 t}$, on a first-order plant. How should we select the pulse parameters with respect to the gain and time constant of the plant?

47. (C‡) It does not take much experience to realize that in most cases we would prefer to work with algebraic equations rather than ordinary differential equations. Therefore it is reasonable to look for a procedure that can be used to evaluate whether or not the accumulation terms of one or more equations in a large set of ordinary differential equations can be neglected. Whenever one of these terms can be dropped, we can replace the original set of equations by an algebraic expression and a smaller set of ordinary differential equations.

There are a number of practical situations where we encounter this kind of behavior. For example, when we consider the response of a thermometer placed in a very large tank, we can easily show that the accumulation of energy within the thermometer is negligible and that we can always assume that the thermometer reading corresponds to the true tank temperature. Of course, any rapidly responding instrument will also behave in this way. Another example of a system of this type is a plate, gas absorption unit, where we expect that the accumulation of material in the gas phase will be negligible in comparison to that in the liquid phase. Similar characteristics are to be expected for most two-phase systems where one phase is a gas.

---

[13] J. O. Hougen and R. A. Walsh, *Chem. Eng. Prog.*, **57**, No. 3, 69 (1961).

Perhaps it should be mentioned that in certain situations it is virtually impossible to solve the set of differential equations unless the simplification is made. As an illustration of the source of the difficulty, let us consider a set of linear differential equations and suppose that one of the eigenvalues is much larger than the rest (or that one of the time constants is very small compared to the others, as is the case with a rapidly responding, measuring instrument). In attempting to use a computer program to determine the eigenvalues, we usually run into trouble and find that the coefficient matrix is essentially singular. Moreover, if we attempt to solve the differential equations numerically, we again run into trouble because when the integration interval is adjusted to be small enough to obtain accurate values of the instrument response, the values of the other variables will hardly change. In other words, it seems as though the solutions will not converge.

Anyone who has programmed an analog computer knows immediately how we can overcome this difficulty. First, the equations are magnitude scaled—that is, we make a change of variables so that the expected changes in every dependent variable cover the range from 0 to 100, -100 to 100, or some other range of interest. In this way we can put all the dependent variables on a common basis, and a composition variation from 0 to 0.01 g mole/liter becomes equivalent to a temperature variation from 480 to $510°R$,

$$x_1 = 10,000A \qquad x_2 = \frac{T - 480}{0.3}$$

Of course, a change of variables of this type will alter the coefficients in the system equations.

Once the magnitude scaling is completed, we time scale the equations. This is accomplished by making a change of variables on the independent variable, $t = \beta\tau$. We select the scale factor β, which will also appear in the coefficient of each term on the right-hand side of the set of equations, so that the largest coefficient is equal to unity. Next, we examine the remaining equations, and if we find that every coefficient in one or more of the set has a value of 10^{-3} or less, we set the derivative term equal to zero and replace the differential equation by an algebraic expression.

Use this procedure, or an appropriate modification, to demonstrate that the dynamics of the thermometer in Exercise 5 can be neglected. Also, use this procedure to evaluate the relative importance of the dynamics of the gas-phase reactions versus the solid-phase reactions in Exercise 45.

48. (C) Develop a solution for the start-up time of the extraction and reaction problem described in Exercise 11, p. 107. Neglect hydraulic effects in the plate column.

49. (C) Acrivos and Amundson[14] used matrix techniques to determine the start-up time for a battery of N isothermal, continuous-stirred-tank reactors. The reactions taking place in each vessel could be described by the mechanism

$$A_1 \underset{k_1'}{\overset{k_1}{\rightleftarrows}} A_2 \underset{k_2'}{\overset{k_2}{\rightleftarrows}} A_3$$

Assuming that the rate of each reaction step is first-order and that the initial concentration of each species in every vessel is zero, use the calculus of finite differences to establish a solution for the start-up time.

[14] A. Acrivos and N. R. Amundson, *Ind. Eng. Chem.*, **47**, 1533 (1955).

50. (C‡) Addition polymerization is a chain reaction where a large number of steps exist—of the order of 1000 reactions. This kind of a model is often used to describe ethylene, vinyl chloride, and styrene polymerizations. The basic approach is to assume mechanisms for an initiation step, a chain propagation process, and a chain termination reaction.[15] As a particular example, we will consider the reactions

$$\text{Initiation} \qquad M_1 \xrightarrow{k_i} P_1$$

$$\text{Propagation} \qquad M_1 + P_n \xrightarrow{k_p} P_{n+1}$$

$$\text{Termination} \qquad M_1 + P_n \xrightarrow{k_t} M_{n+1}$$

where M_1 represents the active monomer, P_n is the active polymer, and M_n is the dead polymer. Also, we will assume that the material balance equations for an isothermal, batch polymerization reaction can be written as

$$\frac{dM_1}{dt} = -k_i M_1^\alpha - (k_p + k_t)M_1 \sum_1^N P_n \qquad M_1(0) = M_0$$

$$\frac{dP_1}{dt} = k_i M_1^\alpha - (k_p + k_t)M_1 P_1 \qquad P_1(0) = 0$$

$$\frac{dP_n}{dt} = k_p M_1 P_{n-1} - (k_p + k_t)M_1 P_n \qquad \text{for } n = 2, 3, \ldots, N \quad P(0) = 0$$

$$\frac{dM_n}{dt} = k_t M_1 P_{n-1} \qquad \text{for } n = 2, 3, \ldots, N, \quad M_n(0) = 0$$

Since these equations include products of the concentration variables, they are nonlinear. Thus we cannot expect to solve the equations analytically. However, we do not relish the thought of developing a numerical solution for a thousand or so equations; therefore we would like to find some procedure for simplifying the calculations.

One simplifying procedure is to define a new variable

$$\phi = \sum_1^N P_n$$

By writing the first few equations for the active polymer

$$\frac{dP_1}{dt} = k_i M_1^\alpha - (k_p + k_t)M_1 P_1$$

$$\frac{dP_2}{dt} = k_p M_1 P_1 - (k_p + k_t)M_1 P_2$$

$$\frac{dP_3}{dt} = k_p M_1 P_2 - (k_p + k_t)M_1 P_3$$

it is easy to see that when we sum all N of the equations, we obtain

$$\frac{d\phi}{dt} = k_i M_1^\alpha - k_t M_1 \phi \qquad \phi(0) = 0$$

Similarly, the expression for the active monomer becomes

$$\frac{dM_1}{dt} = -k_i M_1^\alpha - (k_p + k_t)M_1 \phi \qquad M_1(0) = M_0$$

[15] See S. Liu and N. R. Amundson, *Chem. Eng. Sci.*, **17**, 797 (1962).

It is a simple task to develop a numerical solution for this pair of equations for any reaction order α, and in this way we can find how M_1 and ϕ vary with the batch time.

Once we know M_1 as a function of time, we can define a new independent variable

$$d\tau = M_1 \, dt$$

With this transformation, the equations for the active polymer concentrations become linear. Show that the calculus of finite differences can be used to find an analytical solution for the concentrations of the active polymer species in terms of this new independent variable. Then describe how you could calculate P_n and M_n as a function of time.

51. (B) Find the response of a series of first-order systems described by Eq. 4.6-6 to an impulse, a step, and a sinusoidal change in the control variable.

52. (C‡) We have shown that identical information is obtained from impulse, step, and sinusoidal response measurements on a plant described by a linear differential equation having constant coefficients. Therefore it should be possible to construct Bode diagrams from pulse response data. A number of investigators have developed computer programs for this purpose. In fact, they can be applied even when an arbitrary pulse shape is used as the input signal. Find one of these computer programs in the literature and apply it to a problem of your choice.

Distributed Parameter Systems

5

Although the emphasis is on lumped parameter systems in most books on process dynamics, a majority of plants encountered in industry are actually distributed parameter systems. As mentioned earlier, it is possible to obtain a first estimate of the dynamic characteristics of a distributed parameter process by using a lumped parameter model. However, for some plants it is possible to obtain analytical solutions of the partial differential equations describing the system, and for these cases we can check the validity of the lumped parameter approximations. Also, an investigation of distributed parameter plants hopefully will add further confirmation that most of the dynamic features of linear systems can be interpreted in terms of combinations of simple first- and second-order systems. Although our study will not be a coverage in depth, it will illustrate some of the most important problems and the methods that can be used to solve the partial differential equation models of distributed parameter processes.

SECTION 5.1 PLUG FLOW PROCESSES

One of the most useful assumptions in the analysis of chemical engineering problems is that of plug flow—that is, a fluid moves through a pipe with a flat velocity profile. Even though this assumption can never be precisely

correct, for it implies that there is slip at the wall, it introduces great simplifications into the system equations, thereby frequently making it possible
to obtain analytical solutions.[1] In addition, the comparison of experimental
data with theoretical predictions based on a plug flow assumption often
indicates that the error introduced by the assumption is less than the experimental error. Since the analysis of plug flow processes is the simplest, we will
start our discussion of the dynamics of distributed parameter systems with
some typical plug flow problems.

Dynamic Response of a Tubular Reactor

In Example 3.2-2 we considered a chemical reaction in a tubular reactor
with a constant wall temperature; that is, a system described by a set of
coupled, partial differential equations. If we simplify the problem by restricting
our attention to an isothermal reactor, the equation describing the system
becomes

$$\frac{\partial x}{\partial t} + v\frac{\partial x}{\partial z} = -kx \qquad\qquad 5.1\text{-}1$$

where k is a constant. At steady state conditions, the accumulation term is
equal to zero so that

$$\frac{dx_s}{dz} = -\frac{k}{v}x_s \qquad\qquad 5.1\text{-}2$$

or

$$x_s = x_{fs}\exp\left(-\frac{kz}{v}\right) \qquad\qquad 5.1\text{-}3$$

where x_{fs} is the constant feed composition entering the reactor.

If we are interested in determining the dynamic response for small
changes in feed composition, we first rewrite the equation in terms of perturbation variables. Letting

$$x = x_s + y \qquad x_f = x_{fs} + y_f \qquad\qquad 5.1\text{-}4$$

the state equation becomes

$$\frac{\partial(x_s + y)}{\partial t} + v\frac{\partial(x_s + y)}{\partial z} = -k(x_s + y)$$

or

$$\frac{\partial y}{\partial t} + v\frac{dx_s}{dz} + v\frac{\partial y}{\partial z} = -kx_s - ky$$

or, finally,

$$\frac{\partial y}{\partial t} + v\frac{\partial y}{\partial z} + ky = 0 \qquad\qquad 5.1\text{-}5$$

[1] The relationship between the "plug flow" assumption and "averaging" the general
transport equations is discussed in Exercise 29, p. 300.

Taking the Laplace transform of the left-hand side of this equation, we obtain

$$\mathscr{L}\left[\frac{\partial y}{\partial t} + v\frac{\partial y}{\partial z} + ky\right] = \int_0^\infty e^{-st}\left[\frac{\partial y}{\partial t} + v\frac{\partial y}{\partial z} + ky\right] dt$$

$$= s\tilde{y} + v\frac{\partial \int_0^\infty e^{-st}y \, dt}{\partial z} + k\tilde{y}$$

$$= v\frac{d\tilde{y}}{dz} + (s+k)\tilde{y}$$

Thus the Laplace transform reduces the partial differential equation, Eq. 5.1-5, to an ordinary differential equation

$$\frac{d\tilde{y}}{dz} = -\frac{(s+k)}{v}\tilde{y} \qquad\qquad 5.1\text{-}6$$

The initial conditions, $y(z)$ at $t = 0$, do not appear in this expression, since the state variable y represents deviations from steady state conditions and $y(z) = 0$. At the reactor inlet, $z = 0$, we know that $y = y_f$, and therefore $\tilde{y} = \tilde{y}_f$. Hence the solution of Eq. 5.1-6 in the transform domain is

$$\tilde{y} = \left\{\exp\left[-\frac{(k+s)z}{v}\right]\right\}\tilde{y}_f \qquad\qquad 5.1\text{-}7$$

and the transfer function becomes

$$\frac{\tilde{y}}{\tilde{y}_f} = H(s) = \exp\left[-\frac{(k+s)z}{v}\right] = \exp\left(-\frac{zs}{v}\right)\exp\left(-\frac{kz}{v}\right) \qquad 5.1\text{-}8$$

Remembering that the shift theorem states that

$$\mathscr{L}[f(t - t_0)] = e^{-st_0}\tilde{f}(s) \qquad\qquad 4.2\text{-}10$$

it becomes a simple matter to find the dynamic response of the system for a variety of inputs.

Impulse Response

For an impulse input, $\tilde{y}_f = A$, and

$$\tilde{y} = \left[\exp\left(-\frac{zs}{v}\right)\exp\left(-\frac{kz}{v}\right)\right]A$$

Therefore

$$y(t) = \begin{cases} 0, & \text{if } t \neq \dfrac{z}{v} \\[2mm] Ae^{-kz/v}, & \text{if } t = \dfrac{z}{v} \end{cases} \qquad\qquad 5.1\text{-}9$$

In other words, if we introduce an impulse into the reactor, we see no change in the output until the pulse has had a chance to flow to the outlet (the residence time is z/v) and then we see an impulse whose magnitude can be predicted from the steady state equation describing the system, Eq. 5.1-3. This kind of a process is called a *dead time* system. After some reflection, it is clear that this type of behavior is caused by the assumption that the fluid flows through the reactor as a plug and that the elements at various distances have no effect on each other. Although it is obvious that diffusion will always tend to smooth out such a discontinuity, there are many cases where a dead time response is adequate for engineering purposes.

Step response

For a step input, $\tilde{y}_f = A/s$, and

$$\tilde{y} = \exp\left(-\frac{zs}{v}\right)\exp\left(-\frac{kz}{v}\right)\left(\frac{A}{s}\right)$$

Then

$$y(t) = A\exp\left(-\frac{kz}{v}\right) \qquad t > \frac{z}{v} \qquad\qquad 5.1\text{-}10$$

Thus a step input produces a step output, which appears after one reactor residence time and has a magnitude corresponding to the new steady state output of the system.

Frequency response

Substituting $j\omega$ for s in the transfer function, we obtain

$$H(j\omega) = \exp\left(-\frac{kz}{v}\right)\exp\left(-\frac{j\omega z}{v}\right) \qquad\qquad 5.1\text{-}11$$

Since we know that the polar representation of a complex number $(c_1 + jc_2)$ is simply

$$c_1 + jc_2 = \sqrt{c_1^2 + c_2^2}\, e^{j\phi} \qquad \phi = \tan^{-1}\frac{c_2}{c_1} \qquad\qquad 5.1\text{-}12$$

we see immediately that the amplitude ratio, or system gain $|G|$, is

$$|G| = \exp\left(-\frac{kz}{v}\right) \qquad\qquad 5.1\text{-}13$$

and that the phase angle is

$$\phi = -\frac{\omega z}{v} \qquad\qquad 5.1\text{-}14$$

Hence, on a Bode plot, the gain curve is a constant independent of ω and the phase angle decreases from zero to $-\infty$ as ω increases (see Figure 5.1-1). Since the phase angle curve does not approach an asymptote, this is called a *nonminimum phase* system.

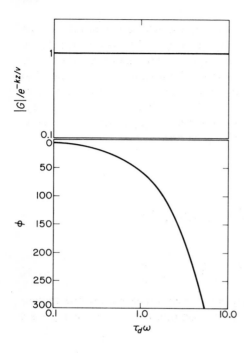

Figure 5.1-1. Bode diagram for a dead-time system.

Finite Difference Approximation of Dead Time Systems

In the previous discussion of partial differential equations, we stated that we could approximate the derivative with respect to length by a finite difference. Then for the isothermal reactor, we would have the set of equations

$$\frac{dy_1}{dt} + \frac{vn}{L}(y_1 - y_f) = -ky_1$$

$$\frac{dy_2}{dt} + \frac{vn}{L}(y_2 - y_1) = -ky_2$$

$$\vdots$$

$$\frac{dy_n}{dt} + \frac{vn}{L}(y_n - y_{n-1}) = ky_n$$

5.1-15

$$\vdots$$

Taking the Laplace transforms of these equations, we obtain

$$\left(s + \frac{vn}{L} + k\right)\tilde{y}_1 = \frac{vn}{L}\tilde{y}_f$$

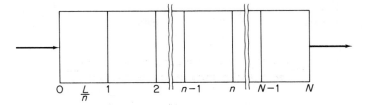

Figure 5.1-2. Difference approximation of tubular reactor.

$$\left(s + \frac{vn}{L} + k\right)\tilde{y}_2 = \frac{vn}{L}\tilde{y}_1$$

$$\cdot$$
$$\cdot \qquad\qquad\qquad\qquad\qquad 5.1\text{-}16$$
$$\cdot$$

$$\left(s + \frac{vn}{L} + k\right)\tilde{y}_n = \frac{vn}{L}\tilde{y}_{n-1}$$

or after eliminating the intermediate variables,

$$\frac{\tilde{y}_n}{\tilde{y}_f} = H(s) = \left(\frac{vn/L}{s + (vn/L) + k}\right)^n = \left\{\frac{1}{[L(s + k)/vn] + 1}\right\}^n \qquad 5.1\text{-}17$$

Since we know that

$$\lim_{n \to \infty}\left(\frac{1}{(\alpha/n) + 1}\right)^n = e^{-\alpha} \qquad\qquad 5.1\text{-}18$$

the transfer function must approach the quantity

$$\lim_{n \to \infty}\left(\frac{\tilde{y}_n}{\tilde{y}_f}\right) = \exp\left[-\frac{L}{v}(s + k)\right] \qquad\qquad 5.1\text{-}19$$

which is the result we obtained by solving the partial differential equation. Thus if we divide the fixed length of the reactor into an infinite number of sections, the two approaches give identical results. For a finite number of sections, the difference-differential equations are merely those for a series of first-order system, and Figures 4.6-2 and 4.6-3 give an indication of the error involved in the difference approximation.

Dynamics of a Steam-Heated Exchanger

The simplest type of exchanger to analyze is one in which the temperature on one side of the walls is constant. This kind of behavior is observed when a pure vapor is condensing on the outside of the tubes, when the shell fluid is perfectly mixed as in a continuous-stirred-tank reactor, or when the flow rate on one side of the exchanger is so high that for practical purposes its temperature can be considered as being independent of the distance from the inlet.

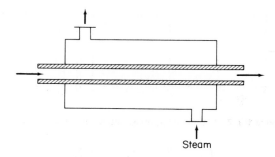

Figure 5.1-3. Steam-heated exchanger.

If we let ρ_f equal the fluid density $(lb)(ft^3)^{-1}$, A equal the cross-sectional area of the tubes, C_{pf} equal heat capacity of the fluid $(Btu)(lb)^{-1}(°F)^{-1}$, h_1 equal the inside film coefficient $(Btu)(hr)^{-1}(ft^2)^{-1}(°F)^{-1}$, A_1 equal the inside heat-transfer area $(ft^2)(ft)^{-1}$, v equal fluid velocity $(ft)(hr)^{-1}$, T equal fluid temperature, T_w equal wall temperature, T_c equal condensate temperature, $M_w C_{pw}$ equal wall heat capacity $(Btu)(ft)^{-1}(°F)^{-1}$, h_2 equal outside film coefficient $(Btu)(hr)^{-1}(ft^2)^{-1}(°F)^{-1}$, and A_2 equal outside heat-transfer area $(ft^2)(ft)^{-1}$, and if we make the assumptions that the fluid motion is plug flow, no axial or radial conduction in the fluid, no resistance of the wall to heat transfer, and no significant accumulation of energy in the condensate film, the energy balances for the fluid and tube wall may be written

$$\rho_f C_{pf} A \frac{\partial T}{\partial t} + \rho_f C_{pf} Av \frac{\partial T}{\partial z} = h_1 A_1 (T_w - T) \qquad \text{5.1-20}$$

$$M_w C_{pw} \frac{\partial T_w}{\partial t} = h_2 A_2 (T_c - T_w) - h_1 A_1 (T_w - T) \qquad \text{5.1-21}$$

Of course, if the accumulation effects in the tube wall were negligible, Eq. 5.1-21 can be used to eliminate the wall temperature T_w from Eq. 5.1-20 and the resulting expression will have the same form, and therefore the same dynamic characteristics, as the tubular reactor model, Eq. 5.1-1.

At steady state conditions, the dynamic equations reduce to

$$\frac{dT_s}{dz} = \frac{h_1 A_1}{\rho_f C_{pf} Av_s}(T_{ws} - T_s) \qquad \text{5.1-22}$$

$$(h_1 A_1 + h_2 A_2)T_{ws} = (h_1 A_1)T_s + h_2 A_2 T_{cs} \qquad \text{5.1-23}$$

For a case where the feed temperature is T_{fs}(at $z = 0$, $T_s = T_{fs}$), the preceding equations can be solved to give the steady state temperature profile for the fluid in the tubes

$$\frac{T_{cs} - T_s}{T_{cs} - T_{fs}} = \exp\left[\frac{-h_1 A_1 h_2 A_2 z}{\rho_f C_{pf} Av_s (h_1 A_1 + h_2 A_2)}\right] \qquad \text{5.1-24}$$

Now we can define a new set of variables that describe the deviations from steady state conditions as

$$\theta = T - T_s, \quad \theta_w = T_w - T_{ws}, \quad \theta_f = T_f - T_{fs},$$
$$\theta_c = T_c - T_{cs}, \quad u = v - v_s \qquad \qquad 5.1\text{-}25$$

Substituting these expressions into the dynamic equations, assuming that the film heat-transfer coefficient h_1 remains constant despite small velocity changes, and neglecting products of small deviations, we obtain

$$\rho_f C_{pf} A \frac{\partial \theta}{\partial t} + \rho_f C_{pf} A v_s \frac{\partial \theta}{\partial z} + h_1 A_1 (\theta - \theta_w) = \rho_f C_{pf} A \frac{\partial T_s}{\partial z} u \qquad 5.1\text{-}26$$

$$M_w C_{pw} \frac{\partial \theta_w}{\partial t} = h_2 A_2 (\theta_c - \theta_w) - h_1 A_1 (\theta_w - \theta) \qquad 5.1\text{-}27$$

with the boundary conditions

$$\text{at } z = 0, \quad \theta = \theta_f \qquad \text{at } t = 0, \quad \theta = \theta_w = 0 \qquad 5.1\text{-}28$$

Letting

$$\tau_1 = \frac{\rho_f C_{pf} A}{h_1 A_1}, \quad \tau_2 = \frac{M_w C_{pw}}{h_2 A_2}, \quad \tau = \frac{M_w C_{pw}}{h_1 A_1} \qquad 5.1\text{-}29$$

taking the Laplace transform of the equations, and rearranging the results somewhat, we find that

$$\tau_1 v_s \frac{\partial \tilde{\theta}}{\partial z} + (1 + \tau_1 s)\tilde{\theta} - \tilde{\theta}_w = -\tau_1 \frac{\partial T_s}{\partial z} \tilde{u} \qquad 5.1\text{-}30$$

$$\left(1 + \frac{\tau_2}{\tau} + \tau_2 s\right) \tilde{\theta}_w = \tilde{\theta}_c + \frac{\tau_2}{\tau} \tilde{\theta} \qquad 5.1\text{-}31$$

or after eliminating $\tilde{\theta}_w$,

$$(\tau_1 v_s)\frac{d\tilde{\theta}}{dz} + \left(1 + \tau_1 s - \frac{\tau_2/\tau}{1 + \tau_2 s + \tau_2/\tau}\right)\tilde{\theta} = \left(\frac{1}{1 + \tau_2 s + \tau_2/\tau}\right)\tilde{\theta}_c - \tau_1 \frac{\partial T_s}{\partial z}\tilde{u}$$
$$5.1\text{-}32$$

Now, letting

$$a = \frac{\tau_1 \tau_2 s^2 + (\tau_1 + \tau_2 + \tau_1 \tau_2/\tau)s + 1}{\tau_1(\tau_2 s + 1 + \tau_2/\tau)}$$

$$\frac{b}{a} = \frac{1}{\tau_1 \tau_2 s^2 + (\tau_1 + \tau_2 + \tau_1 \tau_2/\tau)s + 1} \qquad 5.1\text{-}33$$

$$c = \frac{(T_{cs} - T_{fs})}{v_s \tau_1(1 + \tau_2/\tau)}$$

and substituting the derivative of Eq. 5.1-24, we obtain

$$\frac{v_s}{a}\frac{d\tilde{\theta}}{dz} + \tilde{\theta} = \frac{b}{a}\tilde{\theta}_c - \frac{c}{a}\exp\left[-\frac{z/v_s}{\tau_1(1 + \tau_2/\tau)}\right]\tilde{u} \qquad 5.1\text{-}34$$

Solving this equation with the boundary condition given by Eq. 5.1-28 leads to the result

$$\tilde{\theta} = \tilde{\theta}_f e^{-az/v_s} + \frac{b}{a}(1 - e^{-az/v_s})\tilde{\theta}_c$$

$$+ c\left[\frac{\tau_1(1 + \tau_2/\tau)}{a\tau_1(1 + \tau_2/\tau) - 1}\right]\left\{e^{-az/v_s} - \exp\left[-\frac{z/v_s}{\tau_1(1 + \tau_2/\tau)}\right]\right\}\tilde{u} \qquad 5.1\text{-}35$$

Thus we can define transfer functions for changes in the feed temperature $\tilde{\theta}/\tilde{\theta}_f$, the condensate temperature $\tilde{\theta}/\tilde{\theta}_c$, and the flow rate $\tilde{\theta}/\tilde{u}$. The first of these functions is the simplest and is very similar to the tubular reactor problem considered previously. Both of the other transfer functions are more complex, and we shall limit our attention to the second one

$$\frac{\tilde{\theta}}{\tilde{\theta}_c} = H(s) = \frac{b}{a}(1 - e^{az/v_s}) \qquad 5.1\text{-}36$$

Recognizing that the quantities a and b/a are functions of the Laplace parameter s according to Eq. 5.1-33, it becomes apparent that it will be difficult to invert this transform even for an impulse input. Our previous experience simply does not give us any indication of a procedure that might be appropriate when a ratio of two polynomials in s appears as an exponent. Also, we cannot use the method of partial fractions to break this term up into the product of two exponential functions, for the polynomial in the numerator is of a higher order than that of the denominator. Although the weighting function can be found after an extensive amount of manipulation, the result is almost too complicated to be of much value. Consequently, in the following discussion we will only consider the frequency response of the system, and in a later section we will propose simpler approximate models that can be used to estimate the impulse and step response of the system.

The steady state frequency response is found by substituting $j\omega$ for s in the system transfer function[2]

$$H(j\omega) = \left[\frac{1}{(1 - \tau_1\tau_2\omega^2) + (\tau_1 + \tau_2 + \tau_1\tau_2/\tau)j\omega}\right]$$

$$\times \left\{1 - \exp\left(-\frac{z}{v_s}\right)\left[\frac{(1 - \tau_1\tau_2\omega^2) + (\tau_1 + \tau_2 + \tau_1\tau_2/\tau)j\omega}{(\tau_1/\tau)(\tau + \tau_2 + \tau\tau_2 j\omega)}\right]\right\} \qquad 5.1\text{-}37$$

By rationalizing the complex numbers, this result can be written

$$H(j\omega) = (K_1 + K_2 j)[1 - e^{(K_3 + K_4 j)}] \qquad 5.1\text{-}38$$

where

$$K_1 = \frac{1 - \tau_1\tau_2\omega^2}{(1 - \tau_1\tau_2\omega^2)^2 + (\tau_1 + \tau_2 + \tau_1\tau_2/\tau)^2\omega^2}$$

$$K_2 = \frac{-(\tau_1 + \tau_2 + \tau_1\tau_2/\tau)\omega}{(1 - \tau_1\tau_2\omega^2)^2 + (\tau_1 + \tau_2 + \tau_1\tau_2/\tau)^2\omega^2} \qquad 5.1\text{-}39$$

[2] The validity of this substitution has not been proved for transcendental functions, and we leave this as an exercise for the reader.

$$K_3 = \left[\frac{(1 + \tau_2/\tau)(1 - \tau_1\tau_2\omega^2) + \tau_2\omega^2(\tau_1 + \tau_2 + \tau_1\tau_2/\tau)}{\tau_1[(1 + \tau_2/\tau)^2 + (\tau_2\omega)^2]}\right]\left(-\frac{z}{v_s}\right)$$

$$K_4 = \left[\frac{(1 + \tau_2/\tau)(\tau_1 + \tau_2 + \tau_1\tau_2/\tau) - \tau_2(1 - \tau_1\tau_2\omega^2)}{\tau_1[(1 + \tau_2/\tau)^2 + (\tau_2\omega)^2]}\right]\left(-\frac{z\omega}{v_s}\right)$$

Since

$$e^{K_3 + jK_4} = e^{K_3}(\cos K_4 + j\sin K_4) \qquad 5.1\text{-}40$$

the transfer function can also be written as

$$H(j\omega) = [K_1 - (K_1\cos K_4 + K_2\sin K_4)e^{K_3}]$$
$$+ j[K_2 - (K_2\cos K_4 - K_1\sin K_4)e^{K_3}] \qquad 5.1\text{-}41$$

which is just a single complex number

$$H(j\omega) = K_5 + jK_6 \qquad 5.1\text{-}42$$

where K_5 and K_6 can be determined by inspection. Then the amplitude ratio is

$$|G| = \sqrt{K_5^2 + K_6^2} \qquad 5.1\text{-}43$$

and the phase lag is

$$\tan\phi = \frac{K_6}{K_5} \qquad 5.1\text{-}44$$

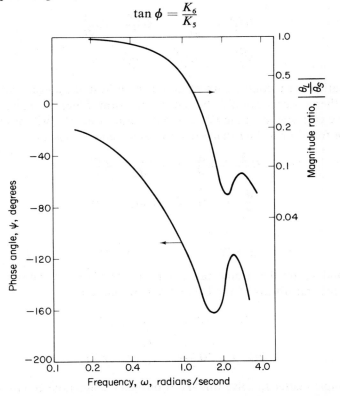

Figure 5.1-4. Frequency response of a steam-heated exchanger. Reproduced from Cohen, W. C. and E. F. Johnson, *Ind. Eng. Chem.*, **48**, 1031 (1956), by permission.

The interesting feature of this result is that both the amplitude ratio and the phase lag contain periodic functions of the forcing frequency. Thus a Bode plot will contain a number of resonant peaks (see Figure 5.1-4).[3] In physical terms, the resonance phenomenon arises because we are forcing the system in a distributed manner; that is, the condensate temperature fluctuates uniformly along the whole length of the exchanger. Depending on the ratio of the residence time in the exchanger to the period of the imposed oscillations, the driving force for heat transfer can be reinforced or partially destroyed as the fluid flows through the exchanger. However, for many processes the resonance phenomenon is not observed experimentally because it occurs at such high frequencies that the data are inaccurate.

Dynamic Response of Countercurrent Equipment

A dynamic model of a packed-bed separation unit was described in Example 3.2-3.

$$H\frac{\partial X}{\partial t} + h\frac{\partial Y}{\partial t} = -G\frac{\partial Y}{\partial z} + L\frac{\partial X}{\partial z} \qquad 3.2\text{-}7$$

$$h\frac{\partial Y}{\partial t} = -G\frac{\partial Y}{\partial z} - k_g a(Y - mX) \qquad 3.2\text{-}9$$

The equation for a countercurrent double-pipe heat exchanger and a number of other systems can also be put in this form. Since the equations are linear, the corresponding model for the deviation variables will have exactly the same form, providing that the flow rates remain constant,

$$H\frac{\partial x}{\partial t} + h\frac{\partial y}{\partial t} = -G\frac{\partial y}{\partial z} + L\frac{\partial x}{\partial z} \qquad 5.1\text{-}45$$

$$h\frac{\partial y}{\partial t} = -G\frac{\partial y}{\partial z} - k_g a(y - mx) \qquad 5.1\text{-}46$$

where

$$x = X - X_s \qquad y = Y - Y_s \qquad 5.1\text{-}47$$

and X_s and Y_s are the steady state compositions of the two phases. Taking the Laplace transforms of these equations, we obtain

$$Hs\tilde{x} + hs\tilde{y} = -G\frac{d\tilde{y}}{dz} + L\frac{d\tilde{x}}{dz} \qquad 5.1\text{-}48$$

$$hs\tilde{y} = -G\frac{d\tilde{y}}{dz} - k_g a(\tilde{y} - m\tilde{x}) \qquad 5.1\text{-}49$$

It is a simple matter to solve this set of ordinary differential equations with

[3] W. C. Cohen and E. F. Johnson, *Ind. Eng. Chem.*, **48**, 1031 (1956).

the boundary conditions

$$\text{at } z = 0, \quad \tilde{y} = \tilde{y}_f = \mathscr{L}(y_f - y_{fs})$$

$$\text{at } z = L, \quad \tilde{x} = \tilde{x}_f = \mathscr{L}(x_f - x_{fs}) \qquad 5.1\text{-}50$$

Thus we can obtain expressions relating the transforms of the outlet composition of each phase, $\tilde{x}$ and $\tilde{y}$, to the inlet disturbances, $\tilde{x}_f$ and $\tilde{y}_f$, and then we can define a number of transfer functions between these inputs and outputs. However, obviously the solution of the preceding ordinary differential equations will include exponential functions and the exponents (the characteristic roots) will depend on the Laplace parameter s. From our analysis of the last problem, we know that it will be a difficult task to invert the transformed expression in order to find the impulse and step response of the system. However, it should be possible to obtain an expression for the steady state frequency response in a relatively simple fashion. Rather than pursue this matter now, we will defer the discussion of countercurrent plug flow systems until a later section, where we develop simpler, approximate models.

Response of a Packed Reactor

Crider and Foss[4] have studied the dynamic response of a nonisothermal tubular reactor containing an inert packing material. This type of system has many of the dynamic features of packed catalytic reactors, but at the same time it avoids some of the experimental and analytical difficulties encountered in catalytic systems. The material and energy balances for the system are

$$\frac{\partial c}{\partial t} + v\frac{\partial c}{\partial z} = -kc \qquad 5.1\text{-}51$$

$$C_{pf}\rho_f\frac{\partial T}{\partial t} + C_{pf}\rho_f v\frac{\partial T}{\partial z} = (-\Delta H)kc + h_p A_p\left(\frac{1-\epsilon}{\epsilon}\right)(T_p - T) \qquad 5.1\text{-}52$$

$$C_{pp}\rho_p\frac{\partial T_p}{\partial t} = h_p A_p(T - T_p) \qquad 5.1\text{-}53$$

where A_p = specific surface area of packing
  h_p = solid-fluid heat-transfer coefficient
  ϵ = void fraction of the bed

and the other symbols are similar to those used previously. Of course, the same equations can be used to describe an empty reactor for a case where the thermal capacity of the wall is significant.

At steady state conditions, the accumulation terms are equal to zero, and the system equations become

$$v\frac{dc}{dz} = -kc \qquad 5.1\text{-}54$$

[4] J. E. Crider and A. S. Foss, *AIChE Journal*, **12**, 514 (1966).

$$C_{pf}\rho_f v \frac{dT}{dz} = (-\Delta H)kc \qquad\qquad 5.1\text{-}55$$

$$T = T_p \qquad\qquad 5.1\text{-}56$$

Douglas and Eagleton[5] showed that an analytical solution can be obtained for this pair of nonlinear equations in terms of exponential integrals. If we linearize the equations about these steady state profiles and let

$$x_1 = \frac{c - c_s}{c_{fs}}, \quad x_2 = \frac{C_{pf}\rho_f(T - T_s)}{(-\Delta H)c_{fs}}, \quad x_p = \frac{C_{pf}\rho_f(T_p - T_{ps})}{(-\Delta H)c_{fs}}$$

$$\tau = k_f t, \quad Z = \frac{k_f z}{v}, \quad r_c = \frac{k_s}{k_f}, \quad r_T = \frac{(-\Delta H)Ek_s c_s}{C_{pf}\rho_f RT_s^2 k_f} \qquad 5.1\text{-}57$$

$$H_p = \frac{h_p A_p(1 - \epsilon)}{\epsilon k_f C_{pf}\rho_f}, \quad B = \frac{C_{pf}\rho_f}{C_{pp}\rho_p(1 - \epsilon)}$$

we obtain

$$\frac{\partial x_1}{\partial \tau} + \frac{\partial x_1}{\partial Z} = -r_c x_1 - r_T x_2 \qquad\qquad 5.1\text{-}58$$

$$\frac{\partial x_2}{\partial \tau} + \frac{\partial x_2}{\partial Z} = r_c x_1 + r_T x_2 + H_p(x_p - x_2) \qquad\qquad 5.1\text{-}59$$

$$\frac{\partial x_p}{\partial \tau} = H_p B(x_2 - x_p) \qquad\qquad 5.1\text{-}60$$

Taking the Laplace transforms of these equations gives

$$\frac{d\tilde{x}_1}{dZ} = -(r_c + s)\tilde{x}_1 - r_T \tilde{x}_2 \qquad\qquad 5.1\text{-}61$$

$$\frac{d\tilde{x}_2}{dZ} = r_c \tilde{x}_1 + (r_T - H_p - s)\tilde{x}_2 + H_p \tilde{x}_p \qquad\qquad 5.1\text{-}62$$

$$0 = H_p B \tilde{x}_2 - (H_p B + s)\tilde{x}_p \qquad\qquad 5.1\text{-}63$$

Solving the last equation for $\tilde{x}_p$ leads to the result

$$\tilde{x}_p = \frac{H_p B}{H_p B + s}\tilde{x}_2 \qquad\qquad 5.1\text{-}64$$

and this expression can be used to eliminate $\tilde{x}_p$ from the energy equation

$$\frac{d\tilde{x}_2}{dZ} = r_c \tilde{x}_1 + \left(r_T - H_p + \frac{H_p^2 B}{H_p B + s} - s\right)\tilde{x}_2$$

$$= r_c \tilde{x}_1 + \left(r_T - \frac{H_p s}{H_p B + s} - s\right)\tilde{x}_2 \qquad\qquad 5.1\text{-}65$$

Thus we have reduced the state equations to a pair of linear, ordinary differential equations with variable coefficients. Since the equations are linear, it is

[5] J. M. Douglas and L. C. Eagleton, *Ind. Eng. Chem. Fundamentals*, **1**, 116 (1962).

possible to eliminate the composition variable and obtain a second-order equation in the temperature deviation, $\tilde{x}_2$.

Perhaps the simplest way of following this somewhat complicated elimination procedure is to redefine some of the terms. If we write the steady state reaction rate as

$$r_s = F_s(T_s)c_s \qquad\qquad 5.1\text{-}66$$

its partial derivatives are

$$r_c = F_s(T_s) \qquad r_T = F_s'(T_s)c_s \qquad\qquad 5.1\text{-}67$$

where

$$F_s'(T_s) = \frac{\partial F_s(T_s)}{\partial T_s} \qquad\qquad 5.1\text{-}67$$

Letting

$$H = \frac{H_p s}{H_p B + s} \qquad\qquad 5.1\text{-}68$$

substituting these expressions into Eqs. 5.1-61 and 5.1-65, and manipulating the results somewhat gives

$$\frac{d\tilde{x}_1}{dZ} + s\tilde{x}_1 = -F_s\tilde{x}_1 - F_s'c_s\tilde{x}_2 \qquad\qquad 5.1\text{-}69$$

$$\frac{d\tilde{x}_2}{dZ} + s\tilde{x}_2 = F_s\tilde{x}_1 - (F_s'c_s - H)\tilde{x}_2 \qquad\qquad 5.1\text{-}70$$

Since

$$\frac{d(e^{sZ}\tilde{x}_1)}{dZ} = e^{sZ}\frac{d\tilde{x}_1}{dZ} + s\tilde{x}_1 e^{sZ} \qquad\qquad 5.1\text{-}71$$

we can rewrite the equations as

$$\frac{d(e^{sZ}\tilde{x}_1)}{dZ} = -F_s\tilde{x}_1 e^{sZ} - F_s'c_s\tilde{x}_2 e^{sZ} \qquad\qquad 5.1\text{-}72$$

$$\frac{d(e^{sZ}\tilde{x}_2)}{dZ} = F_s\tilde{x}_1 e^{sZ} + (F_s'c_s - H)\tilde{x}_2 e^{sZ} \qquad\qquad 5.1\text{-}73$$

which eliminates the second terms on the left-hand sides of the equations. Now, dividing both equations by $r_s = F_s c_s$ and letting

$$dy = r_s dZ, \quad \tilde{X}_1 = e^{sZ}\tilde{x}_1, \quad \tilde{X}_2 = e^{sZ}\tilde{x}_2 \qquad\qquad 5.1\text{-}74$$

we obtain

$$\frac{d\tilde{X}_1}{dy} = -\frac{\tilde{X}_1}{c_s} - \frac{F_s'\tilde{X}_2}{F_s} \qquad\qquad 5.1\text{-}75$$

$$\frac{d\tilde{X}_2}{dy} = \frac{\tilde{X}_1}{c_s} + \left(\frac{F_s'}{F_s} - \frac{H}{F_s c_s}\right)\tilde{x}_2 \qquad\qquad 5.1\text{-}76$$

Differentiating the second equation with respect to y gives

$$\frac{d^2\tilde{X}_2}{dy^2} = \frac{1}{c_s}\frac{d\tilde{X}_1}{dy} - \frac{1}{c_s^2}\tilde{X}_1\frac{dc_s}{dy} + \frac{F_s'}{F_s}\frac{d\tilde{X}_2}{dy} + \tilde{X}_2\left(\frac{F_sF_s'' - F_s'^2}{F_s^2}\right)\frac{dT_s}{dy}$$

$$- \frac{H}{F_s c_s}\frac{d\tilde{X}_2}{dy} - H\tilde{X}_2\left(-\frac{F_s'}{F_s^2 c_s}\frac{dT_s}{dy} - \frac{1}{F_s c_s^2}\frac{dc_s}{dy}\right) \qquad 5.1\text{-}77$$

or after eliminating $d\tilde{X}_1/dy$, using Eq. 5.1-75, and rearranging

$$\frac{d^2\tilde{X}_2}{dy^2} - \frac{F_s'}{F_s}\frac{d\tilde{X}_2}{dy} - \left[\left(\frac{F_s''}{F_s} - \frac{F_s'^2}{F_s^2}\right)\frac{dT_s}{dy} - \frac{F_s'}{F_s c_s}\right]\tilde{X}_2 + \frac{1}{c_s^2}\tilde{X}_1 + \frac{1}{c_s^2}\tilde{X}_1\frac{dc_s}{dy}$$

$$+ \frac{H}{F_s c_s}\left[\frac{d\tilde{X}_2}{dy} - \frac{F_s'}{F_s}\tilde{X}_2\frac{dT_s}{dy} - \frac{1}{c_s}\tilde{X}_2\frac{dc_s}{dy}\right] = 0 \qquad 5.1\text{-}78$$

However, since

$$\frac{dc_s}{dZ} = -r_s \qquad \frac{dT_s}{dZ} = r_s \qquad 5.1\text{-}79$$

or

$$\frac{dc_s}{r_s dZ} = \frac{dc_s}{dy} = -1 \qquad \frac{dT_s}{r_s dZ} = \frac{dT_s}{dy} = 1 \qquad 5.1\text{-}80$$

the equation becomes

$$\left\{\frac{d^2\tilde{X}_2}{dy^2} - \frac{F_s'}{F_s}\frac{d\tilde{X}_2}{dy} - \left[\frac{F_s''}{F_s} - \left(\frac{F_s'}{F_s}\right)^2 - \frac{F_s'}{F_s c_s}\right]\tilde{X}_2\right\}$$

$$+ \frac{H}{F_s c_s}\left\{\frac{d\tilde{X}_2}{dy} - \left(\frac{F_s'}{F_s} - \frac{1}{c_s}\right)\tilde{X}_2\right\} = 0 \qquad 5.1\text{-}81$$

In order for there to be a solution of this equation which is independent of H, it must satisfy both of the expressions enclosed in braces independently. Since the second term only contains a first-order equation, it can be solved to give

$$\tilde{X}_2 = F_s c_s = r_s \qquad 5.1\text{-}82$$

It is a simple matter to show by direct substitution that this expression is also a solution of the first term. Since one solution of the equation is now known, the general solution can be found by reduction of order, and the independent variable can be transformed from y back to Z. The integration constants can be evaluated in terms of the feed disturbances

$$\text{at } Z = 0, \qquad \tilde{x}_1 = \tilde{x}_{1f}, \quad \tilde{x}_2 = \tilde{x}_{2f} \qquad 5.1\text{-}83$$

and the transfer functions can be found. One of the possible results is

$$\frac{\tilde{x}_1}{\tilde{x}_f} = \exp\left[-sZ - \frac{H_p sZ}{s + H_p B}\right] - \left[\frac{1 - H_p sc_s}{r_s(s + H_p B)}\right]r_s\bar{I}(s, Z)\exp(-sZ) \qquad 5.1\text{-}84$$

where

$$\bar{I}(s, Z) = \int_0^Z \frac{1}{c_s(y)}\exp\left(-\frac{H_p sy}{s + H_p B}\right)dy \qquad 5.1\text{-}85$$

Additional information concerning these transfer functions can be found in the original paper.

SECTION 5.2 TRANSPORT PROBLEMS

For some simple problems concerning the transport of mass, energy, and momentum, it is possible to obtain analytical solutions of the rigorous equations describing the system. A thorough study of problems of this nature is of great value, for it makes it possible to gain a very basic understanding of some of the concepts (e.g., film-transfer coefficients) that have been found useful in the design of plants. Then, when it becomes necessary to use these quantities to solve more complicated engineering problems, the designer has a better "feel" for the validity of a particular correlation or the range where an extrapolation of the correlation should give reasonable results.

In addition, situations where an exact analysis of the mass, energy, and momentum transfer are possible provide a method for designing experimental test equipment such as viscometers. Although the major effort in transport studies is directed toward the analysis of steady state systems, it is apparent that the same principles should be applicable for dynamic processes. Therefore in this section we will consider the relationship between transport properties of fluids and film-transfer coefficients, as well as the theories used in certain experimental techniques for measuring transport properties.

Periodic Friction Factors

The analysis of steady state laminar flow in a pipe is perhaps the most common example used to illustrate the fundamental approach to transport phenomenon; it is also generally used to introduce the concept of a friction factor. It is a simple matter to show that an unsteady momentum balance

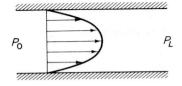

P_0 P_L

Figure 5.2-1. Laminar flow in a pipe.

on a cylindrical shell of fluid leads to the equation[1]

$$\rho \frac{\partial v_z}{\partial t} = -\frac{\partial P}{\partial z} + \eta \left[\frac{1}{r} \frac{\partial}{\partial r} \left(r \frac{\partial v_z}{\partial r} \right) \right] \qquad 5.2\text{-}1$$

Of course, at steady state conditions, the time derivative is equal to zero and

[1] See p. 85 of the reference given in footnote 3 on p. 53, for example.

we obtain

$$-\left(\frac{\partial P}{\partial z}\right)_s = \eta\left[\frac{1}{r}\frac{\partial}{\partial r}\left(r\frac{\partial v_{zs}}{\partial r}\right)\right] \qquad 5.2\text{-}2$$

which is the normal starting point for developing the Hagen-Poiseuille law relating the flow rate through the pipe to the pressure gradient, pipe radius, and fluid viscosity (see Eq. 4.3-67). In order to consider the possibility of imposing a fluctuating pressure gradient on the system,[2,3] we let

$$u = v_z - v_{zs} \qquad P = -\left(\frac{\partial P}{\partial z}\right) + \left(\frac{\partial P}{\partial z}\right)_s \qquad 5.2\text{-}3$$

Substituting these expressions into the dynamic equation and subtracting the steady state expression gives

$$\rho\frac{\partial u(r,t)}{\partial t} = P(t) + \eta\frac{1}{r}\frac{\partial}{\partial r}\left[r\frac{\partial u(r,t)}{\partial r}\right] \qquad 5.2\text{-}4$$

which is identical to the original equation except that now we are considering variables describing the deviation from steady state conditions. Taking the Laplace transform of this expression for a system that is at steady state originally, we obtain

$$\rho s\tilde{u} = \tilde{P} + \eta\frac{\partial}{\partial r}\left(r\frac{\partial \tilde{u}}{\partial r}\right) = \tilde{P} + \eta\left(\frac{d^2\tilde{u}}{dr^2} + \frac{1}{r}\frac{d\tilde{u}}{dr}\right) \qquad 5.2\text{-}5$$

or

$$r^2\frac{d^2\tilde{u}}{dr^2} + r\frac{d\tilde{u}}{dr} - \frac{\rho s}{\eta}r^2\tilde{u} = -\frac{\tilde{P}r^2}{\eta} \qquad 5.2\text{-}6$$

This ordinary differential equation in the transformed variable is a form of Bessel's equation. The complementary solution is

$$\tilde{u}_c = c_1 J_0(\alpha r) + c_2 Y_0(\alpha r) \qquad 5.2\text{-}7$$

where

$$\alpha^2 = -\frac{\rho s}{\eta} \qquad 5.2\text{-}8$$

A particular solution can be determined by inspection

$$\tilde{u}_p = \frac{\tilde{P}}{\rho s} \qquad 5.2\text{-}9$$

so that the complete solution is

$$\tilde{u} = c_1 J_0(\alpha r) + c_2 Y_0(\alpha r) + \frac{\tilde{P}}{\rho s} \qquad 5.2\text{-}10$$

---

[2] Extension of the simple analysis to viscoelastic fluids is described in A. G., Fredrickson, *Principles and Applications of Rheology*, p. 139, Prentice-Hall, Englewood Cliffs, N.J., 1964.

[3] Some experimental data were published by B. Shizgal, H. L. Goldsmith, and S. G. Mason, *Can. J. Chem Eng.*, **43**, 97 (1965).

The boundary conditions for the problem are that the velocity at the pipe wall must be equal to zero and that the velocity at the center of the pipe must be finite, or that the velocity gradient must be equal to zero, in order to maintain the symmetry of the system.

$$\text{at } r = R, \quad u(R, t) = 0 = \tilde{u}(R, s)$$

$$\text{at } r = 0, \quad u(0, t) = \text{finite} = \tilde{u}(0, s) \qquad 5.2\text{-}11$$

Since the function $Y_0(\alpha r)$ becomes unbounded as r approaches zero, the second boundary condition requires that $c_2 = 0$. The first boundary condition leads to the result $c_1 = -\tilde{P}/[\rho s J_0(\alpha R)]$. Thus the complete solution of the system equation is

$$\tilde{u} = \frac{\tilde{P}}{\rho s}\left[1 - \frac{J_0(\alpha r)}{J_0(\alpha R)}\right] \qquad 5.2\text{-}12$$

and the transfer function of the system is

$$\frac{\tilde{u}}{\tilde{P}} = H(s) = \frac{1}{\rho s}\left[1 - \frac{J_0(\alpha r)}{J_0(\alpha R)}\right] \qquad 5.2\text{-}13$$

where the parameter α is a function of the Laplace parameter s (see Eq. 5.2-8).

In order to determine the response of the system to a sinusoidal or cosinusoidal variation in the pressure gradient, we could let $\tilde{P}$ in Eq. 5.2-12 be equal either to $P_i\omega/(s^2 + \omega^2)$ or to $P_i s/(s^2 + \omega^2)$ and then attempt to invert the expression for the transform. The algebra associated with this problem is tedious, however, since the Laplace parameter appears in the argument of the Bessel function. A trick that is commonly used to simplify the algebra in problems of this type is to let

$$P(t) = P_i e^{j\omega t} \qquad 5.2\text{-}14$$

so that

$$\tilde{P} = \frac{P_i}{s - j\omega} \qquad 5.2\text{-}15$$

From our earlier work we know that

$$P(t) = P_i e^{j\omega t} = P_i(\cos \omega t + j \sin \omega t) \qquad 5.2\text{-}16$$

and therefore the input signal will contain both a real and an imaginary component. For this kind of an input, we expect to obtain a velocity $u(r, t)$, which also has a real and imaginary part. Because the system is linear, the real part of the output signal will correspond to the real part of the input, and similar results will be obtained for the imaginary parts. Hence the response of the system to cosine inputs will be the solution of the equation after the imaginary terms have been dropped.

Substituting Eq. 5.2-15 into Eq. 5.2-12 gives

$$\tilde{u} = \frac{P_i}{\rho s(s - j\omega)}\left[1 - \frac{J_0(\alpha r)}{J_0(\alpha R)}\right] \qquad 5.2\text{-}17$$

and now we want to find the inverse transform of this expression, using Heaviside's expansion theorems. The roots of the denominator occur when

$$s = 0, \quad s = j\omega, \quad J_0(\alpha R) = 0 \qquad\qquad 5.2\text{-}18$$

The first root does not lead to any term in the output since $\alpha = 0$ when $s = 0$ and

$$\frac{P_i}{\rho(-j\omega)}\left[1 - \frac{J_0(0)}{J_0(0)}\right] = 0 \qquad\qquad 5.2\text{-}19$$

For the second root we obtain

$$\frac{P_i}{j\omega\rho}\left[1 - \frac{J_0(\beta r)}{J_0(\beta R)}\right]e^{j\omega t} \qquad\qquad 5.2\text{-}20$$

where

$$\beta^2 = -\frac{j\omega\rho}{\eta} \qquad\qquad 5.2\text{-}21$$

An infinite number of roots satisfy the expression $J_0(\alpha R) = 0$. If we let γ_n be the nth root of this equation, we find that

$$-\frac{\rho s R^2}{\eta} = \gamma_n$$

Thus

$$s_n = -\frac{\gamma_n \eta}{\rho R^2} \qquad\qquad 5.2\text{-}22$$

are the values of s corresponding to the roots of the denominator. Since these values are distinct, the contribution of the nth term to the solution will be

$$\frac{P_i}{\rho s_n(s_n - j\omega)}\left|\frac{J_0(\sqrt{\gamma_n}) - J_0(\sqrt{\gamma_n}\,r/R)}{[dJ_0(\alpha R)/ds]|_{s=s_n}}\right|e^{s_n t}$$

and the contribution for all of these terms is just the sum from $n = 1$ to infinity of the preceding expression. However, each value of s_n is negative and real, and therefore every term in the summation will approach zero as time approaches infinity.

Hence the pseudo-steady state frequency response is

$$u(r, t) = \frac{P_i}{j\omega\rho}\left[1 - \frac{J_0(\beta r)}{J_0(\beta R)}\right]e^{j\omega t} \qquad\qquad 5.2\text{-}23$$

The flow rate through the pipe can now be determined:

$$q(t) = \int_0^R 2\pi r u(r, t)\, dr = \frac{\pi P_i R^2}{j\omega\rho}\left[1 - \frac{2J_1(\beta R)}{\beta R J_0(\beta R)}\right]e^{j\omega t} \qquad\qquad 5.2\text{-}24$$

and the average velocity can be shown to be

$$u_{av}(t) = \frac{q}{\pi R^2} = \frac{P_i}{j\omega\rho}\left[1 - \frac{2J_1(\beta R)}{\beta R J_0(\beta R)}\right]e^{j\omega t} \qquad\qquad 5.2\text{-}25$$

Also, the velocity gradient at the wall is

$$\left.\frac{du}{dr}\right|_{r=R} = +\frac{P_i \beta}{j\omega\rho}\frac{J_1(\beta R)}{J_0(\beta R)}e^{j\omega t} \qquad 5.2\text{-}26$$

Although these solutions are exact, they are not convenient to work with, for they involve Bessel functions with complex arguments. For small values of the argument (βR), we can write

$$\frac{J_1(\beta R)}{J_0(\beta R)} \simeq \frac{(\beta R)}{2}\left[1 + \frac{(\beta R)^2}{8} + \frac{(\beta R)^4}{48} + \cdots\right] \qquad 5.2\text{-}27$$

With this approximation the expression for flow rate, Eq. 5.2-24, becomes

$$q(t) = \frac{\pi P_i R^4 j\beta^2}{8\rho\omega}\left[1 + \frac{(\beta R)^2}{6} + \cdots\right]e^{j\omega t}$$

or after substituting Eqs. 5.2-16 and 5.2-21,

$$q(t) = \frac{\pi P_i R^4}{8\eta}\left[1 + \frac{-j\omega\rho R^2}{6\eta} + \cdots\right](\cos\omega t + j\sin\omega t)$$

$$= \frac{\pi P_i R^4}{8\eta}\left(\cos\omega t + \frac{\omega\rho R^2}{6}\sin\omega t\right)$$

$$+ \frac{\pi P_i R^4}{8\eta}\left(\sin\omega t - \frac{\omega\rho R^2}{6}\cos\omega t\right)j \qquad 5.2\text{-}28$$

Thus the real output corresponding to the real input is

$$q(t) = \frac{\pi P_i R^4}{8\eta}\left(\cos\omega t + \frac{\omega\rho R^2}{6}\sin\omega t\right) \qquad 5.2\text{-}29$$

The corresponding result for the average velocity can be shown to be

$$u_{av} = \frac{q}{\pi R^2} = \frac{P_i R^2}{8\eta}\left(\cos\omega t + \frac{\omega\rho R^2}{6}\sin\omega t\right) \qquad 5.2\text{-}30$$

In a similar manner, we find that the shear stress at the wall is

$$\tau_w = -\eta\left.\frac{du}{dr}\right|_{r=R} = \frac{P_i R}{2}\left(\cos\omega t + \frac{\omega\rho R^2}{8\eta}\sin\omega t\right)$$

$$+ \frac{P_i R}{2}\left(\sin\omega t - \frac{\omega\rho R^2}{8\eta}\cos\omega t\right)j$$

so that the real value corresponding to the real input becomes

$$\tau_w = \frac{P_i R}{2}\left(\cos\omega t + \frac{\omega\rho R^2}{8\eta}\sin\omega t\right) \qquad 5.2\text{-}31$$

Since the variables in the foregoing analysis represent deviations from steady state conditions, the total flow rate, average velocity, and shear stress at the wall can be obtained by adding the expressions derived above to the results of the normal steady state analysis. Thus we find that when the pressure gradient fluctuates according to the equation

$$\frac{\Delta P(t)}{L} = \frac{P_0 - P_L}{L}[1 + A\cos\omega t] \qquad 5.2\text{-}32$$

the total flow rate through the pipe is

$$q(t) = \frac{\pi R^4 (P_0 - P_L)}{8\eta L}\left[1 + A\left(\cos \omega t + \frac{\omega \rho R^2}{6} \sin \omega t\right)\right] \qquad 5.2\text{-}33$$

the average velocity is

$$v_{z\,\text{av}}(t) = \frac{(P_0 - P_L)R^2}{8\eta L}\left[1 + A\left(\cos \omega t + \frac{\omega \rho R^2}{6} \sin \omega t\right)\right] \qquad 5.2\text{-}34$$

and the wall shear stress τ_w is

$$\tau_w(t) = \frac{(P_0 - P_L)}{2L}R\left[1 + A\left(\cos \omega t + \frac{\omega \rho R^2}{6} \sin \omega t\right)\right] \qquad 5.2\text{-}35$$

It is common practice to characterize the drag force of a fluid acting on a solid body by a friction factor. This step is accomplished by assuming that the drag force F_D can be written as the product of some characteristic area A_D (normally taken as the projected area of a particle in the direction of its flow through the fluid or as the wetted-surface area of a solid), the kinetic energy per unit volume K (generally taken as $\frac{1}{2}\rho v^2$, where v is the particle velocity or the average fluid velocity v_{av}), and a friction factor f. Thus

$$F_D = A_D K f \qquad 5.2\text{-}36$$

For some simple flow systems, it is possible to find an analytical expression for the friction factor. In fact, for steady state laminar flow, we see that

$$f = \frac{\tau_w (2\pi R L)}{(2\pi R L)\frac{1}{2}\rho v_{\text{av}} v_{\text{av}}} = \frac{(P_0 - P_L)R/2L}{\frac{1}{2}\rho v_{\text{av}}[(P_0 - P_L)R^2/8\eta L]}$$

$$= \frac{8\eta}{\rho v_{\text{av}} R} = \frac{16}{R_e} \qquad 5.2\text{-}37$$

where the Reynolds number Re is given by the expression

$$\text{Re} = \frac{\rho D v_{\text{av}}}{\eta}$$

and D is the pipe diameter. For more complicated steady state flow systems, it is usually necessary to develop empirical relationships between the friction factor and Reynolds number.

However, the analogous results for unsteady state flows are seldom discussed in the literature, and for these cases it is generally assumed that the friction factor remains constant at a value corresponding to the average flow conditions. The fact that the frction factor must be time dependent for our simple example is easily seen by substituting Eqs. 5.2-34 and 5.2-35 into Eq. 5.2-36.

$$f = \frac{[(P_0 - P_L)R/2L]\{1 + A[\cos \omega t + (\omega \rho R^2/6) \sin \omega t]\}}{\frac{1}{2}\rho v_{s\,\text{av}}\{1 + A[\cos \omega t + (\omega \rho R^2/6) \sin \omega t]\} \times} \\ {[(P_0 - P_L)R^2/8\eta L]\{1 + A[\cos \omega t + (\omega \rho R^2/6) \sin \omega t]\}}$$

or

$$f = \frac{16}{Re_s[1 + A(\cos \omega t + (\omega \rho R^2/6) \sin \omega t)]}$$ 5.2-38

where Re_s is the steady state Reynolds number.

Similarly, it is possible to show in some simple cases that film-heat-transfer coefficients and film-mass-transfer coefficients should be considered as time-dependent quantities for dynamic systems. The nature of the time dependence is expected to be a function of the fluctuations in the system input. Therefore we would have to carry out an overwhelming number of experiments in order to develop a set of correlations for complex flow systems that would relate the film coefficients to the normal dimensionless quantities used to describe heat or mass transfer (Reynolds number, Prandtl number, Schmidt number, etc.) and the input fluctuations. At present, it is common practice to assume that the time dependence of the film-transfer coefficients is negligible in comparison with other dynamic effects in a system under investigation, but the reader should be aware that this kind of an inherent assumption is present in the analysis of most dynamic systems.

A Frequency Response Technique for Measuring Viscosity

Whenever the behavior of some system is understood in great detail, that system is frequently used to measure the physical or transport properties of various materials. Although many of the experimental methods involve only steady state systems, there are a great number of cases where the system dynamics are important. As an example of this type, we will consider the behavior of a torsionally oscillating rheometer.[4] For simplicity we will restrict

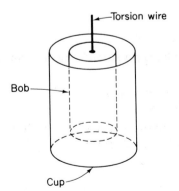

Figure 5.2-2. A torsionally oscillating rheometer.

[4] A more general analysis is given in S. Middleman, *The Flow of High Polymers*, p. 67, Interscience, N. Y., 1968.

our attention to the analysis of Newtonian fluids, but it should be recognized that the theory is not limited to this case.

If we submerge a cylindrical bob of radius R coaxially into a cylindrical cup of radius R_0, fill the gap between the two cylinders to a height L with a fluid of viscosity η, and then oscillate the outer cup cosinusoidally, we expect that the oscillations imposed on the fluid will induce an oscillating torque on the inner cylinder. By using a torsion wire to suspend the bob, we can measure the amplitude ratio and phase lag between the outer and inner cylinders. From these measured frequency response characteristics of the system, we can compute the viscosity of the fluid.

The dynamic equation describing the motion of the bob is derived by using Newton's law:

$$I\frac{\partial^2 x_i}{\partial t^2} = T - Kx_i \qquad\qquad 5.2\text{-}39$$

where $I =$ the moment of inertia of the bob
   $T =$ the torque exerted on the bob by the fluid
   $K =$ the torsion constant of the wire
   $x_i =$ the angular deflection of the inner cylinder.

Similarly, a momentum balance for the fluid gives the result

$$\rho\frac{\partial v_\theta}{\partial t} = -\frac{1}{r^2}\frac{\partial}{\partial r}(r^2\tau_{r\theta}) = -\frac{1}{r^2}\frac{\partial}{\partial r}\left\{r^2\left[-\eta r\frac{\partial}{\partial r}\left(\frac{v_\theta}{r}\right)\right]\right\} \qquad 5.2\text{-}40$$

where $\rho =$ fluid density
   $\eta =$ fluid viscosity
   $r =$ radius
   $v_\theta =$ angular velocity in centimeters per second
   $\tau_{r\theta} =$ momentum flux

Letting

$$v_\theta = \omega r \qquad\qquad 5.2\text{-}41$$

where ω is the angular velocity in radians per second, the momentum balance becomes

$$\rho r\frac{\partial \omega}{\partial t} = \frac{\eta}{r^2}\frac{\partial}{\partial r}\left(r^3\frac{\partial \omega}{\partial r}\right) \qquad\qquad 5.2\text{-}42$$

The equation of motion of the bob and the actual quantities that we measure are in terms of angular displacements; therefore we let

$$\omega = \frac{\partial x}{\partial t} \qquad\qquad 5.2\text{-}43$$

where x is the angular displacement. The system equation becomes

$$\rho r\frac{\partial^2 x}{\partial t^2} = \frac{\eta}{r^2}\frac{\partial}{\partial r}\left(r\frac{\partial^2 x}{\partial r\partial t}\right) \qquad\qquad 5.2\text{-}44$$

This is a linear partial differential equation, so that it would be possible to take the Laplace transform with respect to time and obtain an ordinary differential equation in terms of a transformed displacement variable. By solving the ordinary differential equation with the appropriate boundary conditions and inverting the solution, we could find the time and spatial dependence of the displacement. However, the algebra involved in this approach is quite complex. Also, since we know that the response of a linear system to a cosine input is generally a sine wave plus a cosine wave, we will assume that we can find a solution like

$$x = \theta e^{j\omega_0 t} = \theta(\cos \omega_0 t + j \sin \omega_0 t) \qquad 5.2\text{-}45$$

where ω_0 is the frequency of the oscillation and θ is the amplitude, which depends on the radial position but which is independent of time. Here, again, we are considering both real and imaginary components of the oscillations in an attempt to simplify the algebra, but we must recognize that with this approach the amplitude variable θ may have both real and imaginary parts.

Substituting the assumed solution into the equation of motion for the fluid, we obtain

$$\frac{d^2\theta}{dr^2} + \frac{3}{r}\frac{d\theta}{dr} + \beta^2\theta = 0 \qquad 5.2\text{-}46$$

where

$$\beta^2 = \frac{\rho\omega_0}{j\eta} \qquad 5.2\text{-}47$$

as the equation describing the amplitude of the oscillations. This result is a Bessel equation and has the solution

$$\theta(r) = \frac{1}{r}[c_1 J_1(\beta r) + c_2 Y_1(\beta r)] \qquad 5.2\text{-}48$$

where c_1 and c_2 are integration constants.

Providing that there is no slip between the fluid and the cup wall, the fluid displacement at $r = R_0$ will be

$$x(R_0) = \theta_0 e^{j\omega_0 t} \qquad 5.2\text{-}49$$

where θ_0 is the amplitude imposed on the system. Thus one boundary condition is

$$\text{at } r = R_0, \qquad \theta(R_0) = \theta_0 = \frac{1}{R_0}[c_1 J_1(\beta R_0) + c_2 Y_1(\beta R_0)] \qquad 5.2\text{-}50$$

The second boundary condition is obtained by recognizing that the fluid in contact with the bob must satisfy the equation of motion of the bob, Eq. 5.2-39. The torque can be obtained from the expression

$$T = 2\pi R L(-\tau_{r\theta}|_{r=R} R) = 2\pi R^2 \eta \left(r\frac{d\omega}{\partial r}\right)\Big|_{r=R} \qquad 5.2\text{-}51$$

After substituting Eqs. 5.2-43, 5.2-45, and 5.2-48, this becomes

$$T = 2\pi R^2 L \eta j \omega_0 e^{j\omega_0 t} \beta [c_1 J_2(\beta R) + c_2 Y_2(\beta R)] \qquad \text{5.2-52}$$

Now, if we substitute this result and Eqs. 5.2-45 and 5.2-48 into Eq. 5.2-39, we find that

$$(K - I\omega_0^2)\frac{1}{R}[c_1 J_1(\beta R) + c_2 Y_1(\beta R)] = 2\pi R^2 L \eta \beta j \omega [c_1 J_2(\beta R) + c_2 Y_2(\beta R)]$$

$$\text{5.2-53}$$

The integration constants c_1 and c_2 can be obtained by solving Eqs. 5.2-50 and 5.2-53 simultaneously. Following this procedure, eliminating c_1 and c_2 from the general solution, Eq. 5.2-48, and solving for the amplitude of the inner cylinder, we obtain

$$\theta_i = \frac{[-j2\pi L \eta \omega_0/(I\omega_0^2 - K)]\beta\theta_0 R^2[J_2(\beta R)Y_1(\beta R) - J_1(\beta R)Y_2(\beta R)]}{\{[1 - j2\pi L \eta \omega_0/(I\omega_0^2 - K)]R^2\}[Y_1(\beta R)J_1(\beta R_0) - J_1(\beta R)Y_1(\beta R_0)] + [j2\pi L \eta \omega_0 \beta R^3/(I\omega_0^2 - K)][J_1(\beta R_0)Y_2(\beta R) - J_2(\beta R)Y_1(\beta R_0)]}$$

$$\text{5.2-54}$$

However, this expression is too complex to be of much value so we expand it in a series to obtain

$$\frac{\theta_0}{\theta_i} = 1 + \frac{j}{\eta}\left[\left(\frac{I\omega_0^2 - K}{4\pi L \omega_0}\right)\left(\frac{R_0^2 - R^2}{R_0^2 R^2}\right) + \frac{\omega_0 \rho}{8}\frac{(R_0^2 - R^2)^2}{R^2}\right] + \cdots \qquad \text{5.2-55}$$

or after letting

$$M = \frac{I\omega_0^2 - K}{4\pi L \omega_0}\left(\frac{R_0^2 - R^2}{R_0^2 R^2}\right) + \frac{\omega_0 \rho}{8}\frac{(R_0^2 - R^2)^2}{R^2} \qquad \text{5.2-56}$$

we find that

$$\frac{\theta_0}{\theta_i} = 1 + \frac{j}{\eta}M \qquad \text{5.2-57}$$

From Eq. 5.2-49 we know that the displacement we impose on the system at $r = R_0$ is

$$x_0 = \theta_0 e^{j\omega_0 t} = \theta_0(\cos \omega_0 t + j \sin \omega_0 t) \qquad \text{5.2-58}$$

In an actual experiment, we only consider the real part of this signal

$$x_0 = \theta_0 \cos \omega_0 t \qquad \text{5.2-59}$$

Also, from our assumed solution, Eq. 5.2-45, the displacement at the inner cylinder, at $r = R$, must be

$$x_i = \theta_i e^{j\omega_0 t} = \theta_i(\cos \omega_0 t + j \sin \omega_0 t)$$

Substituting the result we obtained for the amplitude ratio, we see that

$$x_i = \left(\frac{\theta_0}{1 + jM/\eta}\right)(\cos \omega_0 t + j \sin \omega_0 t)$$

$$= \frac{\theta_0}{1 + (M/\eta)^2}\left[\left(\cos \omega_0 t + \frac{M}{\eta} \sin \omega_0 t\right)\right.$$
$$\left. + j\left(\sin \omega_0 t - \frac{M}{\eta} \cos \omega_0 t\right)\right] \qquad 5.2\text{-}60$$

In a real experiment, however, we observe only the real part of this output

$$x_i = \frac{\theta_0}{1 + (M/\eta)^2}\left(\cos \omega_0 t + \frac{M}{\eta} \sin \omega_0 t\right) \qquad 5.2\text{-}61$$

or

$$x_i = \frac{\theta_0}{\sqrt{1 + (M/\eta)^2}} \cos (\omega_0 t + \phi) \qquad 5.2\text{-}62$$

where

$$\tan \phi = -\frac{M}{\eta} \qquad 5.2\text{-}63$$

Thus we find that the ratio of the output signal to the input signal—that is, the displacement of the inner cylinder to the displacement of the outer cylinder—is

$$|G| = \frac{1}{\sqrt{1 + (M/\eta)^2}} \qquad 5.2\text{-}64$$

and the phase angle is

$$\phi = \tan^{-1}\left(-\frac{M}{\eta}\right) \qquad 5.2\text{-}65$$

Once the system geometry (R_0, R, and L) has been fixed and the bob characteristics (I and K) specified, measured values of the amplitude ratio $|G|$ and phase lag ϕ at one or more frequencies can be used to compute the viscosity of the fluid from either Eqs. 5.2-56 and 5.2-64 or 5.2-56 and 5.2-65.

For non-Newtonian fluids, the amplitude ratio and phase angle become independent and the experiment described above provides two pieces of information about the characteristics of the fluid. Often it is important to carry out dynamic experiments on these kinds of fluids, for many exhibit time-dependent properties—for example, stress relaxation—and this can never be measured in a steady state experiment. More information on this subject can be obtained by consulting the books by Fredrickson and Middleman mentioned earlier.

Measurement of the Axial Dispersion Coefficient

In order to ascertain the importance of axial dispersion in a packed-bed reactor, it is necessary to measure experimentally the axial dispersion coefficient D, or its equivalent the Peclet number, $Pe = vL/D$. If it is assumed that

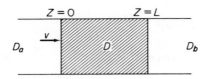

Figure 5.2-3. Packed-bed reactor.

this coefficient depends primarily on the mixing caused by the packing and is independent of any reactions occurring on the packing, then the experiments can be undertaken in an inert system. The equations describing the section before the bed, the packed section, and the after-section are[5]

$$\frac{\partial c_a}{\partial \theta} + \frac{\partial c_a}{\partial Z} - \frac{1}{\text{Pe}_a}\frac{\partial^2 c_a}{\partial Z^2} = 0 \qquad 5.2\text{-}66$$

$$\frac{\partial c}{\partial \theta} + \frac{\partial c}{\partial Z} - \frac{1}{\text{Pe}}\frac{\partial^2 c}{\partial Z^2} = 0 \qquad 5.2\text{-}67$$

$$\frac{\partial c_b}{\partial \theta} + \frac{\partial c_b}{\partial Z} - \frac{1}{\text{Pe}_b}\frac{\partial^2 c_b}{\partial Z^2} = 0 \qquad 5.2\text{-}68$$

where c = concentration in arbitrary units

$$Z = \frac{z}{L}$$

L = length of test section

$$\theta = \frac{vt}{L}$$

v = interstitial velocity

$$\text{Pe} = \frac{vL}{D}$$

D = dispersion coefficient

and the subscripts a and b refer to the inlet and outlet sections. The boundary conditions for the problem are

$$c_a(t, 0) = c_0(t) \qquad\qquad c_a(t, 0^-) = c(t, 0^+)$$

$$c(t, 1^-) = c_b(t, 1^+) \qquad c_b(t, \infty) = \text{finite}$$

$$c_a(0^-) - \frac{1}{\text{Pe}_a}\frac{dc_a(0^-)}{dZ} = c(0^+) - \frac{1}{\text{Pe}}\frac{dc(0^+)}{dZ}$$

$$\qquad\qquad\qquad\qquad\qquad\qquad\qquad\qquad\qquad 5.2\text{-}69$$

$$c(1^-) - \frac{1}{\text{Pe}}\frac{dc(1^-)}{dZ} = c_b(1^+) - \frac{1}{\text{Pe}_b}\frac{dc_b(1^+)}{dZ}$$

which merely state that the composition and flux across the bed inlet and outlet must be continuous and that at time zero we are introducing some input signal $c_0(t)$.

[5] J. F. Wehner and R. H. Wilhelm, *Chem. Eng. Sci.*, **6**, 89 (1956).

For the case where $D_a = D_b = 0$, the problem simplifies to the equation

$$\frac{\partial c}{\partial \theta} + \frac{\partial c}{\partial Z} - \frac{1}{Pe}\frac{\partial^2 c}{\partial Z^2} = 0 \qquad 5.2\text{-}70$$

with the boundary conditions

$$c_a(0^-) = c(0^+) - \frac{1}{Pe}\frac{dc(0^+)}{dZ} = c_0(t) \qquad 5.2\text{-}71$$

$$\frac{dc(1)}{dZ} = 0 \qquad 5.2\text{-}72$$

Now, if we define the Laplace transform in terms of dimensionless time θ as

$$\tilde{c} = \int_0^\infty e^{-s\theta}c(\theta)\, d\theta \qquad 5.2\text{-}73$$

and take the Laplace transform of the system equation and boundary conditions, we obtain

$$s\tilde{c} + \frac{d\tilde{c}}{dZ} - \frac{1}{Pe}\frac{d^2\tilde{c}}{dZ^2} = 0 \qquad 5.2\text{-}74$$

$$\tilde{c}_0 = \tilde{c}(0^+) - \frac{1}{Pe}\frac{d\tilde{c}(0^+)}{dZ} \qquad 5.2\text{-}75$$

$$\frac{d\tilde{c}(1)}{dZ} = 0 \qquad 5.2\text{-}76$$

The characteristic equation of the second-order differential equation is

$$\frac{1}{Pe}\lambda^2 - \lambda - s = 0 \qquad 5.2\text{-}77$$

and the characteristic roots are

$$\lambda = Pe(\tfrac{1}{2} \pm M) \qquad 5.2\text{-}78$$

where

$$M = \left(\frac{s}{Pe} + \frac{1}{4}\right)^{1/2} \qquad 5.2\text{-}79$$

Thus the solution of the differential equation is

$$\tilde{c} = A_1 \exp\left[-Pe(\tfrac{1}{2} + M)Z\right] + A_2 \exp\left[-Pe(\tfrac{1}{2} - M)Z\right] \qquad 5.2\text{-}80$$

where A_1 and A_2 are integration constants. Evaluating these constants, using the boundary conditions, Eqs. 5.2-75 and 5.2-76, we obtain the result

$$\frac{\tilde{c}}{\tilde{c}_0} = \frac{(\tfrac{1}{2} + M)[\exp Pe(\tfrac{1}{2} + M)\exp Pe(\tfrac{1}{2} - M)Z] - (\tfrac{1}{2} - M)[\exp Pe(\tfrac{1}{2} - M)\exp Pe(\tfrac{1}{2} + M)Z]}{(\tfrac{1}{2} + M)^2 \exp Pe(\tfrac{1}{2} + M) - (\tfrac{1}{2} - M)^2 \exp Pe(\tfrac{1}{2} - M)} \qquad 5.2\text{-}81$$

Providing that we measure the output signal at the end of the bed, $Z = 1$, the equation becomes

$$\frac{\bar{c}}{\bar{c}_0} = H_0(s) = \frac{2M \exp \text{Pe}}{(\frac{1}{2} + M)^2 \exp \text{Pe}(\frac{1}{2} + M) - (\frac{1}{2} - M)^2 \exp \text{Pe}(\frac{1}{2} - M)}$$

5.2-82

which is the transfer function of the system.

Once we specify some particular input signal for $c_0(t)$, such as a step function or a sine wave, we can find $\bar{c}_0$ quite simply. However, in order to find an expression for the output time signal $c(t)$, we must invert the product of the transfer function and the input signal. This means that we need to find the roots of the denominator of the transfer function. After looking at Eq. 5.2-82 and remembering that the parameter M depends on both the Laplace parameter s and the Peclet number, which we want to determine, we see that we must find the roots of a transcendental expression. We know that this is not a simple task. Also, even if we find a way to invert the equation for the transform of the output, we expect that this result will depend on some complicated function of the Peclet number. Thus it might still be a very difficult problem to estimate the value of the Peclet number from measured values of the output signal.

The complexities just described can be avoided if we take a completely different approach to the problem. Our main interest is to estimate the Peclet number from measured values of the system input and output. If we can accomplish this goal without obtaining a solution for the output signal, it will be advantageous. The technique we will use is called the method of moments.[6] The theoretical basis for this method is described below, and then its application to the problem of estimating the Peclet number is given.

Method of moments

In Example 4.7-2 we noted that the impulse response of a chemical reactor is called the residence time frequency distribution. Similarly, the step response of the system (i.e., the integral of the impulse response) can be called the cumulative residence time distribution. Since these quantities closely resemble the distribution functions studied in statistics, some of the statistical concepts might prove useful in engineering problems. For example, we might choose to describe a particular distribution function $f(t)$ by its moments.

The moments of a distribution function are defined by the equation

$$m_n = \frac{\int_0^\infty t^n f(t)\, dt}{\int_0^\infty f(t)\, dt}$$

5.2-83

It is clear that the expression $\int_0^\infty f(t)\, dt$ is just the area under the curve. Also, the first moment, m_1, is the mean value of the function, and the second mo-

---

[6] A more complete study of axial dispersion using this approach has been presented by K. B. Bischoff and O. Levenspiel, *Chem. Eng. Sci.*, **17**, 245, 257 (1962).

ment is similar to the variance. When the Laplace transform of the function $f(t)$ is written in terms of an infinite series, it is easy to see that the terms in this series are related to the moments of the function

$$\tilde{f}(s) = \int_0^\infty e^{-st}f(t)\,dt = \int_0^\infty (1 - st + \tfrac{1}{2}s^2t^2 + \cdots)f(t)\,dt$$

$$= \int_0^\infty f(t)\,dt - s\int_0^\infty tf(t)\,dt + \tfrac{1}{2}s^2\int_0^\infty t^2f(t)\,dt + \cdots$$

$$= \left(\int_0^\infty f(t)\,dt\right)[1 - sm_1 + \tfrac{1}{2}s^2m_2 + \cdots] \qquad 5.2\text{-}84$$

Also, it should be recognized that the Laplace transform is proportional to the moment generating function.

$$\lim_{s\to 0}\int_0^\infty e^{-st}f(t)\,dt = \int_0^\infty f(t)\,dt = A_f$$

$$\lim_{s\to 0}\frac{\partial}{\partial s}\left[\int_0^\infty e^{-st}f(t)\,dt\right] = \lim_{s\to 0}\int_0^\infty -te^{-st}f(t)\,dt$$

$$= -\int_0^\infty tf(t)\,dt = -A_fm_1$$

$$\lim_{s\to 0}\frac{\partial^2}{\partial s^2}\left[\int_0^\infty e^{-st}f(t)\,dt\right] = \lim_{s\to 0}\int_0^\infty t^2e^{-st}f(t)\,dt$$

$$= \int_0^\infty t^2f(t)\,dt = A_fm_2 \qquad 5.2\text{-}85$$

and

$$\lim_{s\to 0}\frac{\partial^n}{\partial s^n}\left[\int_0^\infty e^{-st}f(t)\,dt\right] = (-1)^n\int_0^\infty t^nf(t)\,dt$$

$$= (-1)^nA_fm_n$$

Thus the Laplace transform of a frequency distribution function can be expressed as a power series in s, where the coefficients are proportional to the moments of the distribution.

Now we can use these elementary definitions to gain some additional information about the response of systems to arbitrary pulse inputs. The system transfer function is defined by

$$\frac{\tilde{y}}{\tilde{g}} = H(s) \qquad 5.2\text{-}86$$

If the input function $g(t)$ is some pulse, we expect the output $y(t)$ to have a pulse shape. Also, we know that $H(s)$ is merely the Laplace transform of the weighting function $h(t)$ of the system, which has the characteristics of a pulse. Thus $g(t)$, $y(t)$, and $h(t)$ all can be considered as frequency distribution functions, and their Laplace transforms, $\tilde{g}(s)$, $\tilde{y}(s)$, and $H(s)$, can be written in terms of the moments. Letting

$$A_g = \int_0^\infty g(t)\,dt, \quad A_y = \int_0^\infty y(t)\,dt, \quad A_H = \int_0^\infty h(t)\,dt \qquad 5.2\text{-}87$$

we can write

$$\bar{g}(s) = A_g(1 - sm_{g^1} + \tfrac{1}{2}s^2 m_{g^2} + \cdots)$$
$$\bar{y}(s) = A_y(1 - sm_{y^1} + \tfrac{1}{2}s^2 m_{y^2} + \cdots) \qquad\qquad \text{5.2-88}$$
$$H(s) = A_H(1 - sm_{H^1} + \tfrac{1}{2}s^2 m_{H^2} + \cdots)$$

Substituting these expressions into the transfer function, we find that

$$A_y(1 - sm_{y^1} + \tfrac{1}{2}s^2 m_{y^2} + \cdots) = A_H(1 - sm_{H^1} + \tfrac{1}{2}s^2 m_{H^2} + \cdots)$$
$$\times A_g(1 - sm_{g^1} + \tfrac{1}{2}s^2 m_{g^2} + \cdots)$$
$$= A_H A_g[1 - s(m_{H^1} + m_{g^1})$$
$$+ \tfrac{1}{2}s^2(m_{H^2} + 2m_{H^1}m_{g^1} + m_{g^2}) + \cdots]$$

$$\text{5.2-89}$$

Since these infinite series will be equal to each other when the coefficients of like powers of s are equal, we find that

$$A_y = A_H A_g$$
$$m_{y^1} = m_{H^1} + m_{g^1} \qquad\qquad \text{5.2-90}$$
$$m_{y^2} = m_{H^2} + 2m_{H^1}m_{g^1} + m_{g^2}$$

The first relationship in this set states that the product of the area under the weighting function and the area under the input pulse must be equal to the area under the output pulse. The second expression says that the sum of the mean values of the weighting function and input pulse will be equal to the mean value of the output pulse. Rather than consider the third equation directly, we replace the second moment about the origin, m_2, by the second moment about the mean, the variance σ_2^2, where

$$\sigma_2^2 = m_2 - m_1^2 \qquad\qquad \text{5.2-91}$$

to obtain

$$\sigma_{2y}^2 = \sigma_{2H}^2 + \sigma_{2g}^2 \qquad\qquad \text{5.2-92}$$

Hence the variance of the weighting function added to the variance of the input gives the variance of the output signal. By extending this procedure to higher-order moments and recognizing that any frequency distribution curve can be described in terms of its moments, we can predict the output signal from a knowledge of the moments of the input pulse and the weighting function.

Example 5.2-1 Measurement of axial dispersion coefficients, using the method of moments

The transfer function for a packed bed was found to be

$$H(s) = \frac{2M \exp(\text{Pe})}{(\tfrac{1}{2} + M)^2 \exp[\text{Pe}(\tfrac{1}{2} + M)] - (\tfrac{1}{2} - M)^2 \exp[\text{Pe}(\tfrac{1}{2} - M)]} \qquad \text{5.2-82}$$

where

$$M = \left(\frac{s}{\text{Pe}} + \frac{1}{4}\right)^{1/2} \qquad 5.2\text{-}79$$

The area under the weighting function curve is

$$A_H = \lim_{s \to 0} H(s) = \frac{2(\frac{1}{2}) \exp \text{Pe}}{(\frac{1}{2} + \frac{1}{2})^2 \exp [\text{Pe}(\frac{1}{2} + \frac{1}{2})] - (\frac{1}{2} - \frac{1}{2})^2 \exp [\text{Pe}(\frac{1}{2} - \frac{1}{2})]} = 1$$

Also, the mean value of the weighting function, m_{H1}, is

$$-m_{H1} = \lim_{s \to 0} \frac{\partial H(s)}{\partial s} = \lim_{s \to 0} \left(\frac{\partial H(s)}{\partial M} \frac{\partial M}{\partial s}\right) = 1$$

and the second moment of the weighting function is

$$m_{H2} = \lim_{s \to 0} \frac{\partial^2 H(s)}{\partial s^2} = \lim_{s \to 0} \left[\frac{\partial}{\partial M}\left(\frac{\partial H}{\partial M} \frac{\partial M}{\partial s}\right) \frac{\partial M}{\partial s}\right]$$

$$= \frac{1}{\text{Pe}^2}[\text{Pe}^2 - 2 + 2e^{-\text{Pe}} + 2\text{Pe}]$$

Thus for an arbitrary pulse response experiment, the relationship between the moments will be

$$A_y = A_g$$

$$m_{y1} = m_{g1} + 1$$

$$\sigma_{2y}^2 = \sigma_{2g}^2 + \frac{2}{\text{Pe}^2}(\text{Pe} - 1 + e^{\text{Pe}})$$

In other words, the area under the input and output pulses must be equal. The difference between the mean values of the input and output pulses must be

$$\theta = \frac{vt}{L} = 1$$

Also, if the variance of the input and output pulses are measured, the Peclet number, or the axial dispersion coefficient, $\text{Pe} = vL/D$, can be calculated from the equation

$$\sigma_{2y}^2 - \sigma_{2g}^2 = \frac{2}{\text{Pe}^2}(\text{Pe} - 1 + e^{-\text{Pe}})$$

Example 5.2-2 A comparison of lumped and distributed parameter models

In Example 4.3-2 we studied the cooling of an iron sphere in an infinite body of stagnant water and assumed that all the system capacitance was in the iron sphere and that all the resistance to heat transfer was in a stagnant layer of water surrounding the sphere; that is, we treated the problem as a lumped parameter system. A somewhat more rigorous solution of the problem

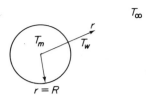

T_∞

Figure 5.2-4. Sphere in a bath.

can be obtained by including the heat conduction and capacitance of the water in the dynamic model, but neglecting the conduction in the sphere. For this case the system equations are

$$\frac{\partial T_w}{\partial t} = \frac{\alpha_w}{r^2} \frac{\partial}{\partial r}\left(r^2 \frac{\partial T_w}{\partial r}\right)$$

$$\rho_M C_{pM} V_M \frac{\partial T_M}{\partial t} = k_w A \frac{\partial T_w}{\partial r}\Big|_{r=R}$$

5.2-93

and the boundary conditions are

at $t = 0$: $T_M = T_0$ for $0 < r < R$, $T_w = T_\infty$ for $R < r < \infty$

at any time t: $T_M = T_w$ at $r = R$, $T_w = T_\infty$ at $r = \infty$

5.2-94

Letting

$$\theta_M = \frac{T_M - T_\infty}{T_0 - T_\infty}, \quad \theta_w = \frac{T_w - T_\infty}{T_0 - T_\infty}, \quad y = \frac{r}{R}$$

$$\tau = \frac{\alpha_w t}{R^2}, \qquad \beta = \frac{3\rho_w C_{pw}}{\rho_M C_{pM}}$$

5.2-95

the equations become

$$\frac{\partial \theta_w}{\theta \tau} = \frac{1}{y^2} \frac{\partial}{\partial y}\left(y^2 \frac{\partial \theta_w}{\partial y}\right) \qquad \frac{\partial \theta_M}{\theta \tau} = \beta \frac{\partial \theta_w}{\partial y}\Big|_{y=1}$$

5.2-96

and the boundary conditions become

at $\tau = 0$: $\theta_M = 1$ for $0 < y < 1$, $\theta_w = 0$ for $1 < y < \infty$

at any τ: $\theta_M = \theta_w$ at $y = 1$, $\theta_w = 0$ at $r = \infty$

5.2-97

Taking the Laplace transforms, we obtain

$$s\tilde{\theta}_w = \frac{1}{y^2} \frac{d}{dy}\left(y^2 \frac{d\tilde{\theta}_w}{dy}\right)$$

5.2-98

$$s\tilde{\theta}_M - 1 = \beta \frac{d\tilde{\theta}_w}{dy}\Big|_{y=1}$$

5.2-99

The solution of Eq. 5.2-98 is

$$\tilde{\theta}_w = \frac{c_1 e^{\sqrt{sy}} + c_2 e^{-\sqrt{sy}}}{y}$$

5.2-100

Since $\tilde{\theta}_w$ must approach zero as y approaches infinity, according to Eq. 5.2-97, we must have $c_1 = 0$. Thus

$$\tilde{\theta}_w = \frac{c_2 \exp\left(-\sqrt{sy}\right)}{y}$$

5.2-101

Then

$$\frac{d\tilde{\theta}_w}{dy} = -c_2 \exp{(-\sqrt{sy})}\left(\frac{\sqrt{sy}+1}{y^2}\right) \qquad 5.2\text{-}102$$

and since $\tilde{\theta}_M = \tilde{\theta}_w$ at $y = 1$, Eq. 5.2-99 becomes

$$sc_2 \exp{(-\sqrt{s})} - 1 = -\beta c_2 \exp{(-\sqrt{s})}(\sqrt{s}+1) \qquad 5.2\text{-}103$$

Hence

$$c_2 = \frac{1}{[s + \beta(\sqrt{s}+1)]}\exp{(\sqrt{s})} \qquad 5.2\text{-}104$$

and we find that

$$\tilde{\theta}_w = \frac{\exp{(-\sqrt{sy})}\exp{(\sqrt{s})}}{y[s + \beta(\sqrt{s}+1)]} \qquad 5.2\text{-}105$$

However, if we again use the result that $\tilde{\theta}_M = \tilde{\theta}_w$ at $y = 1$, we obtain

$$\tilde{\theta}_M = \frac{1}{s + \beta(\sqrt{s}+1)} \qquad 5.2\text{-}106$$

The inverse transform is

$$\tilde{\theta}_M = e^{-\beta\tau/2}I_0(\tfrac{1}{2}\sqrt{\beta^2 - 4\beta\tau}) \qquad 5.2\text{-}107$$

For the problem of interest,

$$\beta = \frac{3C_{pw}\rho_w}{C_{pM}\rho_M} = \frac{3(0.44)62.4}{(0.1178)450} = 1.554$$

$$\frac{\beta\tau}{2} = \frac{\beta\alpha_w t}{2R^2} = \frac{\beta k_w t}{2C_{pw}\rho_w R^2} = \frac{1.554(0.40)t}{2(0.44)62.4(2.18)^2} = 0.002375t$$

$$\frac{1}{2}\sqrt{\beta^2 - 4\beta\tau} = \frac{1}{2}\sqrt{2.42 - 6.216\frac{(0.4)t}{(0.44)(62.4)(2.18)^2}}$$

$$= (0.001411j)t$$

so that

$$\theta_M = \exp{(-0.002375t)}J_0(0.001411t)$$

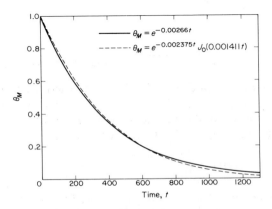

Figure 5.2-5. Cooling of a sphere.

This result and Eq. 4.3-57,

$$\theta = \exp(-0.00266t)$$

are plotted in Figure 5.2-5. It is apparent from the graph that no significant difference exists between the two models for this particular set of parameters.

SECTION 5.3 APPROXIMATE METHODS
FOR DYNAMIC SYSTEMS

Although we have spent a considerable amount of time discussing approximate models for dynamic systems—that is, the use of lumped parameter models to approximate distributed parameter processes and, in particular, the use of matrix differential equations as an approximation of any kind of system, a number of other techniques can be used to develop relatively simple mathematical representations of highly complicated plants. It is obvious that these approximation methods will have certain limitations. However, from a pragmatic standpoint we can say that any procedure that gives us some insight into the dynamic characteristics of a plant in a simple manner is extremely helpful, for generally we can use these results to make a judgment as to whether the system dynamics are negligible, whether we need to develop a very sophisticated dynamic model and conduct a careful study of the dynamic behavior, whether a crude dynamic model will provide sufficient information to serve our purpose, and so forth. Therefore some of the approximation methods that have proven to be very powerful tools in dynamic analysis are discussed below.

Reaction Curve Method

Experience has shown that the transient (step) response of most chemical processes has the general sigmoidal shape shown in Figure 5.3-1. This curve is

y

Time   **Figure 5.3-1.** Step response.

called the *Process Reaction Curve*[1,2] and can be generated either by a simple experiment on an actual operating unit or by a dynamic model describing that unit. Since the shape of the curve is characteristic of high-order systems, it is often convenient to approximate the output by a simpler equation such as

$$y(t) = \begin{cases} AK[1 - e^{-(t-t_d)/\tau}], & t \geq t_d \\ 0, & t < t_d \end{cases} \qquad 5.3\text{-}1$$

[1] J. G. Ziegler and N. B. Nichols, *Trans. ASME*, **64,** 759 (1942).
[2] G. H. Cohen and G. A. Coon, *Trans. ASME*, **75,** 827 (1953).

which corresponds to the transfer function

$$\frac{\tilde{y}}{\tilde{y}_f} = \frac{Ke^{-t_d s}}{\tau s + 1} \qquad\qquad 5.3\text{-}2$$

and a step input $\tilde{y}_f = A/s$.

The constants in this equation are determined by first drawing the tangent to the curve through the inflection point. Then the apparent dead time of the plant, t_d, is found as the intersection of this line and the time axis, and the system time constant τ is the difference between the value where the tangent curve reaches the final output, KA, and the dead time t_d. Our approximate model is simply a dead time plus an exponential rise, as shown in Figure 5.3-2. Although the approximation does not seem to provide a good fit of the actual plant output, the model is often used for the design of control systems.

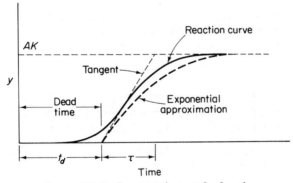

Figure 5.3-2. Constants in transfer function.

An approximate model that matches the actual system output better has the transfer function

$$y = \frac{Ke^{-t_d s}}{(\tau_1 s + 1)(\tau_2 s + 1)}, \qquad \tau_1 > \tau_2 \qquad 5.3\text{-}3$$

The dead time can be read from the graph immediately,[3] and the remainder of the curve is approximated by the transfer function of a second-order overdamped system. The slope-intercept method, the method of Harriott, the method of Oldenbourg and Sartorius, or the method of moments, as described in Section 4.5, can be used to estimate the time constants τ_1 and τ_2. For example, by plotting the tangent to the process reaction curve through the inflection point, we can determine the quantities τ_A and τ_B in Figure 5.3-3. Next we can use the graph prepared by Oldenbourg and Sartorius (Figure 4.5-19) to find τ_1 and τ_2.

---

[3] The asymptotic nature of the curve makes a precise determination of the dead time difficult. Moreover, the noise present in most experimental data adds a significant uncertainty to the measurement.

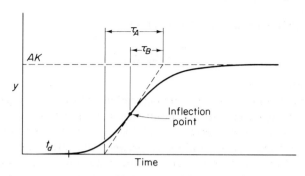

Figure 5.3-3. Constants in transfer function.

Example 5.3-1 Start-up of a plate, gas absorption unit

One start-up procedure for a plate absorption column was described in Example 4.6-3. Use the step response curve to develop an approximate transfer function for the column dynamics.

Solution

The step response of the system originally given in Figure 4.6-5 has been replotted in Figure 5.3-4 and the inflection point is shown. For the dead time plus first-order lag model given by Eq. 5.3-2, we find that

$$K = \frac{\text{final output}}{\text{input change}} = \frac{0.0662}{0.3} = 0.221 \qquad t_d = 3.65$$

$$\tau = 17.80 - 3.65 = 14.15$$

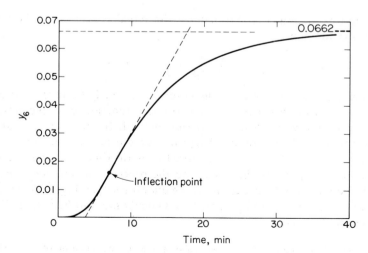

Figure 5.3-4. Step response of gas absorber.

so that

$$\tilde{y} = \frac{0.221e^{-3.65s}}{14.15s + 1}$$

For the second approximate model, given by Eq. 5.3-3, we see that

$$K = 0.221, \quad \tau_A = 14.15, \quad \tau_B = 17.80 - 7.10 = 10.70, \quad t_d = 1.4$$

Then $\tau_B/\tau_A = 0.739$; and plotting a straight line having a slope of minus unity through this point on Figure 4.5-19, we read the values

$$\frac{\tau_1}{\tau_A} = 0.5 \qquad \frac{\tau_2}{\tau_A} = 0.24$$

Hence

$$\tau_1 = 0.5(14.15) = 7.08 \qquad \tau_2 = 0.24(14.15) = 3.40$$

and

$$\tilde{y} = \frac{0.221e^{-1.4s}}{(7.08s + 1)(3.40s + 1)}$$

The Method of Moments

An introduction to the method of moments was presented in Section 5.2. However, in that discussion we were dealing with moments about the origin rather than moments about the mean. Thus we defined the moments as

$$m_n = \frac{\int_0^\infty t^n f(t)\, dt}{\int_0^\infty f(t)\, dt} = \frac{\int_0^\infty t^n f(t)\, dt}{A_f} \qquad \text{5.2-69}$$

and showed that we could express the Laplace transform of any function in terms of a power series in s

$$\tilde{f}(s) = A_f(1 - sm_1 + \tfrac{1}{2}s^2 m_2 - \tfrac{1}{6}s^3 m_3 + \cdots) \qquad \text{5.2-70}$$

The alternate procedure[4] is to define the moments about the mean as

$$\sigma_n = \frac{\int_0^\infty (t - \sigma_1)^n f(t)\, dt}{\int_0^\infty f(t)\, dt}, \qquad n > 2 \qquad \text{5.3-4}$$

where $\sigma_1 = m_1$ and $A_f = \int_0^\infty f(t)\, dt$. In addition, we can show that the Laplace transform of any function can be written

$$\tilde{f}(s) = \exp\left(\sigma_0 - \sigma_1 s + \tfrac{1}{2}\sigma_2 s^2 - \tfrac{1}{6}\sigma_3 s^3 + \cdots\right) = \int_0^\infty e^{-st} f(t)\, dt \qquad \text{5.3-5}$$

[4] H. M. Paynter, and Y. Takahashi, *Trans. ASME*, **78**, 749 (1956).

Since the Laplace transform is proportional to this moment-generating function, we find that

$$\lim_{s \to 0} \tilde{f}(s) = e^{\sigma_0} = \int_0^\infty f(t)\,dt = A_f \qquad 5.3\text{-}6$$

$$\lim_{s \to 0} \frac{d\tilde{f}(s)}{ds} = -\sigma_1 e^{\sigma_0} = -\int_0^\infty t(f)\,dt$$

or

$$\sigma_1 = \frac{\displaystyle\int_0^\infty t f(t)\,dt}{\displaystyle\int_0^\infty f(t)\,dt} = m_1 \qquad 5.3\text{-}7$$

Also

$$\lim_{s \to 0} \frac{d^2 f(s)}{ds^2} = \sigma_1^2 e^{\sigma_0} + \sigma_2 e^{\sigma_0} = \int_0^\infty t^2 f(t)\,dt$$

or

$$\sigma_2 = \frac{\displaystyle\int_0^\infty t^2 f(t)\,dt}{\displaystyle\int_0^\infty f(t)\,dt} - m_1^2 = m_2 - m_1^2 = \frac{\displaystyle\int_0^\infty (t - m_1)^2 f(t)\,dt}{\displaystyle\int_0^\infty f(t)\,dt} \qquad 5.3\text{-}8$$

Similarly,

$$\lim_{s \to 0} \frac{d^3 f(s)}{ds^3} = -\sigma_3 e^{\sigma_0} - \sigma_1^3 e^{\sigma_0} - 3\sigma_1 \sigma_2 e^{\sigma_0}$$

$$= -\int_0^\infty t^3 f(t)\,dt$$

or

$$\sigma_3 = \frac{\displaystyle\int_0^\infty t^3 f(t)\,dt}{\displaystyle\int_0^\infty f(t)\,dt} - 3 m_1 \sigma_2 - m_1^3 = m_3 - 3 m_1 \sigma_2 - m_1^3$$

$$= \frac{\displaystyle\int_0^\infty (t - m_1)^3 f(t)\,dt}{\displaystyle\int_0^\infty f(t)\,dt} \qquad 5.3\text{-}9$$

Thus the values of σ_i in the expansion given by Eq. 5.3-5 are just the moments about the mean.

Now if we consider a case where $\tilde{f}(s)$ is the transfer function of some process (i.e., the Laplace transform of the weighting function of the system), we can always approximate the transfer function by the expression

$$H(s) = \exp\left(\sigma_0 - \sigma_1 s + \tfrac{1}{2}\sigma_2 s - \tfrac{1}{6}\sigma_3 s^3 + \cdots\right) \qquad 5.3\text{-}10$$

We can evaluate the constants in this expression simply by evaluating the derivatives of the transfer function with respect to s and taking the limits of the derivatives as s approaches zero. The frequency response of the system can then be determined by substitution $j\omega$ for s and collecting terms to obtain

$$H(j\omega) = \exp\left[(\sigma_0 - \tfrac{1}{2}\sigma_2\omega^2 + \cdots) + j(-\sigma_1\omega + \tfrac{1}{6}\sigma_3\omega^3 - \cdots)\right] \quad 5.3\text{-}11$$

where the terms in the first group are all the even terms in the series of Eq. 5.3-10 and the second group contains all the odd terms. Remembering that $e^{\alpha + j\beta}$ is a complex number having an amplitude e^α and a phase angle β (see Eq. 5.1-12), we find that the amplitude ratio or system gain $|G|$ must be

$$|G| = \exp(\sigma_0 - \tfrac{1}{2}\sigma_2\omega^2 + \cdots) \qquad 5.3\text{-}12$$

and the phase angle is

$$\phi = -\sigma_1\omega + \tfrac{1}{6}\sigma_3\omega^3 - \cdots \qquad 5.3\text{-}13$$

These results can be plotted on a Bode diagram to give an approximate graphical representation of the system dynamics or substituted into Eq. 4.4-36 to obtain an analytical estimate of the steady state frequency response.

Example 5.3-2 Approximate model for frequency response

Find an approximate description of the frequency response of a series of first-order systems to flow rate variations.

Solution

In Section 4.6 we showed that the system transfer function when $\alpha = 1$ is

$$\tilde{f}(s) = \frac{\tilde{y}(s)}{\tilde{u}(s)} = \frac{\beta \sum_{r=0}^{n-1} (\tau s + 1)^r}{(\tau s + 1)^n} \qquad 4.6\text{-}6$$

Then

$$\lim_{s \to 0} \tilde{f}(s) = \beta n = e^{\sigma_0}$$

$$\lim_{s \to 0} \frac{d\tilde{f}}{ds} = \beta \frac{(n-1)n}{2} - \beta n^2 = -\sigma_1 e^{\sigma_0}$$

$$\lim_{s \to 0} \frac{d^2 \tilde{f}}{ds^2} = \left[\frac{(n-1)n(2n-1)}{6} - \frac{(n-1)n}{2}\right]$$

$$-2\beta n \frac{(n-1)n}{2} + \beta n(n+1)n = (\sigma_1^2 + \sigma_2)e^{\sigma_0}$$

$$\lim_{s \to 0} \frac{d^3 \tilde{f}}{ds^3} = \beta\left\{\left[\frac{(n-1)n}{2}\right]^2 - 3\left[\frac{(n-1)n(2n-1)}{6}\right]\right.$$

$$\left. + 2\left[\frac{(n-1)n}{2}\right]\right\} - 3\beta n\left[\frac{(n-1)n(2n-1)}{6} - \frac{(n-1)n}{2}\right]$$

$$+ 3\beta n(n+1)\frac{(n-1)n}{2} - \beta n(n+1)(n+2)n$$

$$= -(\sigma_3 + \sigma_1^3 + 3\sigma_1\sigma_2)e^{\sigma_0}$$

We can solve these four equations and determine $\sigma_0, \sigma_1, \sigma_2,$ and σ_3. Thus, from Eq. 5.3-12, the system gain is approximately

$$|G| = \exp(\sigma_0 - \tfrac{1}{2}\sigma_2\omega^2) = e^{\sigma_0}\exp(-\tfrac{1}{2}\sigma_2\omega^2)$$

and from Eq. 5.3-13, the phase angle is approximately

$$\phi = -\sigma_1\omega + \tfrac{1}{6}\sigma_3\omega^3$$

Friedly's Method

In the reaction curve technique we attempted to fit very simple models involving one or two time lags and a pure time delay to the transient response of the system. The transient response information is obtained either from an experimental study on the actual plant or by solving the theoretical equations describing the system for the step response. Although step response experiments normally are fairly simple, it is often a difficult task to obtain an analytical expression for the transient output of a complicated plant even for cases where the system transfer function can be developed in a straightfoward fashion. The difficulty arises because fluid flow is frequently the primary means of transport in the system, and the simple assumption of plug flow usually leads to complicated transcedental expressions for the transfer functions. A number of investigators have attempted to develop simple approximations for these complicated transcendental relationships. However, a very general method that appears to have great promise was published by Friedly.[5]

In Friedly's method, a transfer function involving one or two lags plus delays

$$\tilde{G} = \left(\frac{K}{1 + \tau_1 s}\right)[1 - e^{-(\alpha + \beta s)}], \qquad \text{Countercurrent System} \qquad 5.3\text{-}14$$

$$\tilde{G} = \frac{K}{(1 + \tau_1 s)(1 + \tau_2 s)}[1 - e^{-(\alpha + \beta s)}], \qquad \begin{array}{l}\text{Countercurrent or}\\ \text{Complex Flow System}\end{array} \qquad 5.3\text{-}15$$

$$\tilde{G} = \left(\frac{K}{1 + \tau_1 s}\right)(e^{-\alpha s} - e^{-\beta s}), \qquad \text{Cocurrent System} \qquad 5.3\text{-}16$$

is chosen to represent the system, and the constants in these models are selected so that the asymptotic behavior of the actual transfer function and the approximate model are identical. His results show that when the approximation is forced to fit the exact equations both in the high- and low-frequency (short- and long-time) regions, then it generally provides a fairly good fit in the intermediate range. It should be noted, however, that his approach is limited to plug flow systems, where the dynamic behavior can be interpreted in terms of traveling waves, but it gives results that are equally applicable

---

[5] J. C. Friedly, *Proceedings of JACC*, p. 216, Philadelphia, Pa., June 1967.

in the time and frequency domains. Two examples of his procedure are given below.

Example 5.3-3 Response of a countercurrent heat exchanger

The equations describing a countercurrent heat exchanger are

$$\rho_1 C_{P1} A_1 \frac{\partial T_1}{\partial t} + \rho_1 C_{p1} A_1 v_1 \frac{\partial T_1}{\partial x} = UA_H(T_2 - T_1) \qquad \text{5.3-17}$$

$$\rho_2 C_{P2} A_2 \frac{\partial T_2}{\partial t} - \rho_2 C_{p2} A_2 v_2 \frac{\partial T_2}{\partial x} = UA_H(T_1 - T_2) \qquad \text{5.3-18}$$

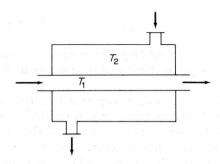

Figure 5.3-5. Countercurrent heat exchanger.

By setting the accumulation terms equal to zero and solving the remaining pair of ordinary differential equations, we can find the steady state profiles, $T_{1s}(x)$ and $T_{2s}(x)$ in the exchanger. Letting

$$\theta_1 = T_1 - T_{1s}(x), \quad \theta_2 = T_2 - T_{2s}(x), \quad a_1 = \frac{UA_H L}{\rho_1 C_{p1} A_1 v_1}$$

$$a_2 = \frac{UA_H L}{\rho_2 C_{p2} A_2 v_2} \qquad r = \frac{v_1}{v_2} \qquad \text{5.3-19}$$

$$\tau = \frac{v_1 t}{L}, \qquad z = \frac{x}{L}$$

the equations can be put into the form

$$\frac{\partial \theta_1}{\partial \tau} + \frac{\partial \theta_1}{\partial z} = a_1(\theta_1 - \theta_2) \qquad \text{5.3-20}$$

$$r \frac{\partial \theta_2}{\partial \tau} - \frac{\partial \theta_2}{\partial z} = a_2(\theta_2 - \theta_1) \qquad \text{5.3-21}$$

The equations describing other countercurrent equipment (packed absorbers, extraction columns, etc.,) can also be put into this form (see Example 3.2-3 and Section 5.1). The system transfer function is obtained by taking the Laplace transforms of the preceding equations, solving the resulting pair of ordinary differential equations, and evaluating the integration constants from a knowledge of the variations of the inputs at the opposite ends of the exchanger, $\theta_1(0, t)$ and $\theta_2(1, t)$. Clearly, we can develop four different transfer functions in this way—that is, the two outlet temperatures in terms of each of the inlet temperatures—but we will only consider a modification of one of these cases:

$$
\begin{aligned}
G(z, s) &= \frac{\tilde{\theta}_1(z, s)}{\tilde{\theta}_2(1, s)} \\
&= \frac{a_1 \exp\{-[(a_1 - a_2)/2 + (1 - r)s/2]\}(z - 1)[e^{q(z-1)} - e^{-q(z+1)}]}{(p + q)\{1 - [(a_1 a_2)/(p + q)^2]e^{-2q}\}}
\end{aligned}
$$

5.3-22

where

$$
p = \frac{1 + r}{2}s + \frac{a_1 + a_2}{2} \quad \text{and} \quad q = \sqrt{p^2 - a_1 a_2}
$$

5.3-23

Thus we have obtained the transfer function relating the change in the tube temperature at any point in the exchanger to changes in the inlet shell temperature.

If we are interested in determining the system-weighting function, we must find some way of inverting the foregoing transform expression. However, doing so will not be a simple task, since we must first find all the roots of the transcendental function in the denominator before applying one or more of the Heaviside expansion theorems. Thus we would like to develop a simpler approximate transfer function that can be inverted more easily. Before undertaking this analysis, however, we will attempt to make some qualitative statements about the nature of the response so that we can ensure that our approximate model will appropriately describe the important dynamic features of the system.

When a step disturbance in the shell inlet temperature enters the exchanger and travels to the left at a relative velocity $1/r$, it will cause a change in the tube temperature as it travels along. After this initial wave front passes out of the exchanger, the entering fluid on the tube side will always encounter shell fluid corresponding to the new inlet condition, and therefore there will be a different driving force for heat transfer. In other words, as the initial wave in the different shell temperature leaves the exchanger, it will seem as if this wave is reflected in the tube temperature. Similarly, after this tube fluid reaches the right-hand side of the exchanger at time $1 + r$, there will appear to be another wave reflection in the shell temperature, for the tube fluid characteristics will be different from the initial case. Hence the effluent tube temperature should indicate the presence of a number of reflected waves inside the exchanger at time intervals of $1 + r$, but the amplitude of these waves should be damped as time increases because of the heat-transfer process. If the damping is large enough, the initial change and the first reflected wave should provide an adequate approximation of the system response. Friedly's method is based on this kind of reflected wave behavior.

It should be noted that this is quite different from a lumped parameter approximation, which can never exhibit wave behavior unless the number of stages approaches infinity.

The asymptotic nature of the system transfer function, Eq. 5.3-22, as s approaches zero is simple to determine. It is clear that in this limit

$$p = \frac{a_1 + a_2}{2}, \quad q = \frac{a_1 - a_2}{2}, \quad p + q = a_1 \qquad 5.3\text{-}24$$

Thus the transfer function approaches the limit

$$G = \frac{a_1 \exp\left[-(a_1 - a_2)(z - 1)/2\right]}{a_1[1 - (a_1 a_2/a_1^2)e^{-(a_1 - a_2)}]}\{\exp\left[(a_1 - a_2)(z - 1)/2\right]$$

$$- \exp\left[-(a_1 - a_2)(z + 1)/2\right]\}$$

or

$$G = a_1 \left[\frac{e^{(a_2 - a_1)z} - 1}{a_2 e^{(a_2 - a_1)} - a_1} \right] \qquad 5.3\text{-}25$$

The upper asymptote is not as apparent. However, we see that as s approaches infinity,

$$p = \frac{1 + r}{2} s \qquad 5.3\text{-}26$$

so that p also approaches infinity. As a result, we find that q is approximately given by the expression

$$q = \sqrt{p^2 - a_1 a_2} \simeq p\left(1 - \frac{a_1 a_2}{2p^2} + \cdots\right) \qquad 5.3\text{-}37$$

and

$$p + q \simeq 2p - \frac{a_1 a_2}{2p} + \cdots \qquad 5.3\text{-}28$$

As p approaches infinity, q must approach infinity, so that the term in braces in the denominator of the transfer function, Eq. 5.3-22, approaches unity—that is, e^{-2q} approaches zero faster than $(p + q)^2$ approaches infinity. Therefore we can approximate the denominator by the expansion

$$\frac{1}{1 - [a_1 a_2/(p + q)^2]e^{-2q}} \simeq 1 + \frac{a_1 a_2}{(p + q)^2}e^{-2q} + \cdots$$

$$= 1 + \frac{a_1 a_2}{(2p + \cdots)^2}e^{-2q} + \cdots \qquad 5.3\text{-}29$$

If we neglect all terms of order $1/p$ and higher and substitute the preceding expressions into the transfer function, we obtain

$$G = \frac{a_1}{2p} \exp\left[-\left(\frac{a_1 - a_2}{2} + \frac{1 - r}{2}s\right)(z - 1)\right][e^{(z-1)p} - e^{-(z+1)p}] \qquad 5.3\text{-}30$$

Now, substituting Eq. 5.3-26 for p in the denominator but inserting the original definition of p (Eq. 5.3-23) in the numerator and manipulating the result somewhat, we find that

$$G = \frac{a_1}{(1+r)s}e^{a_2(z-1)}[e^{-rs(1-z)} - e^{-(a_1+a^2)z}e^{-(z+r)s}] \qquad 5.3\text{-}31$$

The original definition for p is retained in the numerator, for the term of order s^0 provides the scale factor for the effect of the reflected wave.

It now becomes apparent that we should choose our approximate model for the transfer function similar to Eq. 5.3-16, except that α and β correspond to the values in Eq. 5.3-31,

$$G_{approx} = \frac{K}{(1+\tau_1 s)}[e^{-rs(1-z)} - e^{-(a_1+a_2)z}e^{-(z+r)s}] \qquad 5.3\text{-}32$$

At very large values of s, this becomes

$$G_{approx} = \frac{K}{\tau_1 s}[e^{-rs(1-z)} - e^{-(a_1+a_2)z}e^{-(z+r)s}]$$

Therefore, in order for it to match the result of the exact equation, we must have

$$\frac{K}{\tau_1} = \frac{a_1}{(1+r)}e^{a_2(z-1)} \qquad 5.3\text{-}33$$

We can obtain another relationship for the constant K by matching the low-frequency asymptotes. As s approaches zero, the approximate model becomes

$$G_{approx} = K[1 - e^{-(a_1+a_2)z}] = G \qquad 5.3\text{-}34$$

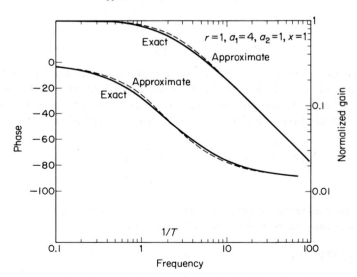

Figure 5.3-6. Frequency response of a heat exchanger. Reproduced from Friedly, J. C., "Proceedings of JACC," p. 216, Philadelphia, Pa., June 1967, by permission.

so that from Eq. 5.3-25 we find that

$$K = \left[\frac{a_1}{1 - e^{-(a_1+a_2)z}}\right]\left[\frac{e^{(a_2-a_1)z} - 1}{a_2 e^{(a_2-a_1)} - a_1}\right] \qquad \text{5.3-35}$$

Thus we can find both the unknown constants K and τ_1 in the approximate transfer function in terms of the known system parameters.

Friedly compared the Bode plots of the exact and approximate transfer functions for a case where $a_1 = 4$, $a_2 = 1$, $r = 1$, and $z = 1$, and found excellent agreement (see Figure 5.3-6). The maximum discrepancy was about

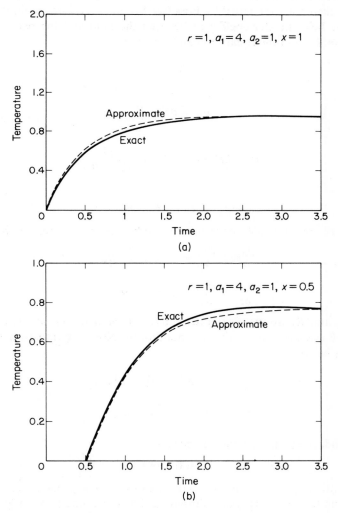

Figure 5.3-7. Step response of a heat exchanger: (a) at $z = 0.5$, (b) at $z = 1.0$. Reproduced from Friedly, J. C., "Proceedings of JACC," p. 216, Philadelphia, Pa., June 1967, by permission.

10 percent in the neighborhood of the break frequency. Also, he calculated the step response for both models at the middle of the exchanger, $z = 0.5$, and at the end, $z = 1$, and again obtained a good agreement (see Figure 5.3-7). When the approach was extended to more complex systems, the aproximate transfer function still provides a much better fit than lumped parameter models.

Example 5.3-4 Response of a packed reactor

Friedly also considered an approximate model for the dynamic response of the packed reactor studied by Foss (see Section 5.1). For the limiting case where the heat-transfer coefficient between the fluid and the packing is infinite, $T = T_p$, and the equations can be written

$$\frac{\partial c}{\partial t} + \frac{\partial c}{\partial z} = -R \qquad\qquad 5.3\text{-}36$$

$$r\frac{\partial \theta}{\partial t} + \frac{\partial \theta}{\partial z} = R \qquad\qquad 5.3\text{-}37$$

where

$$R = 2K_R c(c - \delta) \exp\left(\frac{A\theta}{\theta_0 + \theta}\right) \qquad\qquad 5.3\text{-}38$$

Letting

$$c = c_s + x_1 \qquad \theta = \theta_s + x_2 \qquad\qquad 5.3\text{-}39$$

and linearizing the equations around a steady state profile, we obtain

$$\frac{\partial x_1}{\partial t} + \frac{\partial x_1}{\partial z} = -R_c x_1 - R_\theta x_2 \qquad\qquad 5.3\text{-}40$$

$$r\frac{\partial x_2}{\partial t} + \frac{\partial x_2}{\partial z} = R_c x_1 + R_\theta x_2 \qquad\qquad 5.3\text{-}41$$

where R_c and R_θ are the partial derivatives of the reaction rate with respect to c and θ evaluated at steady state donditions. Now, the transformed equations can be written as

$$\frac{d\tilde{x}_1}{dz} = -(R_c + s)\tilde{x}_1 - R_\theta \tilde{x}_2 \qquad\qquad 5.3\text{-}42$$

$$\frac{d\tilde{x}_2}{dz} = R_c \tilde{x}_1 - (rs - R_\theta)\tilde{x}_2 \qquad\qquad 5.3\text{-}43$$

The boundary conditions considered are

$$\text{at } z = 0, \qquad \tilde{x}_1 = 0 \quad \text{and} \quad \tilde{x}_2 = \tilde{x}_{2f} \qquad\qquad 5.3\text{-}44$$

The low-frequency asymptote—the solution as s approaches zero—must correspond to the steady state solution; and as was noted previously this can be determined analytically in terms of exponential integrals. In order to determine the high-frequency asymptote—the limiting condition as s approaches infinity—we first eliminate the temperature variable from the set of equations. Solving Eq. 5.3-42 for $\tilde{x}_2$.

$$R_\theta \tilde{x}_2 = -\frac{d\tilde{x}_1}{dz} - (R_c + s)\tilde{x}_1 \qquad 5.3\text{-}45$$

and differentiating with respect to z gives

$$R'_\theta \tilde{x}_2 + R_\theta \frac{d\tilde{x}_2}{dz} = -\frac{d^2\tilde{x}_1}{dz^2} - R'_c \tilde{x}_1 - (R_c + s)\frac{d\tilde{x}_1}{dz} \qquad 5.3\text{-}46$$

where R'_θ and R'_c are the derivatives with respect to z. Substituting Eq. 5.3-45 for $\tilde{x}_2$ and rearranging leads to

$$R_\theta \frac{d\tilde{x}_2}{dz} = -\frac{d^2\tilde{x}_1}{dz^2} - \left(s + R_c - \frac{R'_\theta}{R_\theta}\right)\frac{d\tilde{x}_1}{dz} + \left[(s + R_c)\frac{R'_\theta}{R_\theta} - R'_c\right]\tilde{x}_1 \qquad 5.3\text{-}47$$

Now, eliminating $d\tilde{x}_2/dz$ and $\tilde{x}_2$ from 5.4-43, using Eqs. 5.3-45 and 5.3-47, we obtain after some manipulation

$$\frac{d^2\tilde{x}_1}{dz^2} + \left[(1 + r)s + \left(R_c - R_\theta - \frac{R'_\theta}{R_\theta}\right)\right]\frac{d\tilde{x}_1}{dz}$$
$$+ \left[rs^2 + s\left(rR_c - R_\theta - \frac{R'_\theta}{R_\theta}\right) + R'_c\right]\tilde{x}_1 = 0 \qquad 5.3\text{-}48$$

The solution of this equation with the boundary conditions given by Eq. 5.3-44 is the system transfer function. Since the coefficients in the equation depend on the position in the reactor, z, however, normally it is difficult to find a solution of the equation. Fortunately, the asymptotic method is not hampered by this difficulty. Thus, as s approaches infinity, the equation reduces to the form

$$\frac{d^2\tilde{x}_1}{dz^2} + (1 + r)s\frac{d\tilde{x}_1}{dz} + rs^2\tilde{x}_1 = 0 \qquad 5.3\text{-}49$$

which has the solution

$$\tilde{x}_1 = c_1 e^{-sz} + c_2 e^{-rsz} \qquad 5.3\text{-}50$$

Evaluating the integration constants c_1 and c_2 from the boundary conditions, we find that the asymptotic transfer function is

$$\frac{\tilde{x}_1}{\tilde{x}_{2f}} = \frac{R_\theta(0)}{(r - 1)s}(e^{-sz} - e^{-rsz}) \qquad 5.3\text{-}51$$

Then if we select an approximate transfer function having the same form as Eq. 5.3-16 but the delay term matching those in the preceeding expression, we have

$$G_{\text{approx}} = \frac{K}{(1 + \tau_1 s)}(e^{-sz} - e^{-rsz}) \qquad 5.3\text{-}52$$

By matching the high-frequency asymptotes we find that

$$\frac{K}{\tau_1} = \frac{R_\theta(0)}{r - 1} \qquad 5.3\text{-}53$$

The parameter K can be determined by comparing the low-frequency asymptotes or the steady state solutions. For the case where $\tilde{x}_{2f}$ is a step function, x_{2f}/s, the inverse of the approximate transfer function is

$$x_1 = \begin{cases} 0, & t < z \\ -Kx_{2f}\left\{1 - \exp\left[-\frac{(t - z)}{\tau_1}\right]\right\} & z < t < rz \\ -Kx_{2f}\left\{1 - \exp\left[-\frac{(r - 1)z}{\tau_1}\right]\right\} & rz < t \end{cases} \qquad 5.3\text{-}54$$

Hence if x_{1f} is the final steady state concentration following a step change of x_{2f}, K must be

$$K = -\frac{x_{1f}}{x_{2f}}\left\{\frac{1}{1 - \exp\left[-(r - 1)z/\tau_1\right]}\right\} \qquad 5.3\text{-}55$$

Thus we have determined both constants in the approximate transfer function, Eq. 5.3-52.

The Matrizant

In our study of distributed parameter systems in Section 3.3, we found that when we linearized a set of partial differential equations around a steady state profile of interest, we generally obtained a set of linear, partial differential equations with variable coefficients. Next we replaced the spatial derivatives, using a finite-difference approximation, and in this way we obtained a larger set of linear, ordinary differential equations with constant coefficients. However, if we had taken the Laplace transform of the linearized equations instead of introducing the finite-difference approximation, we would have obtained a set of linear, ordinary differential equations with variable coefficients. In some cases it is possible to develop an approximate solution of this set of equations by evaluating the matrizant.

For example, if we consider a pair of equations with variable coefficients

$$\frac{dy_1}{dt} = a_{11}(t)y_1 + a_{12}(t)y_2 \qquad 5.3\text{-}56$$

$$\frac{dy_2}{dt} = a_{21}(t)y_1 + a_{22}(t)y_2 \qquad 5.3\text{-}57$$

where the independent variable t might correspond to time, distance from a reactor inlet, etc., or the more general set of matrix equations

$$\frac{d\mathbf{y}}{dt} = \mathbf{A}(t)\mathbf{y} \qquad 5.3\text{-}58$$

we might attempt to find a solution simply by integrating

$$\mathbf{y} = \mathbf{y}_0 + \int_0^t \mathbf{A}(\tau)\mathbf{y}\, d\tau \qquad \text{5.3-59}$$

If we now substitute this solution into the argument of the integral, we obtain

$$\mathbf{y} = \mathbf{y}_0 + \int_0^t \mathbf{A}\left(\mathbf{y}_0 + \int_0^{\tau_1} \mathbf{A}\mathbf{y}\, d\tau_2\right) d\tau_1 \qquad \text{5.3-60}$$

or

$$y = \left(\mathbf{I} + \int_0^t \mathbf{A}\, d\tau_1 + \int_0^t \mathbf{A} \int_0^{\tau_1} \mathbf{A}\mathbf{y}\, d\tau_2\, d\tau_1\right)\mathbf{y}_0 \qquad \text{5.3-61}$$

where $\mathbf{I}$ is the identity matrix. Continuing this procedure gives

$$\mathbf{y} = \left(\mathbf{I} + \int_0^t \mathbf{A}\, d\tau_1 + \int_0^t \mathbf{A} \int_0^{\tau_1} \mathbf{A}\, d\tau_2\, d\tau_1 + \right.$$
$$\left. \int_0^t \mathbf{A} \int_0^{\tau_1} \mathbf{A} \int_0^{\tau_2} \mathbf{A}\, d\tau_3\, d\tau_2\, d\tau_1 + \cdots\right)\mathbf{y}_0 \qquad \text{5.3-62}$$

or

$$\mathbf{y} = \mathbf{\Omega}_0^t(\mathbf{A})\mathbf{y}_0 \qquad \text{5.3-63}$$

where $\mathbf{\Omega}_0^t$ is called the matrizant.

It is a simple matter to show by direct differentiation that

$$\frac{d\mathbf{\Omega}_0^t(\mathbf{A})}{dt} = \mathbf{A}\mathbf{\Omega}_0^t(\mathbf{A}) \qquad \text{5.3-64}$$

Also, since

$$\mathbf{y}(t_1) = \mathbf{\Omega}_0^{t_1}(\mathbf{A})\mathbf{y}_0 \qquad \text{5.3-65}$$

and

$$\mathbf{y}(t) = \mathbf{\Omega}_{t_1}^t(\mathbf{A})\mathbf{y}(t_1) \qquad \text{5.3-66}$$

then

$$\mathbf{\Omega}_0^t = \mathbf{\Omega}_{t_1}^t(\mathbf{A})\mathbf{\Omega}_0^{t_1}(\mathbf{A}) \qquad \text{5.3-67}$$

In other words, if we divide the interval from zero to t into a number of smaller intervals $(0, t_1), (t_1, t_2), \ldots, (t_{s-1}, t)$, we can write

$$\mathbf{\Omega}_0^t(\mathbf{A}) = \mathbf{\Omega}_{t_{s-1}}^t(\mathbf{A})\mathbf{\Omega}_{t_{s-2}}^{t_{s-1}}(\mathbf{A})\ldots \mathbf{\Omega}_0^{t_1}(\mathbf{A}) \qquad \text{5.3-68}$$

Hopefully, if we choose small enough intervals, we can assume that the coefficient matrix is approximately constant at some average value over each small interval. For the case where $\mathbf{A}$ is identically a constant,

$$\mathbf{\Omega}_0^t(\mathbf{A}) = \mathbf{I} + \mathbf{A}\int_0^t d\tau_1 + \mathbf{A}^2 \int_0^t d\tau_1 \int_0^{\tau_1} d\tau_2$$
$$+ \mathbf{A}^3 \int_0^t d\tau_1 \int_0^{\tau_1} d\tau_2 \int_0^{\tau_2} d\tau_3 + \cdots \qquad \text{5.3-69}$$

or

$$\boldsymbol{\Omega}_0^t(\mathbf{A}) = \mathbf{I} + \mathbf{A}t + \frac{1}{2!}\mathbf{A}^2 t^2 + \frac{1}{3!}\mathbf{A}^3 t^3 + \cdots \qquad 5.3\text{-}70$$

$$= \exp{(\mathbf{A}t)} \qquad 5.3\text{-}71$$

Hence if we let $\mathbf{A}_n$ be the average value of the $\mathbf{A}$ matrix over the nth interval, we can write

$$\boldsymbol{\Omega}_{t_{n-1}}^{t_n}(\mathbf{A}) \simeq \boldsymbol{\Omega}_{t_{n-1}}^{t_n}(A_n) = \exp{(A_n\,\Delta t)} \qquad 5.3\text{-}72$$

where $\Delta t = t_n - t_{n-1}$. Thus the complete solution becomes

$$\mathbf{y} = \boldsymbol{\Omega}_0^t(\mathbf{A})\mathbf{y}_0 \qquad 5.3\text{-}73$$

$$= [\exp \mathbf{A}_n(t_n - t_{n-1})](\exp \mathbf{A}_{n-1}\,\Delta t)\cdots(\exp \mathbf{A}_1\,\Delta t)\mathbf{y}_0 \qquad 5.3\text{-}74$$

Now if we choose the average value of the $\mathbf{A}$ matrix to be an integral average, for each element in the matrix we write

$$(a_n)_{ij} = \frac{1}{\Delta t}\int_{t_{n-1}}^{t_n} a_{ij}(t)\,dt \qquad 5.3\text{-}75$$

and the approximate solution of the system equations becomes

$$\mathbf{y}(t) = \exp\left(\frac{1}{t}\int_0^t \mathbf{A}\,d\tau\right)\mathbf{y}_0 \qquad 5.3\text{-}76$$

Example 5.3-5 Response of a nonisothermal tubular reactor

In order to illustrate the application of the matrizant,[6] we will consider the dynamic model of a nonisothermal tubular reactor derived in Example 3.3-3. The original system equations were

$$\frac{\partial x}{\partial t} + v\frac{\partial x}{\partial z} = -kx \qquad 5.3\text{-}77$$

$$\frac{\partial T}{\partial t} + v\frac{\partial T}{\partial z} = \frac{2h}{C_p\rho r}(T_w - T) + \frac{(-\Delta H)}{C_p\rho}kx \qquad 5.3\text{-}78$$

After linearizing the equations around a steady state profile for the case where the velocity v and the wall temperature T_w are maintained constant, we obtained

$$\frac{\partial y_c}{\partial t} + v\frac{\partial y_c}{\partial z} = -(k_s)y_c - \left(\frac{E}{RT_s^2}k_s x_s\right)y_T \qquad 5.3\text{-}79$$

$$\frac{\partial y_T}{\partial t} + v\frac{\partial y_T}{\partial z} = \left(\frac{(-\Delta H)k_s}{C_p\rho}\right)y_c - \left[\frac{2h}{C_p\rho r} - \frac{(-\Delta H)Ek_s x_s}{C_p\rho RT_s^2}\right]y_T \qquad 5.3\text{-}80$$

where y_c and y_T are the deviations of the composition and temperature from their steady state values. Assuming that the system is at steady state initially,

[6] Following the procedure described by O. Bilous and N. R. Amundson, *AIChE Journal*, **2**, 117 (1956).

but at time zero some disturbance in feed composition y_{cf} and feed temperature y_{Tf} enter the system, the Laplace transforms of the preceding equations can be written

$$\frac{d\tilde{y}_c}{dz} = -\left(\frac{k_s + s}{v}\right)\tilde{y}_c - \left(\frac{Ek_sx_s}{RT_s^2v}\right)\tilde{y}_T \qquad \tilde{y}_c(0) = \tilde{y}_{cf} \qquad 5.3\text{-}81$$

$$\frac{d\tilde{y}_T}{dz} = \left(\frac{(-\Delta H)k_s}{C_p\rho v}\right)\tilde{y}_c - \left[\frac{2h}{C_p\rho rv} - \frac{(-\Delta H)Ek_sx_s}{C_p\rho v_sRT_s^2} + \frac{s}{v}\right]\tilde{y}_T \qquad \tilde{y}_T(0) = \tilde{y}_{Tf}$$

$$5.3\text{-}82$$

which is a pair of linear ordinary differential equations having variable coefficients—that is, x_s, T_s, and k_s, vary with length. Letting

$$\mathbf{A}(z) =$$

$$\begin{pmatrix} -\dfrac{(k+s)}{v} & -\dfrac{Ek_sx_s}{RT_s^2v} \\[3mm] \dfrac{(-\Delta H)k_s}{C_p\rho} & -\left[\dfrac{2h}{C_p\rho rv} - \dfrac{(-\Delta H)Ek_sx_s}{C_p\rho vRT_s^2} + \dfrac{s}{v}\right] \end{pmatrix}, \quad \tilde{\mathbf{y}} = \begin{pmatrix} \tilde{y}_c \\ \tilde{y}_T \end{pmatrix}, \quad \tilde{\mathbf{y}}_f = \begin{pmatrix} \tilde{y}_{cf} \\ \tilde{y}_{Tf} \end{pmatrix}$$

$$5.3\text{-}83$$

we can write the equations in matrix form

$$\frac{d\tilde{\mathbf{y}}}{dz} = \mathbf{A}(z)\tilde{\mathbf{y}} \qquad 5.3\text{-}84$$

The solution of this equation is

$$\tilde{\mathbf{y}} = \mathbf{\Omega}_0^L(\mathbf{A})\tilde{\mathbf{y}}_f \qquad 5.3\text{-}85$$

and an approximate solution may be written as

$$\tilde{\mathbf{y}} = \left(\exp\int_0^L \mathbf{A}(z)\,dz\right)\tilde{\mathbf{y}}_f = e^{\mathbf{M}}\tilde{\mathbf{y}}_f \qquad 5.3\text{-}86$$

where

$$\mathbf{M} = -\begin{pmatrix} \displaystyle\int_0^L \frac{k_s + s}{v}\,dz & \displaystyle\int_0^L \frac{Ek_sx_s}{RT_s^2v} \\[4mm] \displaystyle\int_0^L \frac{(-\Delta H)k_s}{C_p\rho v}\,dz & \displaystyle\int_0^L\left[\frac{2h}{C_p\rho rv} - \frac{(-\Delta H)Ek_sx_s}{C_p\rho vRT_s^2} + \frac{s}{v}\right]dz \end{pmatrix} \qquad 5.3\text{-}87$$

If we partition this matrix and write

$$\mathbf{M} = \frac{sL}{v}\mathbf{I} + \mathbf{N} \qquad 5.3\text{-}88$$

where

$$\mathbf{N} = -\begin{pmatrix} \displaystyle\int_0^L \frac{k_s}{v}\,dz & \displaystyle\int_0^L \frac{Ek_sx_s}{RT_s^2v}\,dz \\[4mm] \displaystyle\int_0^L \frac{(-\Delta H)k_s}{C_p\rho v}\,dz & \displaystyle\int_0^L\left[\frac{2h}{C_p\rho rv} - \frac{(-\Delta H)Ek_sx_s}{C_p\rho vRT_s^2}\right]dz \end{pmatrix} \qquad 5.3\text{-}89$$

we can evaluate all the integrals in the **N** matrix numerically (this step is necessary, for the steady state composition and temperature profiles, $x_s(z)$ and $T_s(z)$, must be determined numerically), and we can write the solution as

$$\tilde{y} = \exp\left[-\frac{sL}{v}\mathbf{I} + \mathbf{N}\right]\tilde{\mathbf{y}}_f \qquad 5.3\text{-}90$$

For this simple problem, it is possible to eliminate the matrix exponentials by using the identity

$$e^{\alpha \mathbf{I}} = e^{\alpha}\mathbf{I} \qquad 5.3\text{-}91$$

and Sylvester's theorem

$$\exp\left[\begin{pmatrix} A & B \\ C & D \end{pmatrix}\right] = \exp\left[\frac{A+D}{2}\right]\left(\frac{\sinh \Delta}{\Delta}\right)\begin{pmatrix} \dfrac{A-D}{2} + \dfrac{\Delta}{\tanh \Delta} & B \\ C & \dfrac{D-A}{2} + \dfrac{\Delta}{\tanh \Delta} \end{pmatrix}$$

$$= \begin{pmatrix} b_{11} & b_{12} \\ b_{21} & b_{22} \end{pmatrix} \qquad 5.3\text{-}92$$

where

$$2\Delta = \sqrt{(A - D)^2 + 4BC} \qquad 5.3\text{-}93$$

Hence the solution can be put into the form

$$\tilde{\mathbf{y}} = e^{-sL/v}\begin{pmatrix} b_{11} & b_{12} \\ b_{21} & b_{22} \end{pmatrix}\tilde{\mathbf{y}}_f = \mathbf{H}(s)\tilde{\mathbf{y}}_f \qquad 5.3\text{-}94$$

where $\mathbf{H}(s)$ is the transfer matrix of the plant.

Bilous and Amundson carried out some numerical calculations for this system for a single irreversible reaction, $A \longrightarrow B$, using the parameters $E = 22{,}500$, $k_0 = 3.94 \times 10^{12}$, $x_{fs} = 0.2$, $T_{fs} = 340\,°K$, $2h/C_p\rho r = 0.20$, $(-\Delta H)/C_p\rho = 7300$, and for two different wall temperatures, $T_{w1} = 335\,°K$ and $T_{w2} = 337.5\,°K$. The steady state profiles were determined numerically and are given in Table 5.3-1. The integrals required in the solution are

$$K = \frac{1}{v}\int_0^L k_s\, dz = -\frac{1}{v}\int_0^L \frac{v}{x_s}\, dx_s = \ln\frac{x_{fs}}{x} \qquad 5.3\text{-}95$$

and

$$J = \frac{1}{v}\int_0^L \frac{k_s E x_s}{RT_s^2}\, dz = \int_x^{x_f} \frac{E\, dx_s}{RT_s^2} \qquad 5.3\text{-}96$$

The values of these integrals for the profiles given in Table 5.3-1 are listed in Table 5.3-2.

With these results, the matrix coefficients can be written as

$$\begin{pmatrix} A & B \\ C & D \end{pmatrix} = \begin{pmatrix} -K & -J \\ \dfrac{(-\Delta H)K}{C_p\rho} & -\left(\dfrac{2hL}{C_p\rho rv} - \dfrac{(-\Delta H)J}{C_p\rho}\right) \end{pmatrix} \qquad 5.3\text{-}97$$

so that the frequency response of the system becomes

Case A

$$\boldsymbol{\Omega}(j\omega) = \mathbf{H}(j\omega) = e^{-40.50j w}\begin{pmatrix} -1.267 & 0.000021 \\ -142 & -1.308 \end{pmatrix}$$

Case B

$$\boldsymbol{\Omega}(j\omega) = H(j\omega) = e^{-15.90j w}\begin{pmatrix} -35.0 & 0.00924 \\ 128{,}000 & 30.4 \end{pmatrix}$$

In addition, Bilous and Amundson calculated the response of the system to 1 percent step changes in the input variables. Their results are given in Table 5.3-3.

TABLE 5.3-1

Steady State Profiles

Case A $T_w = 335\,°\text{K}$

L/v	0	5.04	12.60	21.50	30.20	35.20	40.50
x_s	0.0200	0.0190	0.0151	0.0107	0.0070	0.0065	0.0057
T_s	340	343	346	351	347	343	340

Case B $T_w = 337.5\,°\text{K}$

L/v	0	2.50	5.04	7.55	10.10	12.60	13.82	15.10	15.90
x_s	0.0200	0.0195	0.0190	0.0175	0.0158	0.0130	0.0105	0.0050	0.0012
T_s	340	342	346	350	355	365	377	410	418

TABLE 5.3-2

Evaluation of Integrals

	Case A	Case B
K	1.253	2.820
J	0.00135	0.001486

TABLE 5.3-3

Ultimate Steady State Value

Output variable	Input variable	Case A Change in output	Case A Percent change	Case B Change in output	Case B Percent change
x	x_f	2.53×10^{-4}	4.4	70.0×10^{-4}	580
x	T_f	7.0×10^{-5}	1.2	0.0317	2600
T	x_f	$0.028\,°\text{K}$	0.0	$25.6\,°\text{K}$	6.1
T	T_f	$4.45\,°\text{K}$	1.3	$103\,°\text{K}$	25

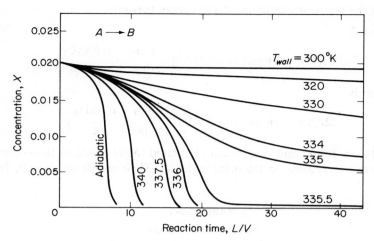

Figure 5.3-8. Composition profiles. Reproduced from Bilous, O. and N. R. Amundson, *AIChE Journal*, **2**, 117 (1956), by permission.

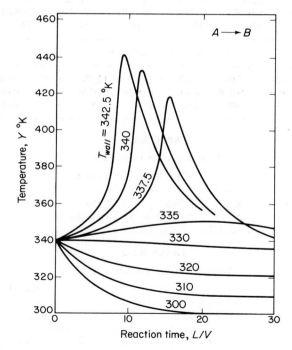

Figure 5.3-9 Temperature profiles. Reproduced from Bilous, O. and N. R. Amundson, *AIChE Journal*, **2**, 117 (1956), by permission.

A study of the results in Table 5.3-3 shows that small changes in the input for case A lead to small output changes, as we would normally expect. However, small input changes for case B cause very large changes in the output. In fact, the output deviations are so large that we cannot expect our linearization of the system equations to be valid, and we must look for another method of estimating the response of the reactor. It seems somewhat surprising that such a small change in one of the system parameters (i.e., the wall temperature in case B is 337.5 °K as compared to 335 °K in case A) can cause the reactor to have such different dynamic characteristics. However, this same kind of behavior is observed in the steady state profiles (see Figures 5.3-8 and 5.3-9). This phenomenon is called *parametric sensitivity*.

Technique of Lamb and Simpkins

After reviewing some of the material presented above, it becomes apparent that we can expect to encounter serious algebraic difficulties as we consider more complex sets of equations. Thus we might also expect that it will be necessary to rely more heavily on numerical, rather than analytical, solutions of the system equations in order to study the dynamic response. Lamb and Simpkins[7] have described a numerical technique that can be used to prepare Bode plots directly from the system equations even for very complicated processes. The method is applicable both to lumped and distributed parameter plants, although only the particular case of the response of a packed reactor is discussed below.

The system equations of interest were given in Section 5.1 as

$$\frac{\partial c}{\partial t} + v\frac{\partial c}{\partial z} = -kc \qquad\qquad 5.1\text{-}51$$

$$C_{pf}\rho_f\frac{\partial T}{\partial t} + C_{pf}\rho_f v\frac{\partial T}{\partial z} = (-\Delta H)kc + h_p A_p\left(\frac{1-\epsilon}{\epsilon}\right)(T_p - T) \qquad 5.1\text{-}52$$

$$C_{pp}\rho_p\frac{\partial T_p}{\partial t} = h_p A_p(T - T_p) \qquad\qquad 5.1\text{-}53$$

At steady state conditions, these equations reduce to

$$v\frac{dc}{dz} = -kc \qquad\qquad 5.1\text{-}54$$

$$C_{pf}\rho_f v\frac{dT}{dz} = (-\Delta H)kc \qquad\qquad 5.1\text{-}55$$

$$T = T_p \qquad\qquad 5.1\text{-}56$$

Previously we noted that the steady state equations can be solved analytically in terms of exponential integrals; but for the purpose of the present discussion

[7] D. E. Lamb and C. R. Simpkins, *Proceedings of JACC*, p. 486, Minneapolis, Minn., June 1963.

we suppose that they must be solved numerically, because numerical calculations will be necessary in more complicated reaction schemes. In order to obtain the linearized dynamic equations, we introduced the definitions given by Eqs. 5.1-57 and found that the desired expressions were

$$\frac{\partial x_1}{\partial \tau} + \frac{\partial x_1}{\partial Z} = -r_c x_1 - r_T x_2 \qquad 5.1\text{-}58$$

$$\frac{\partial x_2}{\partial \tau} + \frac{\partial x_2}{\partial Z} = r_c x_1 + r_T x_2 + H_p(x_p - x_2) \qquad 5.1\text{-}59$$

$$\frac{\partial x_p}{\partial \tau} = H_p B(x_2 - x_p) \qquad 5.1\text{-}60$$

where it is important to note that the coefficients r_c and r_T depend on the solutions of Eqs. 5.1-54 and 5.1-55, so that they also vary with the position in the reactor Z. Taking Laplace transforms of the linearized equations, we obtained

$$s\tilde{x}_1 + \frac{d\tilde{x}_1}{dZ} = -r_c \tilde{x}_1 - r_T \tilde{x}_2 \qquad 5.1\text{-}61$$

$$s\tilde{x}_2 + \frac{d\tilde{x}_2}{dZ} = r_c \tilde{x}_1 + r_T \tilde{x}_2 + H_p(\tilde{x}_p - \tilde{x}_2) \qquad 5.1\text{-}62$$

$$s\tilde{x}_p = H_p B(\tilde{x}_2 - \tilde{x}_p) \qquad 5.1\text{-}63$$

Then after a considerable amount of manipulation, we were finally able to solve these equations analytically for the system transfer functions. Unfortunately, the analytical solutions involved tabulated integrals. Nevertheless, it is possible to use this approach to develop the Bode diagrams for any of the transfer functions of interest.

With Lamb and Simpkin's technique, we attempt to generate the frequency behavior of the process directly from the transformed equations. This is accomplished by substituting $j\omega$ for s and separating the resulting expressions into their real and imaginary parts. Letting

$$\tilde{x}_1 = x_1^R + jx_1^I, \quad \tilde{x}_2 = x_2^R + jx_2^I, \quad \tilde{x}_p = x_3^R + jx_3^I \qquad 5.3\text{-}98$$

we obtain the set of equations

$$\frac{dx_1^R}{dZ} = -r_c x_1^R - r_T x_2^R + \omega x_1^I$$

$$\frac{dx_2^R}{dZ} = r_c x_1^R + (r_T - H_p)x_2^R + H_p x_3^R + \omega x_2^I$$

$$0 = H_p B x_2^R - H_p B x_3^R + \omega x_3^I \qquad 5.3\text{-}99$$

$$\frac{dx_1^I}{dZ} = -\omega x_1^R - r_c x_1^I - r_T x_2^I$$

$$\frac{dx_2^I}{dZ} = -\omega x_2^R + r_c x_1^I + (r_T - H_p)x_2^I + H_p x_3^I$$

$$0 = -\omega x_3^R + H_p B x_2^I - H_p B x_3^I$$

In general, we would have a set of six differential equations in six unknowns, and these equations would have variable coefficients—that is, r_c and r_T depend on Z.

The appropriate boundary conditions for these equations are the disturbances that enter the inlet of the reactor (at $Z = 0$). The simplest way of calculating the plant transfer functions is to consider that only one disturbance enters the process, say the feed composition x_{1f}, and that all the other inputs are maintained constant. After determining the response of all three outputs to this disturbance, we can repeat the analysis for one of the other inputs. The Fourier transform of a unit impulse is the complex number $1 + j0$. Hence the frequency response of the plant outputs can be determined by setting each input in turn equal to this value. Similarly, the desired information for the Bode plots relating reactor composition and temperature and the particle temperature to the feed composition is obtained by evaluating the quantities

$$|G_1| = \sqrt{(x_1^R)^2 + (x_1^I)^2} \qquad \tan \phi_1 = \frac{x_1^I}{x_1^R}$$

$$|G_2| = \sqrt{(x_2^R)^2 + (x_2^I)^2} \qquad \tan \phi_2 = \frac{x_2^I}{x_2^R} \qquad \text{5.3-100}$$

$$|G_3| = \sqrt{(x_3^R)^2 + (x_3^I)^2} \qquad \tan \phi_3 = \frac{x_3^I}{x_3^R}$$

In summary, then, the reactor transfer functions for feed composition changes are computed by numerically by solving Eqs. 5.1-54 through 5.1-56 for the steady state profiles and using these results and the definitions given by Eqs. 5.1-57 to evaluate the coefficients r_c and r_T as a function of Z. Simultaneously, we solve the set of Eqs. 5.3-99 with the initial conditions at $Z = 0$; $x_1^R = 1.0$, $x_1^I = x_2^R = x_2^I = x_3^R = x_3^I = 0$ for some particular value of ω. The values obtained at the end of the reactor, or at any particular position Z of interest, are substituted into Eqs. 5.3-100, and the gain and phase angle at the particular frequency under consideration are calculated. By repeating the procedure for various values of ω, we can plot three Bode diagrams. The procedure is quite simple to implement if an analog computer with repetitive operation is available. By solving the differential equations in the fastest operating time and varying ω in real time, it is possible to plot the Bode diagrams directly.

SECTION 5.4 POPULATION–BALANCE MODELS

In each of the plants discussed previously, the state of the system can be described in terms of one or more dependent (state) variables. The values of these variables may vary from point to point within the system, as well as with time, so that the system may be described by coupled sets of partial differential equations. Nevertheless, at any particular point, we only need a finite number of dependent variables to characterize the plant completely. There are cases, however, where this formulation turns out to be inadequate, and it is more convenient to use an inifinite number of state variables to describe the system. In other words, we consider a function—often called a distribution function—rather than a finite number of dependent variables, as the basis of our mathematical model.

One example of a system of this type is the model of an ideal gas, where we treat the gas molecules as very small, perfectly elastic, spheres that move in random directions at various speeds. The case of greatest interest occurs when there are an extremely large number of molecules present, of the order of 10^{23} molecules/cc. For this case it appears to be impractical to predict the motion of each particle by writing a set of momentum balances describing the collisions between particles or between a particle and the wall of the container. Instead, it is simpler to determine how the velocities are distributed between the molecules at equilibrium conditions. Once we have determined this distribution function, we can picture any small volume element in the box and say that the molecules pass through this element in random directions with various speeds, but, over a sufficiently long time interval, the number of molecules having certain velocities will correspond to the predicted distribution. Of course, any individual particle continually changes direction and speed as it collides with other particles, or the container wall, but the distribution function remains constant at equilibrium conditions. If we suddenly change the temperature of the wall of the box, we could attempt to determine how the velocity distribution would change with time. Similarly, if we impose a small temperature gradient across the box, we could try to determine how the distribution function changed with position. Problems of this nature are treated in texts on the kinetic theory of gases.

Another situation where distribution functions are helpful is in catalyst replacement problems. It is well known that the activity of many catalysts decay with increasing contact times between the catalyst and reactants; for example, in catalytic cracking the deposition of coke on the catalyst surface causes a decrease in the activity. In order to overcome this deficiency, it is common practice to add fresh catalyst continuously and to remove some partially deactivated catalyst from the operating unit. Whenever the new catalyst is thoroughly mixed with the old, we take the chance of removing a fraction of the almost new catalyst along with the old material. Thus we achieve some distribution of catalyst activity within the vessel, and generally it is much easier to predict the behavior of this distribution function rather than the combined behavior of thousands or millions of individual catalyst particles.

Some other problems of possible interest include residence time or age distributions in reactors, crystal-size distribution in continuous crystallizers, the age and size distribution of microorganisms in biochemical processes, and so on. One aspect all these problems have in common is that there is a very large number of countable entities (particles, crystals, microorganisms), and we are interested in how some property, (velocity, catalytic activity, age, size) is distributed between these entities. We often refer to the set of countable entities as a population; therefore an equation describing the behavior of these entities is called a population balance.

The quantitative treatment of population balances can be developed by considering a distribution function $\psi(x, y, z, \alpha, \beta, \gamma, \ldots, t)$, where x, y, and z represent the three spatial coordinate directions; $\alpha, \beta, \gamma, \ldots$ represent certain properties of the entities; and t represents time. To be more specific, α might represent the activity of a catalyst particle, the age of a particle of fluid, or the size of a particular crystal. Similarly, α and β might represent the age and size of a microorganism.

The meaning of the distribution function is that

$$\psi \, \Delta x \, \Delta y \, \Delta z \, \Delta\alpha \, \Delta\beta \, \Delta\gamma \qquad \text{5.4-1}$$

is the fraction of entities in a volume element of size $\Delta x \, \Delta y \, \Delta z$ that have property values in the range $\Delta\alpha, \Delta\beta, \Delta\gamma, \ldots$. We know that all the entities must be in some volume element and in some range of property values, so that

$$\int \psi \, dx \, dy \, dz \, d\alpha \, d\beta \, d\gamma \ldots = 1 \qquad \text{5.4-2}$$

Now we can apply our conservation principles to this distribution function. For example, we can say that the accumulation of entities having properties in the range $\Delta\alpha \, \Delta\beta \, \Delta\gamma \ldots$ within a volume element $\Delta x \, \Delta y \, \Delta z$ is given by

$$[\psi \, |_{t+\Delta t} - \psi \, |_{t}] \, \Delta x \, \Delta y \, \Delta z \, \Delta\alpha \, \Delta\beta \, \Delta\gamma \ldots \qquad \text{5.4-3}$$

The net transport of entities having property values in the range $\Delta\alpha \, \Delta\beta \, \Delta\gamma \ldots$ into the volume element in the x direction by convective flow within a time interval Δt is

$$[(v_x\psi) \, |_x - (v_x\psi) \, |_{x+\Delta x}] \, \Delta y \, \Delta z \, \Delta\alpha \, \Delta\beta \, \Delta\gamma \ldots \Delta t \qquad \text{5.4-4}$$

The corresponding expressions for the y and z directions are

$$[(v_y\psi) \, |_y - (v_y\psi) \, |_{y+\Delta y}] \, \Delta x \, \Delta z \, \Delta\alpha \, \Delta\beta \, \Delta\gamma \ldots \Delta t \qquad \text{5.4-5}$$

and

$$[(v_z\psi) \, |_z - (v_z\psi) \, |_{z+\Delta z}] \, \Delta x \, \Delta y \, \Delta\alpha \, \Delta\beta \, \Delta\gamma \ldots \Delta t \qquad \text{5.4-6}$$

We expect the properties of the various entities to be different, and in most cases we recognize that the property of a particular entity might be a continuously changing quantity. For example, the activity of a catalyst particle might continue to decline with time, or the size of a particular crystal might continue to increase with time. Hence the variation in properties can be treated as a rate process, and we will need some kind of a kinetic mechanism to describe these changes. We let r_α be the rate per unit volume at which the property α increases for entities having properties in the range $\Delta\beta \, \Delta\gamma \ldots$, and then we write that the net change in the entities having properties in the range $\Delta\alpha$, within the volume element $\Delta x \, \Delta y \, \Delta z$ during the time interval Δt, is given by

$$[(r_\alpha\psi)|_\alpha - (r_\alpha\psi)|_{\alpha+\Delta\alpha}] \, \Delta x \, \Delta y \, \Delta z \, \Delta\beta \, \Delta\gamma \ldots \Delta t \qquad \text{5.4-7}$$

Similarly, for the properties β and γ we have

$$[(r_\beta\psi)|_\beta - (r_\beta\psi)|_{\beta+\Delta\beta}] \, \Delta x \, \Delta y \, \Delta z \, \Delta\alpha \, \Delta\gamma \ldots \Delta t \qquad \text{5.4-8}$$

and

$$[(r_\gamma\psi)|_\gamma - (r_\gamma\psi)|_{\gamma+\Delta\gamma}] \, \Delta x \, \Delta y \, \Delta z \, \Delta\alpha \, \Delta\beta \ldots \Delta t \qquad \text{5.4-9}$$

Another possibility we must consider is the birth, or death, of entities within the volume element having properties in the range of interest during the time interval Δt. Letting

$$Q_B = \frac{\text{birth of entities}}{\text{(volume)(unit property change)(time)}} \qquad \text{5.4-10}$$

$$Q_D = \frac{\text{death of entities}}{\text{(volume)(unit property change)(time)}} \qquad \text{5.4-11}$$

the net creation of entities will be

$$(Q_B - Q_D) \, \Delta x \, \Delta y \, \Delta z \, \Delta\alpha \, \Delta\beta \, \Delta\gamma \ldots \Delta t \qquad \text{5.4-12}$$

As a first attempt to develop a population–balance model for a process, we might limit our attention to the phenomena discussed above. The balance is obtained by equating the accumulation term to the sum of the net transfer into the element, the change of properties of entities within the element, and the net creation of entities having the desired properties. After dividing each term in this equation by $\Delta x \, \Delta y \, \Delta z \, \Delta\alpha \, \Delta\beta \, \Delta\gamma \ldots \Delta t$ and taking the limit as each small increment approaches zero, we obtain

$$\frac{\partial\psi}{\partial t} + \frac{\partial}{\partial x}(v_x\psi) + \frac{\partial}{\partial y}(v_y\psi) + \frac{\partial}{\partial z}(v_z\psi) + \frac{\partial}{\partial\alpha}(r_\alpha\psi)$$

$$+ \frac{\partial}{\partial\beta}(r_\beta\psi) + \frac{\partial}{\partial\gamma}(r_\gamma\psi) + \cdots + Q_B - Q_D = 0 \qquad \text{5.4-13}$$

An illustration of the use of a population–balance model to describe the behavior of a continuous crystallizer is presented below. The analysis is limited to steady state operation, although numerous treatments of crystallizer dynamics have been published.[1,2,3,4] Also, a number of other applica-

---

[1] A. D. Randolph and M. A. Larson, *AIChE Journal*, **8**, 639 (1962).

[2] A. D. Randolph, *AIChE Journal*, **11**, 424 (1965).

[3] M. B. Sherwin, R. Shinnar, and S. Katz, *AIChE Journal*, **13**, 1141, (1967).

[4] M. B. Sherwin, R. Shinnar, and S. Katz, *Chem Eng. Prog. Symposium Ser.*, **65**, No. 95, 75 (1969).

tions of the population-balance approach, in addition to the one described below, are presented in Himmelblau and Bischoff's book.[5]

Example 5.4-1 Crystal-size distribution

Let us consider a perfectly mixed vessel in which crystallization is taking place. This kind of system is frequently called a mixed suspension, mixed product removal (MSMPR) crystallizer, providing that we actually remove product containing all crystal sizes present in the vessel. We assume that we can characterize the crystallization phenomenon by a single property: namely, the size of a crystal L. Thus in the population–balance we let the property α correspond to L and neglect the properties $\beta, \gamma, \ldots$. Similarly, we do not expect the conditions within the perfectly mixed vessel to vary from one point to another, so that we must modify the transport terms in the general population–balance equation. After setting the birth and death terms equal to zero, we write the equation as

$$V\frac{\partial \psi}{\partial t} + V\frac{\partial}{\partial L}(r_L \psi) = q(\psi_f - \psi) \qquad 5.4\text{-}14$$

where the right-hand side represents the net flow of crystals through the vessel.

To find the steady state crystal-size distribution, we set the accumulation term equal to zero. Also, for a case where the feed stream does not contain any crystals, $\psi_f = 0$. Letting

$$\tau = \frac{V}{q} \qquad 5.4\text{-}15$$

the equation becomes

$$\tau \frac{\partial}{\partial L}(r_L \psi) = -\psi \qquad 5.4\text{-}16$$

Before progressing, we must have some information about the crystal growth rate r_L. We would expect this rate to depend on both the present crystal size and the composition of the dissolved crystalline material, c (sometimes called the substrate concentration). Empirical evidence indicates that a useful expression for the kinetics of crystal growth is

$$r_L = r_0(c)L^b \qquad 5.4\text{-}17$$

where r_0 is called the specific growth rate, which depends on the substrate concentration, and b is a constant, which depends on the material under consideration. However, for some materials, the data indicate that b should be set equal to zero. This means that the growth rate of a crystal is independent of its size, a result that is called the "McCabe ΔL law." For simplicity, we consider only this last case. Then our system equation becomes

$$\tau r_0 \frac{d\psi}{dL} = -\psi \qquad 5.4\text{-}18$$

[5] See Chaps. 4 and 6 of the reference given in footnote 5 on p. 53.

so that the distribution function (i.e., the number of crystals having a size between L and $L + \Delta L$ per unit volume) is

$$\psi = \psi_0 e^{-L/\tau r_0} \qquad\qquad 5.4\text{-}19$$

where ψ_0 represents the number of nuclei per unit volume—that is, the crystals of zero size.

In many cases it is more convenient to describe the behavior of a crystallizer in terms of the weight fraction distribution rather than the number distribution given above. This can be calculated by recognizing that the mass of crystals of dimension less than L per unit volume can be written as

$$M_c = k_L \rho_c \int_0^L L^3 \psi \, dL = k_L \rho_c \int_0^L L^3 \psi_0 e^{-L/\tau r_0} \, dL \qquad 5.4\text{-}20$$

where ρ_c is the density of the crystals and k_L is a shape factor relating the volume of a crystal to a cube having a side of length L. The cumulative weight fraction distribution is obtained by dividing this expression by the total mass of crystals

$$W(L) = \frac{\displaystyle\int_0^L L^3 e^{-L/\tau r_0} dL}{\displaystyle\int_0^\infty L^3 e^{-L/\tau r_0} dL} \qquad\qquad 5.4\text{-}21$$

or

$$W(L) = \frac{1}{6} \int_0^{L/\tau r_0} x^3 e^{-x} \, dx \qquad\qquad 5.4\text{-}22$$

The corresponding frequency distribution can be obtained by differentiation

$$w(L) = \frac{1}{6\tau r_0} \left(\frac{L}{\tau r_0}\right)^3 e^{-L/\tau r_0} \qquad\qquad 5.4\text{-}23$$

The preceding analysis includes a large number of assumptions, and as we have mentioned many times before, it is necessary to verify the model experimentally before it is used as a design tool. Fortunately, a number of experimental studies of this type have been undertaken, and they demonstrate that there is a good correspondence between theory and experiment. An illustration of the quality of the fit is shown in Figure 5.4-1 for a set of data taken in a laboratory-scale MSMPR crystallizer.

The distribution functions just developed characterize the behavior of the solid phase in the crystallizer. However, in order to obtain a complete description of the unit, we must also write a material balance for the dissolved solids. At steady state conditions, a total material balance gives the expression

$$q_f c_f = q(c + M_T) \qquad\qquad 5.4\text{-}24$$

where q_f and q represent the feed and effluent, volumetric flow rates, c_f and c are the feed and effluent substrate concentrations, and M_T is the total mass

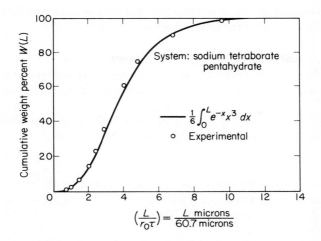

Figure 5.4-1. Crystal size distribution. Reproduced from Randolph, A. D., *AIChE Journal*, **11**, 424 (1965), by permission.

concentration of solid crystals. The value of M_T can be obtained by integrating M_c over all possible crystal sizes.

$$M_T = k_L\rho_c \int_0^\infty L^3\psi \, dL = k_L\rho_c \int_0^\infty L^3\psi_0 e^{-L/\tau r_0} \, dL$$
$$= 6k_L\rho_c\psi_0\tau r_0 \qquad\qquad 5.4\text{-}25$$

We mentioned earlier that we expect the specific growth rate r_0 to depend on the substrate concentration c. Moreover, we might expect that the number of nuclei per unit volume also depends on this quantity. Thus it is necessary to make additional measurements in order to develop these relationships. The results of many experiments indicate that both r_0 and ψ_0 can often be written in terms of powers of the supersaturation.

Our purpose here is to describe some of the features of population–balance models rather than to develop a complete model of a crystallization system. For this reason we will limit our attention to the simple problem where we attempt to find the effect of the flow rate on the most predominant particle size. We expect that the product quality specifications would be based on this size factor, and possibly some measure of the spread of the distribution function around this value—for example, the variance of the distribution.

It is apparent from the frequency distribution function

$$w(L) = \frac{1}{6\tau r_0}\left(\frac{L}{\tau r_0}\right)^3 e^{-L/\tau r_0} \qquad\qquad 5.4\text{-}23$$

that there are almost no particles having very small or very large sizes. The size corresponding to the maximum number of particles can be found by setting the derivative of the function equal to zero

$$L_M = 3\tau r_0 \qquad\qquad 5.4\text{-}26$$

It is interesting to note that this result is merely the second derivative of the cumulative distribution function, so that it corresponds to the inflection point on Figure 5.4-1.

If we assume that both ψ_0 and r_0 can be written as powers of the supersaturation, we can find a relationship between them:

$$\psi_0 = k_0 r_0^{m-1} \qquad\qquad 5.4\text{-}27$$

where k_0 is some specific rate constant and m represents the relative kinetic orders of the nucleation rate to the growth rate when both are written in terms of power of the supersaturation. Then, combining Eqs. 5.4-24 through 5.4-27, we find that

$$L_M = 3\left[\frac{1}{6k_0 k_L \rho_c}\left(\frac{q_f c_f}{q} - c\right)\right]^{1/(m+3)} \tau^{(m-1)/(m+3)} \qquad\qquad 5.4\text{-}28$$

or, for a case where $q_f = q$ and we crystallize all the dissolved solids, $c = 0$,

$$L_M = 3\left(\frac{c_f}{6k_0 k_L \rho_c}\right)^{1/(m+3)} \tau^{(m-1)/(m+3)} \qquad\qquad 5.4\text{-}29$$

Thus the crystal size will either increase or decrease with τ, which is inversely proportional to the flow rate q, depending on whether or not the value of m is greater or less than unity. Both kinds of behavior have been observed in practice.

SECTION 5.5 SUMMARY

In this chapter we reviewed some of the techniques that can be used to determine the dynamic response of distributed parameter systems. The transfer functions describing the process could be obtained in a straightforward fashion, but they were much more complex than the expressions for lumped parameter plants, for they generally involved transcendental functions. Because finding the inverse transforms and obtaining explicit expressions for the time dependence of the system output is difficult, it is common practice to search for simple approximations of the plant transfer function. Of course, different approximations will lead to somewhat different results; therefore the particular approximation employed will depend on the desired use of the model. The main difference between distributed parameter systems and lumped parameter processes is that the former exhibit dead time, nonminimum phase characteristics and wave behavior.

QUESTIONS FOR DISCUSSION

1. What is a viscoelastic fluid? How do you measure its inherent properties? What kind of results would you obtain with a steady state experiment? When are viscoelastic properties important in plant design?

2. What is the relationship between the "time-smoothed" conservation equations used to describe turbulent flow and a steady state conservation equation? In other words, does turbulent flow ever correspond to a steady state situation?

EXERCISES

1. (A) Schnelle[1] measured the response of a thermometer when it was plunged into a bath having a temperature of 139.8 °F. Use several procedures to develop an approximate dynamic model for the thermometer from the data given below:

Time (sec)	0	10	20	30	40	50	60	70	80	90	100	120
Temperature (°F)	78	88	102	113	120	126	130	132	135	136	138	139

2. (A) Cohen and Johnson[2] measured the step response of a steam-heated exchanger. The data are shown in Figure 5.5-1. Use several techniques to develop an approximate dynamic model for the exchanger.

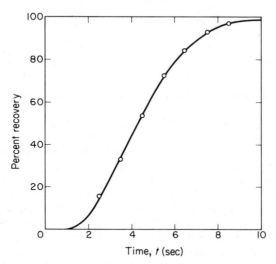

Figure 5.5-1. Step response of a heat exhanger. Reproduced from Cohen, W. C. and E. F. Johnson, *Ind. Eng. Chem.*, **48**, 1031 (1956), by permission.

[1] K. B. Schnelle Jr., "Experiments to Illustrate the Dynamic Testing of Chemical Process Systems," Report to National Science Foundation under Grant Number G-22952.
  [2] See footnote 3, p. 238.

3. (B‡) A thermistor provides a simple device for making continuous measurements of the composition of some binary mixtures of gases. However, it is seldom used to study axial dispersion in packed beds because the time constant associated with the thermistor dynamics is roughly the same as the "effective" time constant of the bed. Hence, the measuring device causes approximately the same amount of distortion in the test signal as the process itself.

As a particular illustration of measurement error, consider a situation where we are attempting to use a pulse-response experiment to determine the axial dispersion coefficient of a packed bed (see Example 5.2-1), and we use a pair of identical thermistors to measure the input and output pulses (see Figure 5.5-2). If the transfer functions of the thermistor are

$$H_m = \frac{K_m}{\tau_m s + 1}$$

find the relationship between the moments of the measured distribution functions, $C_2(t)$ and $C_4(t)$. Discuss in detail any physical, or computational, factors not included in your analysis that might make the configuration shown in the sketch more advantageous, or disadvantageous, than suggested by your analysis.

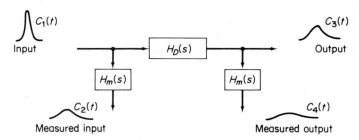

Figure 5.5-2. The effect of measurement dynamics.

4. (C‡) We have used the simplified equations describing a first-order, irreversible, exothermic reaction in a continuous-stirred-tank reactor containing a heating coil to illustrate many of the principles of process dynamics. However, these equations neglected the possibility of dynamic effects in the heating coil. Assuming plug flow in the coil and no wall capacitance or resistance of the wall, derive an appropriate set of equations that includes the coil dynamics. Linearize the equations and obtain the transfer functions relating changes in the hot-fluid inlet temperature and velocity to the reactor composition. Discuss the effect of this modification on the nature of the characteristic equation and the stability of the system.

5. (B) Consider a steam-heated exchanger for a case where the feed temperature is 61 °F, the condensate temperature is 272 °F, the exchanger length is 17 ft, $UA_H/C_p\rho A_t = 0.725$, the velocity through the tubes is 5.2 ft/sec, and the wall capacitance is negligible. Prepare a Bode plot relating the tube effluent temperature to the tube velocity. Estimate the outlet temperature response when $v = 5.2 + 0.522 \sin \omega t$ and $\omega = 22$.

6. (C‡) Stermole and Larson[3] present some data for the step response and the

[3] F. J. Stermole and M. A. Larson, *Ind. Eng. Chem. Fundamentals*, **2**, 62 (1963).

frequency response of a steam-heated exchanger. The model they use to describe the system neglects wall capacitance and the effect of velocity on the heat-transfer coefficient. See if you can demonstrate quantitatively that these effects should be negligible for their system.

7. (B) Derive the transfer function for countercurrent flow in a double-pipe heat exchanger given by Eq. 5.3-22. Also, obtain the transfer function relating the tube outlet temperature to the tube inlet temperature.

8. (B‡) Notice that the Bode diagram for a double-pipe, countercurrent heat exchanger shown in Figure 5.3-6 asymptotically approaches a phase lag that is less than 180 deg. However, phase lags of 180 deg often can be observed experimentally. What would you guess is the major source of error in the simplified exchanger model? Can you develop a quantitative criterion to verify your conclusion?

9. (B‡) A numerical study of heat-exchanger dynamics was undertaken by Lee.[4] One case he considered was *cocurrent* flow in a double-pipe exchanger. The inner and outer pipes corresponded to $\frac{3}{4}$-in. and $1\frac{1}{4}$-in. copper water tubing, so that the wall thickness of the inner tube was 0.035 in. and that of the outer tube was 0.055 in. Hot water entered the tubes at 150 °F at a rate of 1 gpm, while cold water was fed to the shell at 68 °F and at a rate of 10 gpm. The exchanger was 20 ft long, and the film-heat-transfer coefficients were estimated using the Dittus-Boelter equation. Using this information, prepare a Bode plot relating the tube outlet temperature to the shell flow rate.

10. (C) Derive a transfer function relating the tube outlet temperature to the shell inlet temperature for a two-tube-pass, single-shell-pass heat exchanger.

11. (B) To be sure, there is a "generation gap." One problem students of my generation always attacked with gusto was the calculation of how long it would take a can of beer to cool, without letting it freeze, when it was set on an outside window-sill on a cold winter's evening. The modern generation seems to be more concerned with problems in radiation heat transfer to accelerate the rate of growth of grass. Nevertheless, please picture yourself in a bygone era and develop an *engineering correlation* for the beer can problem.

12. (B‡) Many of the techniques used by chemical engineers are also helpful in the food-processing industry. For example, consider the problem of sterilizing food after it has been placed in cylindrical cans. Normally it is assumed that all harmful bacteria will be killed if the food temperature is raised to some value T_1. The heating process is accomplished by placing the can in a sterilization bath that is maintained at a high temperature T_0. Develop an equation for the sterilization time. Also, by selecting various values for the system parameters, see if you can determine whether or not it is necessary to consider the resistance to heat transfer in a stagnant film surrounding the can.

13. (B‡) Most college coeds are lousy cooks because they do not take science courses and therefore do not understand the *fundamental* principles of cooking. On the other hand, virtually all chemical engineers have the potential of being superb cooks because that is the very nature of their training. Perhaps this is the reason that one survey showed that chemical engineers make the best husbands.

In any event, when I shared a house (fondly known as the Pig Pen) with five

---

[4] E. T. Lee, *Dynamics and Feedforward Control of Parametrically Forced Distributed Parameter Processes*, M.S. Thesis, University of Massachusetts, July 1969.

others in graduate school, we cooked our rolled roasts of beef $\frac{1}{2}$ hour per pound at 350° and they always turned out delicious. Can you estimate the temperature at the center of the roast, which we could have measured to see if the roast was done —if we had had a thermometer.

14. (C‡) The culinary experitise we developed in graduate school was modestly mentioned in the previous problem. The greatest challenge we faced in that field occurred when we invited several coeds over for a home-cooked meal. Normally we started with cocktails, and then sat down to a turkey dinner, with stuffing, mashed potatoes, a vegetable, and cranberry sauce. However, the high point would be Baked Alaska for dessert. For those unfamiliar with the term, we note that Baked Alaska is prepared by placing a block of ice cream, which has been frozen very hard so that its uniform temperature is about 15 deg, on a piece of sponge cake. The ice cream is covered by a layer of meringue, about $\frac{3}{8}$ in. thick, and then placed in a hot oven, preheated to 450 deg, for 5 minutes. If we assume that after 5.5 minutes the surface temperature of the ice cream would be sufficiently high so as to melt, see if you can estimate the thermal diffusivity of meringue. Next, try to convince a home economics major to verify your analysis experimentally.

15. (C) Design a frequency response experiment that would make it possible for you to measure the thermal diffusivity of long cylindrical rods of new plastic materials.

16. (C) In Examples 4.3-3 and 5.2-2 we used two different kinds of models to determine the cooling time for a sphere. For the given system parameters, which were completely idiotic, we found that both models gave essentially the same prediction. As a third approach for estimating the cooling time, let us assume that the resistance and capacitance within the sphere are significant and that the water is perfectly mixed (so that there are no temperature gradients in the water). With these assumptions, the system equations become

$$\frac{\partial T_m}{\partial t} = \frac{\alpha_m}{r^2}\frac{\partial}{\partial r}\left(r^2\frac{\partial T_m}{\partial r}\right)$$

$$\rho_w C_{pw} V_w \frac{\partial T_w}{\partial t} = -k_m A \frac{\partial T_m}{\partial r}\bigg|_{r=R}$$

What are the appropriate boundary conditions for these equations? Determine the cooling curve for a case where the volume of the cooling water is 100 times that of the sphere. Compare your solution to Figure 5.2-5.

17. (C) Jury[5] studied unsteady state diffusion into stagnant drops of fluid or spherical particles. He solved the equation

$$\frac{\partial c}{\partial t} = D\frac{1}{r^2}\frac{\partial}{\partial r}\left(r^2\frac{\partial c}{\partial r}\right)$$

with the boundary conditions

$c(r, 0) = c_i$, a constant $c(0, t) =$ a maximum or a minimum

Then he used the solutions to calculate the flux at the surface and the time dependence of the average composition within the sphere. In this way he was able to predict exact values for the film-mass-transfer coefficient that would be applicable

[5] S. H. Jury, *AIChE Journal*, **13**, 1124 (1967).

after an initial transient period. However, he found that he arrived at a different estimate of the Sherwood number when he applied the final value theorem to the Laplace transform of the conservation equation than when he truncated the infinite series for the response in the time domain. See if you can reproduce his results.

Also, note that this model provides an additional way of estimating the cooling time for the sphere described in Examples 4.3-3 and 5.2-2. Calculate the cooling curve and compare your result to Fig. 5.2-5, as well as to the curve obtained in the previous exercise.

18. (C) The "effectiveness" of a fin is defined as the ratio of the heat transferred from the wall through the base of the fin to the heat transferred from the same area when the fin is not present. Consider a rectangular fin having a length L, a width W, and a thickness H, attached to a wall having a temperature T_w and surrounded by a fluid having a temperature T_f. If the wall temperature suddenly changes to a new value, $T_w(1 + A)$, how does the fin effectiveness change with time?

19. (C) The effectiveness factor[6] of a porous, spherical, catalyst particle is defined as the ratio of the total molar flow rate of reactant entering the particle by diffusion divided by the product of the surface reaction rate and the total available surface. How would the effectiveness factor change with time if the surface concentration undergoes a step change?

20. (C‡) After introducing as many simplifying assumptions as you can, find a solution that describes the behavior of the heat regenerator discussed in Exercise 3.5-26.

21. (C‡) A parallel reaction scheme was described in Exercise 37, p. 221, and the parameters for the reaction were given as case III in Table 4.8-3. Show how you would use this information to develop the composition and temperature profiles in a packed tubular reactor. Then linearize the dynamic equations around these steady state profiles and develop an approximate model for the reactor dynamics. Carefully list the assumptions you use in your analysis.

22. (C) Hougen and Watson[7] describe the operation of a bed packed with silica gel that is being used to remove benzene from an air stream in a solvent recovery plant. At their operating conditions, 70 °F and atmospheric pressure, 90 percent of the benzene is adsorbed from the inlet stream, which enters at 7.5 (lb)/(ft²)(min) on a solvent-free basis and contains 0.9 percent benzene by volume. The silica gel is 4 to 6 mesh size ($d_p = 0.0128$ ft, $a_v = 202$ ft²/ft³), the external void fraction is 50 percent, the bulk density of the bed is 39 lb/ft³, and the bed is 3.22 ft long. Over the range of interest, the weight of benzene adsorbed is related to the partial pressure of benzene in the air stream by the expression $w = 1.67P^*/P_s$, where the vapor pressure of benzene, P_s, at 70 °F is 0.125 atm. Correlations were used to estimate the constants

$$a = \frac{1}{H_d} = 0.703, \quad a_v\left(\frac{d_p G}{\eta}\right)^{-0.51} = 11.8, \quad c = \frac{y^*}{w} = \frac{P_s(78)}{\pi(29)(1.67)} = 1.60\frac{P_s}{\pi}$$

$$b = \frac{1.60 G P_s}{\rho_B H_{dG}\pi} = \frac{1.60(7.5)(0.125)}{39(1/11.8)1} = 0.455 \text{ min}^{-1}$$

---

[6] See p. 542 of the reference given in footnote 3 on p. 32.

[7] O. A. Hougen and K. M. Watson, "Chemical Process Principles," Part III *Kinetics and Catalysis*, p. 1091, Wiley, N.Y., 1947.

which appear in the conservation equations

$$\frac{\partial w}{\partial \tau} = b\left(\frac{y}{c} - w\right)$$

$$-\frac{\partial y}{\partial z} = a(y - cw)$$

where y = lb benzene/lb air, w = lb benzene/lb silica gel, z = height from top (ft), and τ = time (hr). The desired removal was accomplished in 1 hour.

Suppose that new air-pollution standards require a 95 percent removal of the benzene. How will this affect the operation? What about a 99.9 percent removal?

23. (C) In an ion-exchange unit used for water softening, the metallic ions from the water are exchanged for sodium ions in the resin. Then, in the regeneration process, the metal ions are exchanged again with a sodium salt solution. Hougen and Watson[8] assume that the exchange rate is proportional to the product of the calcium ion concentration in the water, u, and the square of the concentration of available sodium in the exchanger, v. Thus they write the conservation equations for the ion-exchange unit as

$$\frac{\partial u}{\partial x} = auv^2$$

$$\frac{\partial v}{\partial \tau} = buv^2$$

Show that the solution to these equations is

$$\ln z + z = \ln s + s - r$$

$$\frac{v}{v_0} = \frac{1}{1 + z}$$

$$\frac{u}{u_0} = \frac{1 + 1/s}{1 + 1/z}$$

where

$$r = -av_0^2 x, \quad s = -bu_0 v_0 \tau, \quad z = \frac{v_0}{v} - 1$$

24. (C) McHenry and Wilhelm[9] made frequency response measurements on gas mixtures flowing through packed beds and found that the axial Peclet number was about 1.88. They noted that the experimental value was in good agreement with a theoretically predicted value of 2.0. They obtained this theoretical estimate by first solving the diffusion equation,

$$\frac{\partial c}{\partial t} = -v\frac{\partial c}{\partial z} + D\frac{\partial^2 c}{\partial z^2}$$

with the boundary conditions,

at $z = 0$, $c = c_s + A_0 \sin \omega t$

at $z = \infty$, $c = c_s$

[8] See p. 1074 of reference 7 on p. 297.
[9] K. W. McHenry Jr. and R. H. Wildhelm, *AIChE Journal*, **3**, 83 (1957).

and then approximating the exact solution by the expression

$$c(L) = c_s + A_0 \exp\left(-\frac{N_f^2 L}{\text{Pe}}\right) \cos\left(\omega t - N_f L\right)$$

where

$$N_f = \frac{d_p \omega}{v}, \quad \text{Pe} = \frac{d_p v}{D}, \quad L = \frac{z}{d_p}$$

Next they developed a solution based on the assumption that the bed acted as a series of N perfect mixers, where N is the number of particles. The system equation for this case is

$$\frac{\lambda}{v}\frac{dc_n}{dt} + c_n = c_{n-1}$$

where λ is the length corresponding to a perfect mixer; and for N mixers in series, they developed the approximate solution

$$\ln\frac{A_N}{A_0} \approx -\frac{N\omega^2\lambda^2}{2v^2} \quad \text{for } \frac{\omega\lambda}{v} \ll 1$$

Recognizing that

$$\lambda = \frac{z}{N}$$

and putting the expression in dimensionless form, the result is

$$\ln\frac{A_N}{A_0} \approx \frac{N_f^2 L}{2}, \quad N_f \ll 1$$

By comparing this equation with the solution of the diffusion equation, we find that

$$\text{Pe} \approx 2$$

Develop the complete solutions of the problems described above. Also, determine the effect that changing the boundary conditions on the diffusion equation has on the predicted value of the Peclet number.

25. (C) Ebach and White[10] used a frequency response technique to study the flow of liquids through packed beds. They assumed that the composition of the dye tracer in the fluid could be described by the equation

$$\frac{\partial c}{\partial t} = -v\frac{\partial c}{\partial z} + D\frac{\partial^2 c}{\partial z^2}$$

and that an appropriate set of boundary conditions were

$$\text{at } z = 0, \quad c = c_s + A_0 \cos \omega t$$
$$\text{at } z = \infty, \quad c = c_s$$

Some of their data for a bed 3.01 ft long packed with particles 0.039 in. in diameter, so that the bed porosity was 0.34, are listed below. Use the data to estimate the axial dispersion coefficient. Also, determine the effect of changing the system boundary conditions on the axial dispersion coefficient.

[10] E. A. Ebach and R. R. White, *AIChE Journal*, **4**, 161 (1958).

Interstitial Velocity	Input Amplitude	Output Amplitude	Frequency	Reynolds Number
0.270	0.836	0.680	0.785	29.54
0.146	0.761	0.561	0.561	15.35
0.0788	0.723	0.499	0.313	8.21
0.0326	0.706	0.501	0.126	3.65
0.0150	0.513	0.266	0.0785	1.60

26. (C‡) Bruley and Prados[11] measured the frequency response of a wetted-wall, adiabatic humidifier for cases of laminar and turbulent flow. They used a simple plug flow model to describe the turbulent flow system and found that this gave adequate agreement with their data. Determine the effect that adding an effective axial dispersion term to the model would have on the predicted values.

27. (C‡) Write a critical evaluation of the paper, "The Dynamics of a Packed Gas Absorber by Frequency Response Analysis," by R. I. Gray and J. W. Prados, *AIChE Journal*, **9**, 211 (1963).

28. (C) Consider the reaction mechanism

$$A \underset{k_2}{\overset{k_1}{\rightleftharpoons}} B \xrightarrow{k_3} C,$$

where each reaction is first-order, taking place in an isothermal batch reactor. Show how measured values of the moments of the composition versus time curves can be used to estimate the values of the rate constants.[12]

29. (C‡) Himmelblau and Bischoff[13] present an extensive discussion of the relationship between "transport type" models and "plug flow" models. In particular, they consider a general form of the diffusion equation

$$\frac{\partial c}{\partial t} + \frac{\partial (v_z c)}{\partial z} = D_z \frac{\partial^2 c}{\partial z^2} + D_r \frac{1}{r} \frac{\partial}{\partial r}\left(r \frac{\partial c}{\partial r}\right) + R_c$$

where the concentration c depends on both the radial and axial position in a tube, as well as time, and R_c represents a reaction rate term that is some function of c. They multiply each term in this equation by $2\pi r \, dr$, integrate from 0 to R (the tube radius), and divide by the cross-sectional area of the tube A to obtain

$$\frac{1}{A} \int_0^R \frac{\partial c(r, z)}{\partial t} 2\pi r \, dr + \frac{1}{A} \int_0^R \frac{\partial v_z(r, z)c(r, z)}{\partial z} 2\pi r \, dr$$

$$= \frac{D_z}{A} \int_0^R \frac{\partial^2 c(r, z)}{\partial z^2} 2\pi r \, dr + \frac{D_r}{A} \int_0^R \frac{1}{r} \frac{\partial}{\partial r}\left[r \frac{\partial c(r, z)}{\partial r}\right] 2\pi r \, dr$$

$$+ \frac{1}{A} \int_0^R R_c 2\pi r \, dr$$

By defining an average radial concentration, an average radial molar flux, and an average radial reaction rate as

$$c_{av}(z) = \frac{1}{A} \int_0^R c(r, z) 2\pi r \, dr \qquad [v_z(z)c(z)]_{av} = \frac{1}{A} \int_0^R v_z(r, z)c(r, z) 2\pi r \, dr$$

[11] D. F. Bruley and J. W. Prados, *AIChE Journal*, **10**, 612 (1964).
[12] This problem is similar to one discussed by J. B. Butt, *AIChE Journal*, **8**, 553, (1962).
[13] See footnote 5, p. 32

$$R_{c\,\text{av}} = \frac{1}{A}\int_0^R R_c(r, z)2\pi r\, dr$$

and by assuming that all the functions are well behaved, we can write the conservation equation as

$$\frac{\partial c_{\text{av}}(z)}{\partial t} + \frac{\partial [v_z(z)c(z)]_{\text{av}}}{\partial z} = D_z\frac{\partial^2 c_{\text{av}}(z)}{\partial z^2} + D_r\left(2\pi r\frac{\partial c(r, z)}{\partial r}\right)\Big|_0^R + R_{c\,\text{av}}$$

For most engineering problems of interest, we know that the radial concentration gradient will be equal to zero at the center of the pipe. Moreover, normally we assume that the flux at the wall can be described in terms of a mass-transfer coefficient (or possibly a reaction rate term if the wall is a reactant). Thus we write

$$D_r\left((2\pi r\frac{\partial c(r, z)}{\partial r})\right)\Big|_0^R = k_g a[c_{\text{av}}(z) - c_{\text{eq}}(z)]$$

Similarly, for many engineering systems, axial dispersion effects will be negligible in comparison to convective transport in the axial direction. With these simplifications, the conservation equation becomes

$$\frac{\partial c_{\text{av}}(z)}{\partial t} + \frac{\partial [v_z(z)c(z)]_{\text{av}}}{\partial z} = k_g a[c_{\text{av}}(z) - c_{\text{eq}}(z)] + R_{c\,\text{av}}$$

Now if we can assume that

$$[v_z(z)c(z)]_{\text{av}} = v_{z\,\text{av}}c_{\text{av}}(z)$$

and consider a case of fully developed flow, so that v_z does not change with axial position, our equation will be identical to a plug flow model. Select several functions that might be "reasonable" representations of a radial velocity profile and a radial concentration gradient and use these to calculate the quantities

$$\int_0^R v_z(r)c(r)2\pi r\, dr$$

and

$$\left[\int_0^R v_z(r)2\pi r\, dr\right]\left[\int_0^R c(r)2\pi r\, dr\right]$$

Also, consider the concentration and velocity fluctuations that are expected to occur at turbulent flow conditions and develop the "time-smoothed"[14] conservation equation. Average this expression over the radial coordinate and discuss its relationship to the normal plug flow equation.

After completing the work above, look up the derivations of the Taylor diffusion coefficient[15,16] and discuss the relationship of that quantity to your previous results.

30. (C) If we consider laminar flow in a circular pipe having a constant wall temperature, and neglect heat conduction in the axial direction, we would write the system equation as

$$\rho C_p\left[\frac{\partial T}{\partial t} + v(r)\frac{\partial T}{\partial z}\right] = \frac{k}{r}\frac{\partial}{\partial r}\left(r\frac{\partial T}{\partial r}\right)$$

[14] See p. 626 of the reference given in footnote 3 on p. 53.
[15] G. I. Taylor, *Proc. Roy. Soc.*, **A219**, 186 (1952).
[16] R. Aris, *Proc. Roy Soc.*, **A 235**, 67 (1956).

However, whenever possible we would like to replace this model by the simpler expression

$$\rho C_p \left[\frac{\partial T}{\partial t} + v \frac{\partial T}{\partial z} \right] = h(T_w - T)$$

which is based on the assumptions of plug flow and no radial temperature gradients. Can you develop a criterion that will indicate when the second approach is valid and the time dependence of the film-heat-transfer coefficient is negligible? There are two possible cases of interest: the first is an entrance region problem and the second is with fully developed flow.[17]

31. (C) It is readily apparent from Example 4.6-3 that it is a tedious task to calculate the dynamic response of a plate separation column, even for the case of a gas absorber with a linear equilibrium relationship. Hence it would be highly desirable to find a simple procedure for estimating an "effective" time constant for the unit. Fortunately, Pigford[18] has developed a technique of this type and has applied the method to the complete set of equations describing a distillation column. For simplicity, we will limit our attention to the absorber problem studied earlier.

First, show that the absorber equations can be written in the form

$$\frac{dx_n}{d\theta} = x_{n+1} - (1 + \beta)x_n + \beta x_{n-1}$$

With the equations in this form, Pigford suggested that if there are a large number of trays, we can replace the discrete variable n by a continuous variable. This step is accomplished by making the substitution

$$x_{n+j} = x_n + j \frac{\partial x_n}{\partial n} + \frac{j^2}{2} \frac{\partial^2 x_n}{\partial n^2}$$

Consider a situation where the column equations have been linearized around the final steady state operating point, so that the system proceeds from some initial condition on each plate to the final steady state, and show that an approximate model for the column is

$$\frac{\partial x}{\partial t} = \frac{1}{2}(1 + \beta)\frac{\partial^2 x}{\partial n^2} + (1 - \beta)\frac{\partial x}{\partial n}$$

with the boundary conditions

$$\text{at } n = 0, \quad \frac{\partial x}{\partial n} - x = 0 \quad \text{at } n = N, \quad \frac{\partial x}{\partial n} + x = 0$$

$$\text{at } t = 0, \quad x = x_i(n)$$

Separate the variables and use the spatial boundary conditions to establish a relationship for the eigenvalues. Next, substitute the system parameters given in Example 4.6-3 and calculate the smallest eigenvalue. Compare the time constant

[17] A comprehensive discussion of this problem was presented by J. C. Friedly, *Proceedings of the Joint Automatic Control Conference*, p. 296, University of Colorado, Boulder, Col., August 1969.

[18] See R. L. Pigford, *Dechema Monographien*, **53**, 217 (1965) or R. F. Jackson, and R. L. Pigford, *Ind. Eng. Chem.*, **48**, No. 6, 1020 (1956).

obtained in this way to the values listed in the problem and the response curve shown in Figure 4.6-5.

32. (C) Of course, Pigford's procedure, described above, for replacing difference-differential equations by partial differential equations is also applicable to sets of first-order equations. For example, in the polymerization problem described in Exercise 50, p. 226, we can first numerically solve for the functions $M_1(t)$ and $\phi(t)$, next we can introduce the transformation $d\tau = M_1\, dt$ to linearize the set of difference-differential equations, then we can use Pigford's method for replacing the discrete variable by a continuous one, and, finally, we can solve the partial differential equation describing the polymer distribution function. Find this solution and discuss in detail the significance of the dead time that appears in the result.[19]

33. (C) A common trick used by engineering faculty when writing textbooks is to present to their class as homework assignments some of the problems they encounter during consulting work. An interesting illustration of this type, which indicates the level of chemical engineering consulting activities, has been published by Levenspiel.[20] He cleverly attempts to hide behind an alias in the affair, but to any experienced author his association is apparent. The problem is reproduced below so that you can judge for yourself.

> The Lavender Hill Philanthropic Society is a worthy organization dedicated to the preservation of important historical monuments and buildings.
>
> Like all the other members of this close-knit group, you expect, in compensation for your efforts, only the satisfaction of seeing a job well done, plus a small token commission on each project. You have worked up steadily in the organization and at present have charge of the disposal team.
>
> Now the latest venture, and by far the boldest and grandest one ever undertaken by this group, is to save Fort Knox from sinking completely out of sight because of the excessive and unplanned-for overload on the foundations. The solution to this weighty problem is obvious: eliminate the overload.
>
> Advanced design and planning estimate that the foundation overload can be sufficiently lightened by removing 20 tons of long, cylindrical 1-in. diameter gold bars worth about $40 million. These will be delivered to you at 8:00 P.M. on *the* day. It is up to you to dispose of them as soon as possible, but certainly before 8:00 A.M. the next day, when visiting dignitaries of the various constabularies may be expected. After weighing the various alternatives, you hit on the ingenious plan of dumping the bars in the employees' swimming pool, which will be filled with aqua regia for that occasion.
>
> A literature search produces no useful rate data for this reaction, so an experiment is devised with the only sample of gold available, a $\frac{1}{2}$-in. diameter gold marble. The following results are obtained, using the same fluid as in the pool.

Size of marble (in.)	0.5	0.4	0.3	0.2	0.1	0
Time (min)	0	42	87	130	172	216

[19] See R. J. Zeeman and N. R. Amundson, *AIChE Journal*, **9**, 297 (1963) and *Chem. Eng. Sci.*, **20**, 231 (1965).

[20] See p. 383 of the reference given in footnote 2 on p. 12.

(a) What time can the bars be expected to disappear, and can the 8: 00 A.M. deadline be met?

(b) Certainly the earlier the bars dissolve, the safer the project is from unforeseen contingencies. With the thought that agitation may speed up the reaction, the project director helpfully suggests that the group's phychologist, and not too-reliable member, Harry, with a slight push or prod, may volunteer his services in agitating the pool. Would Harry's services be needed?

Note: Naturally the employees' swimming pool is large enough so that the acid strength is not appreciably lowered during reaction.

Try to use two different models to estimate the time for the sphere to dissolve. First, assume that there are no concentration gradients in the fluid phase and that the surface reaction rate is constant. Next, assume that the surface reaction rate has a first-order dependence on the acid concentration at the surface. Once you have compared these models for the sphere, solve Levenspiel's problem for the cylinders.

34. (C‡) In Exercise 4.8-23 we estimated the time to cool a granular solid in a rotating drum. How does the estimate change if we consider that the air passes through the drum in plug flow, rather than being perfectly mixed? What other kinds of models could you use to describe the cooling of the solid phase? Which of these can be handled without the aid of a computer?

35. (C‡) Our study of distributed parameter plants revealed that in some cases it was possible to observe a resonance-type phenomenon (see Figure 5.1-4). However, in our earlier discussion of the use of matrix techniques to estimate the dynamic response of a process (see Section 3.3), we replaced all partial differential equations by lumped parameter models. Therefore an obvious question we must answer is, Does discretization of the partial differential equations ever destroy the resonance behavior? For simplicity, compare the frequency response of the lumped and distributed parameter models describing a steam-heated exchanger with a sinusoidally varying condensate temperature but no wall capacitance. Then attempt to generalize your results.

Analysis of Nonlinear Systems: Perturbation Theory

6

Our previous analysis has shown that the mathematical models used to describe most chemical plants are nonlinear and often have variable coefficients. It is common practice to linearize these models around some steady state operating condition of interest and to use this approximate linear model to obtain a first estimate of the dynamic response of the system. The linearization is based on a Taylor series expansion; therefore it is possible to show that if the input fluctuations are small enough, the estimates obtained by using the linearized equations will be valid.

However, the problem of establishing the range over which the linear approximation leads to satisfactory results has received little attention in the literature. Most investigators make the assumption that linear models will be applicable over the entire range of engineering interest, and then they attempt to confirm this assumption by comparing experimental data with the predictions obtained from a linear analysis. Of course, this approach runs into difficulty when an exact confirmation of the theoretical prediction is not observed, for one is never certain whether the linearization or some other assumption introduced to simplify the model is the cause of the discrepancy.

Actually, a straightforward extension of simple perturbation analysis can be used to test the validity of the linearization technique. In addition, the method gives a qualitative insight into the different kinds of behavior of linear and nonlinear systems. Specifically, the results show that the frequency response of nonlinear plants contains higher harmonics, in addition to the fundamental component, and the time average value of the system output will be different than that predicted by a steady state analysis. Of course, this last result implies that dynamic operation of nonlinear systems might lead to a better performance than that corresponding to the optimum steady state design.

SECTION 6.1 LUMPED PARAMETER SYSTEMS

The original developments of perturbation theory were presented by Poincaré and others in the early 1900s to solve problems in celestial mechanics. In order to outline the application of the method to dynamically stable chemical processes,[1] we will begin by considering a lumped parameter plant that is described by the set of equations.

$$\frac{dx_i}{dt} = f_i(x_1, \ldots, x_n; u_1, \ldots, u_p; v_1, \ldots, v_r), \qquad i = 1, 2, \ldots, n \qquad 6.1\text{-}1$$

where $x_1, \ldots, x_n$ are the n state variables; $u_1, \ldots, u_p$ are the p inputs which can be manipulated (the control variables); $v_1, \ldots, v_r$ are the r inputs which cannot be manipulated (the disturbances); and f_i represents some arbitrary function of the quantities. The actual dynamic response of the plant to variations in any of the inputs can be obtained by solving this set of differential equations. The task is difficult, however, if any of the functions f_i are nonlinear or if some of the inputs appear as variable coefficients. Thus we hope that we can at least estimate the response for nonlinear systems by considering the simpler problem where we approximate the functions f_i by their Taylor series expansions about some steady state operating point of interest.

$$\frac{dx_i}{dt} = (f_i|_s) + (x_j - x_{js})\left(\frac{\partial f_i}{\partial x_j}\Big|_s\right) + (u_j - u_{js})\left(\frac{\partial f_i}{\partial u_j}\Big|_s\right) + (v_j - v_{js})\left(\frac{\partial f_i}{\partial v_j}\Big|_s\right).$$

$$+ \frac{1}{2}(x_j - x_{js})(x_k - x_{ks})\left(\frac{\partial^2 f_i}{\partial x_j \partial x_k}\Big|_s\right) + \frac{1}{2}(u_j - u_{js})(u_k - u_{ks})$$

$$\times \left(\frac{\partial^2 f_i}{\partial u_j \partial u_k}\Big|_s\right) + \frac{1}{2}(v_j - v_{js})(v_k - v_{ks})\left(\frac{\partial^2 f_i}{\partial v_j \partial v_k}\Big|_s\right) + (x_j - x_{js})$$

$$\times (u_k - u_{ks})\left(\frac{\partial^2 f_i}{\partial x_j \partial u_k}\Big|_s\right) \qquad\qquad 6.1\text{-}2$$

$$+ (u_j - u_{js})(v_k - v_{ks})\left(\frac{\partial^2 f_i}{\partial u_j \partial v_k}\Big|_s\right)$$

[1] The analysis described below is taken from A. B. Ritter and J. M. Douglas, *Ind. Eng. Chem. Fundamentals*, **9**, 21 (1970).

$$+ (x_j - x_{js})(v_k - v_{ks})\left(\frac{\partial^2 f_i}{\partial x_j \partial v_k}\bigg|_s\right) + \cdots$$

where a repeated index indicates summation. The argument proceeds that if we consider small enough changes in $(u_j - u_{js})$ and $(v_j - v_{js})$, the deviation from steady state conditions of the control and disturbance variables, then $(x_j - x_{js})$, the deviation of the state variables from their steady state values, will be small, and, furthermore, all the quadratic and higher-order terms in the expansion will be negligible in comparison with these linear terms. With this assumption, the system equations become a set of linear differential equations having constant coefficients, and we know that it is always possible to find analytical solutions for the dynamic response.

As an additional consideration, however, we would like to be able to ascertain the range of system parameters and the magnitudes of the input changes where the quadratic terms begin to become appreciable. This process can be accomplished by associating an artificial parameter μ with the quadratic and higher-order terms. For the simple case where the disturbances are equal to zero, Eq. 6.1-2 becomes

$$\frac{d(x_i - x_{is})}{dt} = a_{ij}(x_j - x_{js}) + b_{ij}(u_j - u_{js}) + \mu\left[\frac{1}{2}a_{ijk}(x_j - x_{js})\right.$$

$$\times (x_k - x_{ks}) + \frac{1}{2}b_{ijk}(u_j - u_{js})(u_k - u_{ks}) + c_{ijk}(x_j - x_{js})$$

$$\left. \times (u_k - u_{ks}) + \cdots\right] \qquad \text{6.1-3}$$

where

$$a_{ij} = \left(\frac{\partial f_i}{\partial x_j}\bigg|_s\right), \quad b_{ij} = \left(\frac{\partial f_i}{\partial u_j}\bigg|_s\right), \quad a_{ijk} = \frac{\partial^2 f_i}{\partial x_j \partial x_k}\bigg|_s, \quad \text{etc.} \qquad \text{6.1-4}$$

and the term $(f_i|_s)$ in Eq. 6.1-2 has been set equal to zero, for it is just the steady state system equation. Now we assume that we can find a solution of this set of equations, having the form

$$(x_i - x_{is}) = y_{is} + \mu y_{i1} + \mu^2 y_{i2} + \cdots \qquad \text{6.1-5}$$

where $y_{i0}, y_{i1}, \ldots$ are functions of time still to be determined. Substituting this solution into the system equation gives

$$\frac{dy_{i0}}{dt} + \mu\frac{dy_{i1}}{dt} + \mu^2\frac{dy_{i2}}{dt} + \cdots = a_{ij}(y_{j0} + \mu y_{j1} + \mu^2 y_{j2} + \cdots)$$

$$+ b_{ij}(u_j - u_{js}) + \mu\left[\frac{1}{2}a_{ijk}(y_{j0} + \mu y_{j1} + \mu^2 y_{j2} + \cdots)\right.$$

$$\times (y_{k0} + \mu y_{k1} + \mu^2 y_{k2} + \cdots) + \frac{1}{2}b_{ijk}(u_j - u_{js})(u_k - u_{ks})$$

$$\left. + c_{ijk}(u_k - u_{ks})(y_{j0} + \mu y_{j1} + \mu^2 y_{j2} + \cdots) + \cdots + \cdots\right]$$

$$\text{6.1-6}$$

If we equate terms having like coefficients of μ, we obtain the set of equations

$$\frac{dy_{i0}}{dt} = a_{ij}y_{j0} + b_{ij}(u_j - u_{js}) \tag{6.1-7}$$

$$\frac{dy_{i1}}{dt} = a_{ij}y_{j1} + \left[\frac{1}{2}a_{ijk}y_{j0}y_{k0} + \frac{1}{2}b_{ijk}(u_j - u_{js})(u_k - u_{ks})\right.$$

$$\left. + c_{ijk}(u_k - u_{ks})y_{j0} + \cdots\right] \tag{6.1-8}$$

$$\frac{dy_{i2}}{dt} = a_{ij}y_{j2} + \left[\frac{1}{2}a_{ijk}(y_{j0}y_{k1} + y_{k0}y_{j1}) + c_{ijk}(u_k - u_{ks})y_{j1} + \cdots\right] \tag{6.1-9}$$

This set of differential equations defines the functions $y_{i0}, y_{i1}, \ldots$, providing that we also specify a set of boundary conditions for the equations. However, it is immediately apparent that we have replaced a set of nonlinear ordinary differential equations which might have variable coefficients by a larger set of nonhomogeneous, linear differential equations having constant coefficients. This is a tremendous advantage, since we can always find solutions for this latter case, whereas we can obtain only numerical solutions for the former case. Of course, we pay the price of having to solve two or three times as many equations in order to obtain this simplification, but for most engineering systems the investment is worthwhile.

After a closer examination of Eqs. 6.1-7 through 6.1-9, we find that the first equation, which is called the *generating solution*, is identical to the linearized equation normally used to describe the system (assuming that we let $y_{i0} = x_i - x_{is}$ at $t = 0$). Thus if we impose some change on the system by manipulating one of the control variables, such as a sinusoidal variation, we will obtain the usual results for the linear frequency response from the first set of equations. The equations for the first-order correction functions, y_{i1}, given by Eq. 6.1-8 must have exactly the same complementary solutions as were obtained in the linear frequency response problem, but there will be a new particular integral that depends on both the solution of the linear response problem and the input signal (the terms in brackets in Eq. 6.1-8). We take the initial conditions for this set of equations as $y_{i1} = 0$ at $t = 0$, since we have previously matched the initial conditions for $(x_1 - x_{is})$ at $t = 0$ with y_{i0}. A similar result is obtained for the equation for the second-order correction function, y_{i2}. Hence the problem of finding solutions of the expanded set of equations actually requires only the determination of a number of particular integrals.

Once the correction functions have been determined, we can find the system outputs by substituting the results into Eq. 6.1-5. The parameter μ was introduced in an artificial manner (see Eqs. 6.1-2 and 6.1-3), and therefore we must set μ equal to unity in the final equation. Although this procedure seems somewhat contradictory, what we are actually doing is using this parameter to keep track of the order of magnitude of various terms. We know that in some sufficiently small neighborhood of the steady state operating condition, the linearized equations will provide a valid description of the system dynamics. In this region the quadratic and higher-order terms in the state equations, Eq. 6.1-3, will be negligible in comparison with the linear terms; and the first- and second-order correction functions in the solution, which have coefficients of μ and μ^2, are negligible in comparison with the generating solution. As we increase the size of the input fluctuations, there will be some point where the quadratic and higher-order terms in the state equations start to have a significant influence on the output. In terms of our assumed solution, Eq. 6.1-5, this means that the first-order correction functions, y_1 start to become appreciable. Similarly, as we continue to increase the size of the inputs, we expect that the second-order correction functions will increase in importance, and so forth.

Of course, these transitions will take place in a continuous manner, and the significance of various terms is a matter of engineering judgment. However, if the system parameters and amplitude of the input signal are such that the first-order correction function is always less than 10 percent of the generating solution, then for most practical purposes we would be willing to state that the linearized model provides a valid description of the plant. Several examples of the method are given below.

Example 6.1-1 Step response of a nonlinear, isothermal reactor problem

In Example 4.3-1 we studied the effect of a pair of rectangular pulses on the dynamic response of an isothermal, stirred-tank reactor for the case of a second-order reaction. The material balance was

$$V\frac{dA}{dt} = q(A_f - A) - kVA^2 \qquad \text{3.1-5}$$

or after making the transformation of variables,

$$y = \frac{A - A_s}{A_{fs}}, \quad y_f = \frac{A_f - A_{fs}}{A_{fs}}, \quad \tau = \frac{1}{2kA_s + q/V}, \quad \theta = \frac{t}{\tau}$$

$$\beta = \frac{k}{2kA_s + q/V} \qquad \gamma = \frac{q/V}{2kA_s + q/V}$$

we obtained the state equation

$$\tau\frac{dy}{dt} = -(y + \beta y^2) + \gamma y_f \qquad \text{4.3-34}$$

When we considered a simple step change in the feed composition, $y_f = a$, the equation could be solved analytically, and we found that the result was

$$y = \frac{\sqrt{1 + 4a\beta\gamma} - 1}{2\beta}$$
$$\times \left\{ \frac{1 - \exp\left[-\sqrt{1 + 4a\beta\gamma}(t/\tau)\right]}{1 - [(1 - \sqrt{1 + 4a\beta\gamma})/(1 + \sqrt{1 + 4a\beta\gamma})] \exp\left[-\sqrt{1 + 4a\beta\gamma}(t/\tau)\right]} \right\}$$

4.3-45

Find the perturbation solution for the step response of this system.

Solution

The state equation is

$$\tau\frac{dy}{dt} = -(y + \beta y^2) + \gamma y_f$$

6.1-10

and if we expand the nonlinear function on the right-hand side of the equation in a Taylor series around the steady state operating point, $y = 0$ and $y_f = 0$, we obtain the same equation. Then we introduce the artificial parameter μ in front of the nonlinear term

$$\tau\frac{dy}{dt} = -y + \gamma y_f - \mu\beta y^2$$

6.1-11

and assume a solution

$$y = y_0 + \mu y_1 + \mu^2 y_2 + \cdots$$

6.1-12

Substituting this solution into the state equation and equating the terms having like coefficients of μ, we obtain

$$\tau\frac{dy_0}{dt} + y_0 = \gamma y_f$$

6.1-13

$$\tau\frac{dy_1}{dt} + y_1 = -\beta y_0^2$$

6.1-14

$$\tau\frac{dy_2}{dt} + y_2 = -2\beta y_0 y_1$$

6.1-15

The first equation in this set is identical to the linearized system equation, Eq. 4.3-35. For a step input, $y_f = a$; and taking the boundary conditions as $y_0 = 0$ at $t = 0$, the solution of this equation is

$$y_0 = \gamma a(1 - e^{-t/\tau})$$

6.1-16

Now we substitute this expression into the right-hand side of Eq. 6.1-14

$$\tau\frac{dy_1}{dt} + y_1 = -\beta\gamma^2 a^2(1 - e^{-t/\tau})^2$$

6.1-17

Since we imposed the system boundary conditions on the generating solution, y_0, we take the boundary conditions for this equation as $y_1 = 0$ at $t = 0$. The solution is

$$y_1 = -\beta\gamma^2 a^2\left(1 - 2\frac{t}{\tau}e^{-t/\tau} - e^{-2t/\tau}\right)$$

6.1-18

Similarly, if we substitute this result into the right-hand side of Eq. 6.1-15 and solve the equation with a zero initial condition, we obtain

$$y_2 = 2\beta^2\gamma^3 a^3 \left\{ 1 + \left[\frac{5}{2} - \frac{t}{\tau} - \left(\frac{t}{\tau} \right)^2 \right] e^{-t/\tau} - \left(3 + \frac{2t}{\tau} \right) e^{-2t/\tau} - \frac{1}{2} e^{-3t/\tau} \right\}$$

Hence, according to Eq. 6.1-12, an approximate solution of the equation is

$$y = \gamma a (1 - e^{-t/\tau}) - \mu \beta \gamma^2 a^2 \left[1 - 2\frac{t}{\tau} e^{-t/\tau} - e^{-2t/\tau} \right]$$

$$+ \mu^2 2 \beta^2 \gamma^3 a^3 \left\{ 1 + \frac{5}{2} - \frac{t}{\tau} - \left(\frac{t}{\tau} \right)^2 \right] e^{-t/\tau}$$

$$- \left(3 + \frac{2t}{\tau} \right) e^{-2t/\tau} - \frac{1}{2} e^{-3t/\tau} \right\} + \cdots \qquad\qquad 6.1\text{-}20$$

We must set μ equal to unity in this result, for this small parameter was introduced in an artificial manner to keep track of terms having different orders of magnitude. However, if the quantity $\beta\gamma a$ is small, the coefficients of the correction functions become small in the same way as was assumed for μ. In fact, if we consider that the group $\mu\beta\gamma a$ is a small parameter, rather than just μ, all the arguments of classical perturbation theory become valid for our problem. This fact becomes more apparent if we define a new set of variables as

$$z = \frac{y}{\gamma a}, \quad \theta = \frac{t}{\tau}, \quad \mu = \beta\gamma a, \quad y_f = a \qquad\qquad 6.1\text{-}21$$

and write the original equation, Eq. 6.1-11, in the form

$$\frac{dz}{d\theta} = -z + 1 + \mu z^2 \qquad\qquad 6.1\text{-}22$$

In order to assess the value of perturbation analysis, we first compare the solution of the linearized equation, Eq. 6.1-16, which often is the only result considered in a dynamic study, with the exact analytical solution, Eq. 4.3-45. The first thing we notice is that the linear equation does not predict the correct value of the final steady state; that is,

$$\frac{\sqrt{1 + 4\beta\gamma a} - 1}{2\beta} \neq \gamma a \qquad \text{unless } \beta = 0 \qquad\qquad 6.1\text{-}23$$

This is a somewhat trivial result, for we know that if we set the accumulation term in Eq. 4.3-34 equal to zero and solve the steady state equations for both the linear and nonlinear cases, we cannot possibly obtain the same solutions. Nevertheless, this point is often overlooked in dynamic studies. The other apparent differences in the equations are that the time constant in the exponent and the functional dependence on time do not agree.

It is interesting to note, however, that if we expand the exact solution in a power series in β, we obtain the result given by Eq. 6.1-20. Hence the linear solution will provide an adequate engineering approximation whenever the term $\beta\gamma a$ is small. Although the analytical solution is not available for most cases, we can obtain essentially the same information by merely comparing the first correction function y_1, Eq. 6.1-18, with the linear response, Eq. 6.1-16. In other words, for engineering purposes we can consider that the linearized system equations provide an adequate approximation of the nonlinear model whenever the first-order correction function is negligible (less than 10 percent) compared with the linear response.

Example 6.1-2 Response of an isothermal reactor to sinusoidal flow rate variations

Exactly the same procedure can be used to estimate the effects of variable coefficients in a system equation. For example, if we consider sinusoidal variations in the flow rate to a CSTR with a first-order reaction, the system equation is

$$V\frac{dx}{dt} = q(x_f - x) - kVx = q_s(1 + a \sin \omega_0 t)(x_f - x) - kVx \qquad 6.1\text{-}24$$

Letting

$$x = x_s + y, \quad \tau = \frac{q_s t}{V}, \quad \alpha = \frac{kV}{q_s}, \quad \beta = x_f - x_s, \quad \omega = \frac{\omega_0 V}{q_s} \qquad 6.1\text{-}25$$

the system equation becomes

$$\frac{dy}{d\tau} + (1 + \alpha + a \sin \omega\tau)y = \beta a \sin \omega\tau \qquad 6.1\text{-}26$$

which is slightly different from the standard form we have been using. This is a first-order, linear differential equation; thus we can write its solution as

$$y = ce^{-(1+\alpha)\tau} e^{(a/\omega) \cos \omega\tau} + e^{-(1+\alpha)\tau} e^{(a/\omega) \cos \omega\tau}$$
$$\times \int \beta a \sin \omega\tau \, e^{(1+\alpha)\tau} e^{-(a/\omega) \cos \omega\tau} \, d\tau \qquad 6.1\text{-}27$$

where c is an integration constant. Using the identity[2]

$$e^{(a/\omega) \cos \omega\tau} = I_0\left(\frac{a}{\omega}\right) + 2\sum_{n=1}^{\infty} I_n\left(\frac{a}{\omega}\right) \cos n\omega\tau \qquad 6.1\text{-}28$$

where I_0 and I_n are modified Bessel functions of the first kind, the equation can be written

$$y = ce^{-(1+\alpha)\tau}\left[I_0\left(\frac{a}{\omega}\right) + 2\sum_{n=1}^{\infty} I_n\left(\frac{a}{\omega}\right) \cos n\omega\tau\right]$$
$$+ \beta ae^{-(1+\alpha)\tau}\left[I_0\left(\frac{a}{\omega}\right) + \sum_{n=1}^{\infty} I_n\left(\frac{a}{\omega}\right) \cos n\omega\tau\right]\int \sin \omega\tau \, e^{(1+\alpha)\tau}$$
$$\times \left[I_0\left(\frac{-a}{\omega}\right) + 2\sum_{p=1}^{\infty} I_p\left(\frac{-a}{\omega}\right) \cos p\omega\tau\right] d\tau \qquad 6.1\text{-}29$$

[2] See p. 370 of the reference given in footnote 6 on p. 86.

After carrying out the integration, introducing the relationships for even and odd functions for the Bessel functions, dropping the term involving the integration constant so that we consider only the pseudo-steady state output, we obtain

$$y = \left[I_0\left(\frac{a}{\omega}\right) + 2 \sum_{n=0}^{\infty} I_n\left(\frac{a}{\omega}\right) \cos n\omega\tau \right] \left\{ \sum_{p=0}^{\infty} \left[\frac{(-1)^p \beta a [I_p(a/\omega) - I_{p+2}(a/\omega)]}{(1+\alpha)^2 + (1+p)^2 \omega^2} \right] \right.$$
$$\times \left[(1+\alpha) \sin (p+1)\omega\tau - (p+1)\omega \cos (p+1)\omega\tau\right] \qquad \text{6.1-30}$$

It is apparent that this exact solution will contain an infinite number of higher harmonic components in addition to the fundamental and will have a non-zero mean value, for it contains cosine terms raised to even powers.

It is a fairly difficult task to find the analytical solution even for this very simple linear differential equation with a periodic coefficient; therefore we would like to see if we can obtain essentially the same results using perturbation theory. If this process is possible, we might also expect to be successful when the state equation was nonlinear and an analytical solution is not available or if we had sets of coupled state equations with periodic coefficients. The procedure used to develop an approximate solution is essentially the same as described previously. We rewrite the state equation, Eq. 6.1-26, as

$$\frac{dy}{d\tau} + (1+\alpha)y = \beta a \sin \omega\tau - \mu a \sin \omega\tau y \qquad \text{6.1-31}$$

where a small parameter μ has been artificially introduced in front of the terms which make an analytical solution difficult, and then we assume a solution having the form

$$y = y_0 + \mu y_1 + \mu^2 y_2 + \cdots \qquad \text{6.1-32}$$

Substituting this assumed solution into the state equation and equating terms having like coefficients of μ leads to the set of equations

$$\frac{dy_0}{d\tau} + (1+\alpha)y_0 = \beta a \sin \omega\tau \qquad \text{6.1-33}$$

$$\frac{dy_1}{d\tau} + (1+\alpha)y_1 = -a \sin \omega\tau \, y_0 \qquad \text{6.1-34}$$

$$\frac{dy_2}{d\tau} + (1+\alpha)y_2 = -a \sin \omega\tau \, y_1 \qquad \text{6.1-35}$$

$$\vdots$$

The generating solution—the solution of Eq. 6.1-33—is

$$y_0 = c_0 e^{-(1+\alpha)\tau} + \left[\frac{\beta a}{(1+\alpha)^2 + \omega^2} \right] [(1+\alpha) \sin \omega\tau - \omega \cos \omega\tau]$$

We are interested only in the pseudo-steady state response, so we set $c_0 = 0$

to eliminate the transient term. Thus

$$y_0 = \left[\frac{\beta a}{(1 + \alpha)^2 + \omega^2}\right][(1 + \alpha)\sin\omega\tau - \omega\cos\omega\tau] \qquad 6.1\text{-}36$$

This result is simply the linear frequency response of the system.

Now if we use this expression to eliminate y_0 from Eq. 6.1-31, we obtain

$$\frac{dy_1}{d\tau} + (1 + \alpha)y_1 = \left[\frac{-\beta a^2/2}{(1 + \alpha)^2 + \omega^2}\right]$$

$$\times [(1 + \alpha) - (1 + \alpha)\cos 2\omega\tau - \omega\sin 2\omega\tau] \qquad 6.1\text{-}37$$

where the trigonometric identities

$$\sin^2\omega\tau = \tfrac{1}{2}(1 - \cos 2\omega\tau) \qquad \cos\omega\tau\sin\omega\tau = \tfrac{1}{2}\sin 2\omega\tau \qquad 6.1\text{-}38$$

have been used to simplify the right-hand side of the equation. The pseudo-steady state solution for the first correction function, y_1, can be shown to be

$$y_1 = \frac{-\beta a^2/2}{(1 + \alpha)^2 + \omega^2} + \frac{\beta a^2/2}{[(1 + \alpha)^2 + \omega^2][(1 + \alpha)^2 + 4\omega^2]}$$

$$\times \{3\omega(1 + \alpha)\sin 2\omega\tau + [(1 + \alpha)^2 - 2\omega^2]\cos 2\omega\tau\} \qquad 6.1\text{-}39$$

Thus the first correction function term introduces the second harmonic and has a nonzero mean value—that is, a "DC" component.

It is possible to use this result to eliminte y_1 from Eq. 6.1-35 and then to solve for the second correction term y_2. This procedure would lead to third harmonic terms and an additional correction to the linear frequency response. However, we are mainly interested in developing a technique for use in esti-mating the error in a linear frequency response analysis, and thus we need consider only the first correction term. The approximate solution becomes

$$y = y_0 + \mu y_1 + \cdots \qquad 6.1\text{-}32$$

or

$$y = \left[\frac{\beta\alpha}{(1 + \alpha)^2 + \omega^2}\right][(1 + \alpha)\sin\omega\tau - \omega\cos\omega\tau] - \frac{\mu\beta a^2/2}{(1 + \alpha)^2 + \omega^2}$$

$$+ \frac{\mu\beta a^2/2}{[(1 + \alpha)^2 + \omega^2][(1 + \alpha)^2 + 4\omega^2]}$$

$$\times \{3\omega(1 + \alpha)\sin 2\omega\tau + [(1 + \alpha)^2 - 2\omega^2]\cos 2\omega\tau\} + \cdots \qquad 6.1\text{-}40$$

If the quantity μa was a small number, we expect that our approximate solution will coverage rapidly. This condition can be met if the amplitude of the forcing signal, a, is small even if we set μ equal to unity, which is re-quired because μ was introduced arbitrarily. Of course, in this problem it was not really necessary to use a parameter μ, for we could carry out the perturbation analysis in terms of the amplitude a.

In order to compare the approximate solution with the exact analytical solution, we can first write the explicit form of the first few terms of the analytical solution.

$$y = \left[I_0\left(\frac{a}{\omega}\right) + 2I_1\left(\frac{a}{\omega}\right)\cos\omega\tau + \cdots\right]\left\{\frac{\beta a[I_0(a/\omega) - I_2(a/\omega)]}{(1+\alpha)^2 + \omega^2}[(1+\alpha)\sin\omega t\right.$$

$$- \omega\cos\omega\tau] - \frac{\beta a[I_1(a/\omega) - I_3(a/\omega)]}{(1+\alpha)^2 + 4\omega^2}$$

$$\times [(1+\alpha)\sin 2\omega\tau - 2\omega\cos 2\omega\tau] + \cdots\}$$

Expanding this result and using trigonometric identities to replace products of trigonometric functions by harmonic terms, we obtain

$$y = \frac{\beta aI_0(a/\omega)[I_0(a/\omega) - I_2(a/\omega)]}{(1+\alpha)^2 + \omega^2}[(1+\alpha)\sin\omega\tau - \omega\cos\omega\tau]$$

$$+ \frac{2\beta aI_1(a/\omega)[I_0(a/\omega) - I_2(a/\omega)]}{(1+\alpha)^2 + \omega^2}\left(\frac{-\omega}{2}\right)$$

$$+ \frac{-\beta aI_0(a/\omega)[I_1(a/\omega) - I_3(a/\omega)]}{(1+\alpha)^2 + 4\omega^2}[(1+\alpha)\sin 2\omega\tau - 2\omega\cos 2\omega\tau]$$

$$+ \frac{2\beta aI_1(a/\omega)[I_0(a/\omega) - I_2(a/\omega)]}{(1+\alpha)^2 + \omega^2}\frac{1}{2}[1+\alpha)\sin 2\omega\tau - \omega\cos 2\omega\tau]$$

$$+ \cdots \qquad\qquad\qquad\qquad\qquad\qquad\qquad\qquad\qquad 6.1\text{-}41$$

Now if we use the small argument approximations for Bessel functions, that is,

$$I_p(x) \simeq \frac{x^p}{2^p p!} \qquad \text{for small } x \qquad\qquad 6.1\text{-}42$$

neglect all terms of higher order than a^2, and collect terms, we obtain

$$y = \left[\frac{\beta a}{(1+\alpha)^2 + \omega^2}\right][(1+\alpha)\sin\omega\tau - \omega\cos\omega\tau] - \frac{\beta a^2/2}{(1+\alpha)^2 + \omega^2}$$

$$+ \frac{3\beta a^2\omega(1+\alpha)/2}{[(1+\alpha)^2 + \omega^2][(1+\alpha)^2 + 4\omega^2]}\sin 2\omega\tau$$

$$+ \frac{(\beta a^2/2)[1+\alpha)^2 - 2\omega^2]}{[(1+\alpha)^2 + \omega^2][(1+\alpha)^2 + 4\omega^2]}\cos 2\omega\tau + \cdots \qquad 6.1\text{-}43$$

It is a simple matter to show that this expansion of the analytical solution is identical to the perturbation solution, Eq. 6.1-40, when μ is set equal to unity. Since the perturbation analysis gives us a valid approximate solution in this case, we hope that it will also yield fruitful results when exact analytical solutions are not available.

Example 6.1-3 Frequency response of a nonisothermal,
continuous-stirred-tank reactor

The equations describing the dynamic behavior of a nonisothermal CSTR were developed in Example 3.1-3, and the procedure used to linearize the equations was presented in Example 3.3-1. Then an illustration of the linear system response was discussed in Example 4.5-2. In order to estimate the response of the nonlinear system,[3] or to determine the region where the linearized equations are valid, we want to develop a solution using perturbation theory. However, rather than treat the original set of nonlinear equations, we will expand the nonlinear terms in a Taylor series and keep all the terms up to second-order. Although this approach will give us only an approximation of the nonlinear system response, it greatly simplifies the analysis.

Thus if we consider the original set of equations

$$V\frac{dA}{dt} = q(A_f - A) - kVA \qquad 3.1\text{-}23$$

$$VC_p\rho\frac{dt}{dt} = qC_p\rho(T_f - T) + (-\Delta H)kVA + \frac{UA_H Kq_H}{1 + Kq_H}(T_H - T) \qquad 3.1\text{-}24$$

where

$$K = \frac{2C_{pH}\rho_H}{UA_H} \qquad k = k_0 \, e^{-E/RT} \qquad 3.1\text{-}25$$

and expand the right-hand sides in a Taylor series for a case where the feed composition is the only time-variable input, we obtain

$$\frac{dx_1}{d\tau} = a_{11}x_1 + a_{12}x_2 + c_{11}x_f + m_{11}x_1 x_2 + m_{12}x_2^2 + \cdots \qquad 6.1\text{-}44$$

$$\frac{dx_2}{d\tau} = a_{21}x_1 + a_{22}x_2 + m_{21}x_1 x_2 + m_{22}x_2^2 + \cdots \qquad 6.1\text{-}45$$

where

$$x_1 = \frac{A - A_s}{A_s}, \quad x_2 = \frac{T - T_s}{T_s}, \quad x_f = \frac{A_f - A_{fs}}{A_s}, \quad \tau = \frac{qt}{V}$$

$$a_{11} = -\left(1 + \frac{k_s V}{q}\right), \quad a_{12} = -\left(\frac{k_s V}{q}\right)\left(\frac{E}{RT_s}\right), \quad a_{21} = \left(\frac{k_s V}{q}\right)\left[\frac{(-\Delta H)A_s}{C_p\rho T_s}\right]$$

$$a_{22} = -\left[1 + \left(\frac{Kq_H}{1 + Kq_H}\right)\frac{UA_H}{q_s C_p\rho} - \left(\frac{k_s V}{q}\right)\left(\frac{E}{RT_s}\right)\frac{(-\Delta H)A_s}{C_p\rho T_s}\right] \qquad 6.1\text{-}46$$

$$m_{11} = \left(\frac{k_s V}{q}\right)\left(\frac{E}{RT_s}\right), \quad m_{12} = -\left(\frac{k_s V}{q}\right)\left(\frac{E}{RT_s}\right)\left(\frac{E}{RT_s} - 2\right),$$

$$m_{21} = \left(\frac{k_s V}{q}\right)\left[\frac{(-\Delta H)A_s}{C_p\rho T_s}\right]\left(\frac{E}{RT_s}\right),$$

$$m_{22} = \left(\frac{k_s V}{2q}\right)\left[\frac{(-\Delta H)A_s}{C_p\rho T_s}\right]\left(\frac{E}{RT_s}\right)\left(\frac{E}{RT_s} - 2\right)$$

Although this terminology is slightly different from that used previously, it is a simple matter to convert from one set of definitions to the other.

Now if we introduce an artificial small parameter μ before the nonlinear

[3] See footnote 1, p. 306.

terms

$$\frac{dx_1}{d\tau} = a_{11}x_1 + a_{12}x_2 + c_{11}x_f + \mu(m_{11}x_1x_2 + m_{12}x_2^2) \qquad 6.1\text{-}47$$

$$\frac{dx_2}{d\tau} = a_{21}x_1 + a_{22}x_2 + \mu(m_{21}x_1x_2 + m_{22}x_2^2) \qquad 6.1\text{-}48$$

assume a solution having the form

$$x_1 = y_0 + \mu y_1 + \mu^2 y_2 + \cdots \qquad 6.1\text{-}49$$

$$x_2 = z_0 + \mu z_1 + \mu^2 z_2 + \cdots \qquad 6.1\text{-}50$$

substitute these expressions into Eqs. 6.1-47 and 6.1-48, and equate terms having like coefficients of μ, we find that

$$\frac{dy_0}{d\tau} = a_{11}y_0 + a_{12}z_0 + x_f \qquad 6.1\text{-}51$$

$$\frac{dz_0}{d\tau} = a_{21}y_0 + a_{22}z_0 \qquad 6.1\text{-}52$$

$$\frac{dy_1}{d\tau} = a_{11}y_1 + a_{12}z_1 + m_{11}y_0z_0 + m_{12}z_0^2 \qquad 6.1\text{-}53$$

$$\frac{dy_2}{d\tau} = a_{21}y_1 + a_{22}z_1 + m_{21}y_0z_0 + m_{22}z_0^2 \qquad 6.1\text{-}54$$

$$\vdots$$

For sinusoidal variations in A_f,

$$x_f = A_0 \sin \omega\tau \qquad 6.1\text{-}55$$

the solutions of Eqs. 6.1-51 and 6.1-52 for the case of an underdamped system give the linear frequency response

$$y_0 = \frac{(A_0/A_s)\tau^2 \sqrt{a_{22}^2 + \omega^2}}{\sqrt{(1 - \tau^2\omega^2)^2 + (2\gamma\tau\omega)^2}} \sin(\omega\tau + \phi_1) \qquad 6.1\text{-}56$$

$$z_0 = \frac{a_{21}(A_0/A_s)\tau^2}{\sqrt{(1 - \tau^2\omega^2)^2 + (2\gamma\tau\omega)^2}} \sin(\omega\tau + \phi_2) \qquad 6.1\text{-}57$$

where

$$\frac{1}{\tau^2} = a_{11}a_{22} - a_{12}a_{21} \qquad \gamma = -\frac{\tau}{2}(a_{11} + a_{12}) \qquad 6.1\text{-}58$$

$$\phi_1 = \tan^{-1}\left(\frac{-2\gamma\tau\omega}{1 - \tau^2\omega^2}\right) \qquad \phi_2 = \tan^{-1}\left(\frac{-2\gamma\tau\omega}{1 - \tau^2\omega^2}\right) + \tan^{-1}\left(\frac{-\omega}{a_{22}}\right)$$

After substituting these results into Eqs. 6.1-53 and 6.1-54 and using trigonometric identities to replace the powers of trigonometric functions by constants and second harmonics, we obtain a set of equations having the same complementary solution as before, an exponentially damped sine wave, and a particular integral containing constants and second harmonics. For example, the pseudo-steady state solution for y_1 is

$$y_1 = y_{1 \text{ av}} + K_1 \sin 2\omega\tau + K_2 \cos 2\omega\tau \qquad \text{6.1-59}$$

where

$$y_{1 \text{ av}} = \frac{a_{21}(A_0/A_s)^2\tau^6}{2}$$

$$\times \left[\frac{(a_{12}m_{21} - a_{22}m_{11})\sqrt{a_{22}^2 + \omega^2}\cos(\phi_2 - \phi_1) + (a_{12}m_{22} - a_{22}m_{12})a_{21}}{(1 - \tau^2\omega^2)^2 + (2\gamma\tau\omega)^2} \right]$$

$$\text{6.1-60}$$

and K_1 and K_2 are constants that can be evaluated in terms of the system parameters after some algebraic manipulation.[4] A similar result can be obtained for z_1. Hence the complete solution for the effluent composition is

$$x_1 = y_0 + \mu y_1 + \cdots = \frac{(A_0/A_s)\tau^2\sqrt{a_{22}^2 + \omega^2}}{\sqrt{(1 - \tau^2\omega^2)^2 + (2\gamma\tau\omega)^2}} \sin(\omega\tau + \phi) + y_{1 \text{ av}}$$

$$+ K_1 \sin 2\omega\tau + K_2 \cos 2\omega\tau \qquad \text{6.1-61}$$

Actually, the last three terms in this result should be multiplied by μ, but since we introduced this parameter in an artificial manner, we must set μ equal to unity. However, we still expect that the approximate solution will be valid, providing that the coefficients of the last three terms are small in comparison to the linear system response. Of course, if this is not the case, either more correction terms must be determined to improve the accuracy of the approximate solution or the equations have to be solved numerically. In any case, the procedure does make it possible to assess the validity of the linearization technique. A comparison between the perturbation solution and a numerical solution is given in Section 10.1 of Volume 2.

SECTION 6.2 DISTRIBUTED PARAMETER SYSTEMS

Exactly the same technique can be used to estimate the response of distributed parameter systems described by equations containing nonlinear terms or time-variable coefficients. Again we try to rearrange the equations and introduce an artificial, small parameter μ such that when μ is set equal to zero, we obtain the normal linearized response of the system. Then by assuming a solution as a power series in μ, we develop a set of equations for the generating solution and correction functions. Many times the complementary solutions of all these equations will be the same, but the particular integrals will all

---

[4] The complete expressions are given in A. B. Ritter, *Frequency Response of Nonlinear Systems*, M.S. Thesis, University of Rochester, Rochester, N.Y., 1968.

be different and will depend on the lower-order terms in the solution. Thus we replace nonlinear equations, or ones with variable coefficients, by a larger set of linear equations with constant coefficients, but that are nonhomogeneous. Although, in theory, it is always easier to solve this expanded set of equations, in some cases the algebra becomes very tedious.

Example 6.2-1 Response of a steam-heated exchanger to sinusoidal flow rate variations

The equations describing the dynamic response of a steam-heated exchanger were discussed in Section 5.1. For the case where the thermal capacity of the metal walls is negligible, the system equation becomes

$$C_p \rho A \frac{\partial T}{\partial t} + C_p \rho A v \frac{\partial T}{\partial z} = h A_H (T_c - T) \qquad 6.2\text{-}1$$

At steady state conditions, the accumulation term $\partial T/\partial t$ is equal to zero, and it is a simple matter to show that the steady state temperature profile is

$$T_s = T_{cs} + (T_f - T_{cs})e^{-Bz/v_s} \qquad 6.2\text{-}2$$

where T_f is the feed temperature and

$$B = \frac{U A_H}{C_p \rho A} \qquad 6.2\text{-}3$$

In order to study the dynamic response of the system to flow rate changes, we introduce the deviation variables

$$T = T_s + \theta \qquad v = v_s + u \qquad 6.2\text{-}4$$

Substituting these expressions into the unsteady state energy balance and subtracting the steady state equation gives the result

$$\frac{\partial \theta}{\partial t} + v_s \frac{\partial \theta}{\partial z} + u \frac{\partial T_s}{\partial z} + u \frac{\partial \theta}{\partial z} = -B\theta \qquad 6.2\text{-}5$$

This is a linear equation with a time-variable coefficient. Actually, the overall heat-transfer coefficient U, and therefore the term B, also depends on the flow velocity, but we will assume that the variation is small and can be neglected. Although it is possible to find an analytical solution for the system response,[1,2,3] the result is cumbersome in use because, in general, it cannot be written in an explicit form. Therefore we will attempt to develop a simple approximate solution, using perturbation theory.[4]

It is clear that it would be an easy task to determine the system response, providing that the last term on the left-hand side of the equation was missing.

[1] L. B. Koppel, *Ind. Eng. Chem. Fundamentals*, **1**, 131 (1962).
[2] W. H. Ray, *Ind. Eng. Chem. Fundamentals*, **5**, 138 (1966).
[3] D. D. Penrod, and E. D. Crandall, *Ind. Eng. Chem. Fundamentals*, **5**, 581 (1966).
[4] See footnote 1, p. 306.

Therefore we multiply this term by the parameter μ

$$\frac{\partial \theta}{\partial t} + v_s \frac{\partial \theta}{\partial z} + B\theta = \frac{\partial T_s}{\partial z} u - \mu u \frac{\partial \theta}{\partial z} \qquad 6.2\text{-}6$$

and assume a solution having the form

$$\theta = \theta_0 + \mu \theta_1 + \mu^2 \theta_2 + \cdots \qquad 6.2\text{-}7$$

where $\theta_0, \theta_1, \theta_2, \ldots$ are unknown functions of length z and time t. Substituting this solution into the system equation and equating terms having like coefficients of μ leads to the set of equations

$$\frac{\partial \theta_0}{\partial t} + v_s \frac{\partial \theta_0}{\partial z} + B\theta_0 = \frac{\partial T_s}{\partial z} u \qquad 6.2\text{-}8$$

$$\frac{\partial \theta_1}{\partial t} + v_s \frac{\partial \theta_1}{\partial z} + B\theta_1 = -\frac{\partial \theta_0}{\partial z} u \qquad 6.2\text{-}9$$

.
.
.

The first equation in this set, the generating equation, is identical to the linearized equation normally used to estimate the frequency response. Letting

$$u = A \sin \omega\tau \qquad 6.2\text{-}10$$

eliminating $\partial T_s/\partial z$ by means of Eq. 6.2-2, taking the Laplace transform, and solving the resulting ordinary differential equation with the boundary conditions

$$\text{at } z = 0, \qquad \bar{\theta}_0(s) = 0 \qquad 6.2\text{-}11$$

we obtain the solution for the transform of θ_0

$$\bar{\theta}_0 = \frac{(T_f - T_{cs})AB\omega}{v_s s(s^2 + \omega^2)} e^{-Bz/v_s}[1 - e^{-sz/v_s}] \qquad 6.2\text{-}12$$

This expression can be inverted to give the linear frequency response for $t > z/v_s$, the dead time,

$$\theta_0 = \frac{(T_f - T_{cs})AB}{v_s \omega} e^{-Bz/v_s} \left(\cos \omega \frac{t-z}{v_s} - \cos \omega t \right) \qquad 6.2\text{-}13$$

or after using a trigonometric identity to simplify the result,

$$\theta_0 = \frac{(T_f - T_{cs})AB}{v_s \omega} e^{-Bz/v_s} \left[\left(\cos \frac{\omega z}{v_s} - 1 \right) \cos \omega t + \sin \frac{\omega z}{v_s} \sin \omega t \right]$$

$$6.2\text{-}14$$

Now that the linear frequency response is known, the term $\partial \theta_0/\partial z$ in the equation for the first correction function, Eq. 6.2-9, can be determined. After the application of some trigonometric identities and some manipulation, the equation can be written as

$$\frac{\partial \theta_1}{\partial t} + v_s \frac{\partial \theta_1}{\partial z} + B\theta_1 = K_1[K_2 \sin \omega t + K_3(1 - \cos 2\omega t)] \qquad 6.2\text{-}15$$

where

$$K_1 = -\frac{(T_f - T_{cs})A^2 B^2}{2 v_s^2 \omega} e^{-Bz/v_s}$$

$$K_2 = 1 - \frac{\omega}{B} \sin \frac{\omega z}{v_s} - \cos \frac{\omega z}{v_s} \qquad 6.2\text{-}16$$

$$K_3 = \frac{\omega}{B} \cos \frac{\omega z}{v_s} - \sin \frac{\omega z}{v_s}$$

Taking the Laplace transform of this equation leads to an ordinary differential equation, which can be solved in a straightforward manner with the boundary condition

$$\text{at } z = 0, \qquad \theta_1(t) = \tilde{\theta}_1(s) = 0 \qquad 6.2\text{-}17$$

Inverting the transform then gives the solution for the first correction function for $t > 2\omega z/v_s$

$$\theta_1 = \left[\frac{(T_f - T_{cs})A^2 B^2}{2 v_s^2 \omega^2} e^{-Bz/v_s} \right] \left[\left(1 - \frac{\omega}{B} \sin \frac{\omega z}{v_s} - \cos \frac{\omega z}{v_s} \right) \right.$$

$$+ \left(\frac{\omega}{B} \cos \frac{\omega z}{v_s} - \sin \frac{\omega z}{v_s} - \frac{\omega}{B} \cos \frac{2\omega z}{v_s} + \frac{1}{2} \sin \frac{2\omega z}{v_s} \right) \sin 2\omega t$$

$$+ \left. \left(\frac{1}{2} - \frac{\omega}{B} \sin \frac{\omega z}{v_s} - \cos \frac{\omega z}{v_s} + \frac{\omega}{B} \sin \frac{2\omega z}{v_s} + \frac{1}{2} \cos \frac{2\omega z}{v_s} \right) \cos 2\omega t \right]$$

$$6.2\text{-}18$$

The approximate solution for the exchanger response is

$$\theta = \theta_0 + \mu \theta_1 + \cdots \qquad 6.2\text{-}7$$

where, of course, $\mu = 1$, for it was introduced in an arbitrary manner in order to keep track of terms having the same order of magnitude. Therefore, substituting Eqs. 6.2-14 and 6.2-18, we find that the approximate frequency response of the system must be

$$\theta = \left[\frac{(T_f - T_{cs})AB}{v_s \omega} e^{-Bz/v_s} \right] \left[\left(\cos \frac{z}{v_s} - 1 \right) \cos \omega t + \left(\sin \frac{\omega z}{v_s} \right) \sin \omega t \right]$$

$$+ \left[\frac{(T_f - T_{cs})A^2 B^2}{2 v_s^2 \omega^2} e^{-Bz/v_s} \right] \left[\left(1 - \frac{\omega}{B} \sin \frac{\omega z}{v_s} - \cos \frac{\omega z}{v_s} \right) \right.$$

$$6.2\text{-}19$$

$$+ \left(\frac{\omega}{B} \cos \frac{\omega z}{v_s} - \sin \frac{\omega z}{v} - \frac{\omega}{B} \cos \frac{2\omega z}{v_s} + \frac{1}{2} \sin \frac{2\omega z}{v_s} \right) \sin 2\omega t$$

$$+ \left. \left(\frac{1}{2} - \frac{\omega}{B} \sin \frac{\omega z}{v_s} - \cos \frac{\omega z}{v_s} + \frac{\omega}{B} \sin \frac{2\omega z}{v_s} + \frac{1}{2} \cos \frac{2\omega z}{v_s} \right) \cos 2\omega t \right.$$

This result shows that if

$$\frac{AB}{v_s\omega} \ll 1 \qquad\qquad 6.2\text{-}20$$

we can expect that the first correction function will be negligible in comparison to the generating solution, or that the linear frequency response will adequately describe the output. On the other hand, if the preceding criterion is not satisfied, we expect the time average output to be different from the steady state value and that higher harmonics will be present in the output; that is, the sinusoidal output signal will be distorted. It is interesting to note that exactly the same kind of distortion was obtained for the case of a first-order, linear ordinary differential equation with a periodic coefficient (see Example 6.1-2).

SECTION 6.3 SUMMARY

Most of the equations describing chemical processes are nonlinear or the inputs appear as time-variable coefficients. In order to obtain a first estimate of the dynamic response of the plant, so that the feasibility of an optimum steady state control system can be assessed or a dynamic experiment can be planned to measure a system parameter, the standard practice is to linearize the equations about a steady state operating condition and then determine the output of the linearized system. If it becomes necessary to establish the range of system parameters and the amplitude of the inputs where the linearization provides a valid approximation of the system equations, a perturbation solution can be developed. This process requires only slightly more effort than the determination of the linearized system response.

The results of applying perturbation theory to some simple dynamic problems indicate that, in general, the linearized equations predict the wrong steady state gain for the system (which agrees with the results from a simple steady state analysis), the system nonlinearities or the presence of periodic coefficients cause higher harmonics to be produced in the output (sinusoidal signals are distorted), and the time average value of the output of a periodic nonlinear process is different from the normal steady state output. Although the error in the steady state gain and the presence of distortion have been recognized for many years, the difference in the time-average operating levels was not realized until recently. This phenomenon is of possible importance because it implies that sometimes it should be possible to achieve a periodic performance which is superior to the optimum steady state design. The subject of periodic process operation will be discussed in additional detail in Chapter 10, Volume 2.

EXERCISES

1. (A) In Examples 4.3-1 and 6.1-1 we studied the step response of a second-order irreversible reaction in an isothermal, continuous-stirred-tank reactor. Estimate the frequency response of the nonlinear system if the amplitude of the feed composition oscillations is 10 percent of the steady state value. How large does the input amplitude have to be before we observe significant distortion of the output signal?

2. (A) Estimate the response of the plant described in the previous exercise if the feed rate to the reactor oscillates sinusoidally with an amplitude equal to 15 percent of the steady state value. How large an input fluctuation can we tolerate before it becomes necessary to account for the nonlinear nature of the system?

3. (B‡) If we consider a sphere that is losing heat to its surroundings by radiation, such as would be the case for the cannon ball described in Example 4.3-3 during its flight through the air, we could attempt to describe the behavior of the system by the simple model

$$VC_p\rho \frac{dT}{dt} = \sigma\epsilon A(T^4 - T_a^4)$$

Although it is possible to obtain an exact solution to this equation by recognizing that $T^4 - T_a^4 = (T^2 - T_a^2)(T^2 + T_a^2)$ and using partial fraction expansions, a more common approach might be to introduce a radiation heat-transfer coefficient and write the equation in the form

$$VC_p\rho \frac{dT}{dt} = h_r A(T - T_a)$$

Of course, the radiation heat-transfer coefficient must depend on temperature, according to the relation

$$h_r = \sigma\epsilon \left(\frac{T^4 - T_a^4}{T - T_a}\right)$$

See if you can use perturbation theory to develop a criterion for when we can neglect the temperature dependence of this quantity.

Using the parameters given in Example 4.3-3, as well as making any *reasonable* assumptions for other values you may need, calculate the cooling curve for the sphere both by the exact radiation equation and by the first two terms in the perturbation solution. Plot your results and discuss them in detail.

4. (B) Hairless Joe really misses Lonesome Polecat. He took your suggestions and, after some trial-and-error manipulations, got the Kickapoo Joy Juice process described in Exercise 4, p. 210, started up OK, but now he finds that he must console himself for his "friend's" absence by imbibing his wonderful brew. Of course, he does this in typical Lower-Slobovian fashion by lifting the whole vat and drinking directly from it. Man, this is really hard on the piping, although it has held up so far. However, he is curious as to how this continual variation in the kettle volume will affect the output. Therefore he asks you to estimate the response to sinusoidal variations in the volume of the vessel for a case where he drinks 20 percent of the contents once an hour.

5. (B) Coughanowr and Koppel[1] present some values describing the steady state operation of a tank 10 ft high and 5 ft in diameter. Use the data to develop a relationship between liquid level and the effluent flow rate.

If the tank is initially operating at steady state conditions with an input flow rate of 20,000 gal/hr and suddenly one of the two feed pumps fails so that the feed rate is cut in half, calculate the response of the level in the tank. Plot the generating solution and first correction functions separately.

Inlet flow rate (gal/hr)	0	5,000	10,000	15,000	20,000	25,000	30,000
Steady state level (ft)	0	0.7	1.1	2.3	3.9	6.3	8.8

6. (B*) Horn and Lin[2] studied the behavior of the pair of parallel reactions, $A \longrightarrow B$ and $A \longrightarrow C$, in an isothermal, stirred tank reactor. The dynamic equations describing the system were taken to be

$$V\frac{dA}{dt} = q(A_f - A) - k_1 VA^2 - k_2 VA$$

$$V\frac{dB}{dt} = q(B_f - B) + k_1 VA^2$$

The equations were written in dimensionless form by letting

$$x_1 = \frac{A}{A_f}, \quad x_2 = \frac{B}{A_f}, \quad u = \frac{Vk_{10}A_f \exp(-E_1/RT)}{q}, \quad \rho = \frac{E_2}{E_1},$$

$$B_f = 0, \quad a = \frac{Vk_{20}}{q}\left(\frac{Vk_{10}A_f}{q}\right)^{-\rho}, \quad \tau = \frac{qt}{V}$$

so that they become

$$\frac{dx_1}{d\tau} = 1 - x_1 - ux_1^2 - au^\rho x_1$$

$$\frac{dx_2}{d\tau} = -x_2 + ux_1^2$$

When they chose the system parameters as $\rho = \frac{3}{4}$ and $a = 1.0$, they found that the value of the dimensionless reactor temperature $u = 2.52$ gave the maximum possible dimensionless composition of component B, $x_2 = 0.185$, and that the corresponding value of A was $x_1 = 0.2718$.

Use perturbation theory to estimate the response of the reactor to sinusoidal changes in the temperature u. In particular, make certain that you calculate the time-average composition of component B and discuss how you would choose the amplitude and the frequency of the oscillations in order to maximize this composition.

7. (B*) In Exercise 25, p. 216, we described a dynamic model for a plate, gas absorption unit with two plates. Coughanowr and Koppel[3] originally solved these equations numerically using an analog computer, whereas we linearized them and estimated the response. Use perturbation theory to develop correction functions

[1] See p. 72 of the reference given in footnote 2 on p. 213.
[2] F. J. M. Horn and R. C. Lin, *Ind. Eng. Chem. Proc Design Develop.*, **6**, 21, (1967).
[3] See footnote 2, p. 213.

that can be used to improve the linear response results. Plot both the linear response curves and the correction functions.

8. (B*) Suppose that the vapor-liquid equilibrium relationship in the gas absorption unit described in the previous exercise was nonlinear. Show how it would be possible to estimate the dynamic response of the unit if the nonlinear equilibrium curve was approximated by a quadratic expression.

9. (B*) A perturbation solution for the frequency response of a steam-heated exchanger was described in Example 6.2-1. However, the effect of velocity on the overall heat-transfer coefficient was not considered in the analysis. Using conventional correlations, estimate the error caused by neglecting this variation.

10. (C) Find the response of a first-order isothermal reaction in a tubular reactor for sinusoidal variations in the velocity. Include an axial dispersion term in your conservation equation.

Appendix A
Matrix Operations

Most engineers have had a brief introduction to vector analysis in a physics course. A vector was defined as a quantity having both magnitude and direction, such as force or velocity. It was common practice to write the vector in terms of its components, which are merely the projections of the vector on the x, y, and z axes. Thus we write

$$\mathbf{P} = p_1\mathbf{i} + p_2\mathbf{j} + p_3\mathbf{k} \qquad\qquad 1.$$

where $\mathbf{i}$, $\mathbf{j}$, and $\mathbf{k}$ are unit vectors in the x, y, and z directions. The procedure for adding two vectors or determining the length of a vector is usually presented in terms of the components. The reader is expected to have been familiar with these concepts, even if he does not remember the details of the operations.

Our main interest in vector analysis is the development of a simple notation we can use to describe very complex problems. For example, we know that engineers are often called on to solve sets of simultaneous equations, such as

$$
\begin{aligned}
a_{11}x + a_{12}y + a_{13}z &= b_1 \\
a_{21}x + a_{22}y + a_{23}z &= b_2 \qquad\qquad 2. \\
a_{31}x + a_{32}y + a_{33}z &= b_3
\end{aligned}
$$

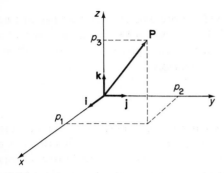

Figure A-1. Vector components.

These can be any three simultaneous equations in three unknowns, and they do not necessarily have anything to do with real vector quantities. To be specific, they may represent two material balances and an energy balance used to characterize the behavior of a pair of parallel reactions in a noniso-thermal, continuous-stirred-tank reactor. Alternately, they may represent the composition of the most volatile component on each of the three trays in a laboratory, plate, gas absorption column. Of course, if we consider dis-tillation of a 5-component mixture in a commercial-scale unit having 20 plates, we may need $5(20) = 100$ material balance expressions, plus 20 energy balance equations, to describe the system completely. Obviously it will be a tedious matter to write down these 120 equations, and it is for this reason that we want to find a simple notation.

Let us return to our set of three equations and suppose that we multiply the first expression by a unit vector $\mathbf{i}$, the second by a unit vector $\mathbf{j}$, and the third by a unit vector $\mathbf{k}$. When we add the equations together, we obtain

$$(a_{11}x + a_{12}y + a_{13}z)\mathbf{i} + (a_{21}x + a_{22}y + a_{23}z)\mathbf{j}$$
$$+ (a_{31}x + a_{32}y + a_{33}z)\mathbf{k} = b_1\mathbf{i} + b_2\mathbf{j} + b_3\mathbf{k} \quad 3.$$

This result is a vector equation, and it satisfies the normal rule that two vectors are equal if and only if their respective components are equal. Thus we can define a vector $\mathbf{b}$, having as its components

$$\mathbf{b} = b_1\mathbf{i} + b_2\mathbf{j} + b_3\mathbf{k} \quad\quad 4.$$

and a vector $\mathbf{r}$, having as its components

$$\mathbf{r} = r_1\mathbf{i} + r_2\mathbf{j} + r_3\mathbf{k} = (a_{11}x + a_{12}y + a_{13}z)\mathbf{i}$$
$$+ (a_{21}x + a_{22}y + a_{23}z)\mathbf{j} + (a_{31}x + a_{32}y + a_{33}z)\mathbf{k} \quad 5.$$

and write our vector equation as

$$\mathbf{r} = \mathbf{b} \quad\quad 6.$$

So far we have done nothing new except to show that a set of three simul-taneous equations can be interpreted as a single vector equation. We have not simplified the procedure for determining the value of x, y, and z that satisfy these equations, but we have developed a more efficient notation for writing our original set of three equations. What we would like to do now is find a way of extending this approach so that we can represent the 120 equations for the distillation column as a single equation.

Generalized Vectors

At first glance it appears as if we cannot accomplish this task because there are only three unit vectors corresponding to the three directions in a cartesian coordinate system. However, it is also clear that these three vectors really had nothing to do with the problem, except to keep track of which terms went together so that the components of the vectors would be equal. In other words, we could just as easily have defined a *vector* as a *set of ordered quantities* and written that

$$\mathbf{b} = (b_1 \; b_2 \; b_3) \qquad \mathbf{r} = (r_1 \; r_2 \; r_3) \qquad\qquad 7.$$

and

$$\mathbf{b} = \mathbf{r} \qquad\qquad 8.$$

Then, for the set of five simultaneous equations

$$
\begin{aligned}
a_{11}x + a_{12}y + a_{13}z + a_{14}u + a_{15}v &= b_1 \\
a_{21}x + a_{22}y + a_{23}z + a_{24}u + a_{25}v &= b_2 \\
a_{31}x + a_{32}y + a_{33}z + a_{34}u + a_{35}v &= b_3 \\
a_{41}x + a_{42}y + a_{43}z + a_{44}u + a_{45}v &= b_4 \\
a_{51}x + a_{52}y + a_{53}z + a_{54}u + a_{55}v &= b_5
\end{aligned}
\qquad 9.
$$

we can define **b** and **r** as

$$\mathbf{b} = (b_1 \; b_2 \; b_3 \; b_4 \; b_5) \qquad \mathbf{r} = (r_1 \; r_2 \; r_3 \; r_4 \; r_5) \qquad 10.$$

and write the total set of equations as

$$\mathbf{b} = \mathbf{r} \qquad\qquad 11.$$

Perhaps it should be emphasized that Eqs. 8 and 11 are different vector equations because the vectors contain a different number of components.

Of course, it is not necessary to order the components in a row, such as we did in Eqs. 7 and 10. Instead, we could order them vertically and define **b** and **r** as

$$
\mathbf{b} = \begin{pmatrix} b_1 \\ b_2 \\ b_3 \\ b_4 \\ b_5 \end{pmatrix}
\qquad
\mathbf{r} = \begin{pmatrix} r_1 \\ r_2 \\ r_3 \\ r_4 \\ r_5 \end{pmatrix}
\qquad 12.
$$

and again write

$$\mathbf{b} = \mathbf{r} \qquad\qquad 13.$$

In both cases the final equation merely states that the vector components are equal in the two vectors. Not surprisingly, the first case, Eq. 10, is called a row vector and the second case, Eq. 12, is called a column vector.

Vector Multiplication

If we now examine the components of the vector $\mathbf{r}$ (see Eq. 5), it seems as if we should be able to write them as the product of vectors or as some other quantity multiplied by a vector. Consequently, it becomes necessary to define what we mean by the product of our generalized vectors. Certainly we want this to agree with the conventional results if our ordered arrays contain only three terms and therefore can be considered as the classical type of vector. In particular, we know that the dot product of two vectors is defined as

$$\mathbf{m} \cdot \mathbf{n} = (m)(n) \cos \theta \qquad\qquad 14.$$

where (m) and (n) are the magnitudes of the vectors and θ is the angle between them. When written in terms of the components, the dot product becomes

$$(m_1\mathbf{i} + m_2\mathbf{j} + m_3\mathbf{k}) \cdot (n_1\mathbf{i} + n_2\mathbf{j} + n_3\mathbf{k}) = m_1n_1 + m_2n_2 + m_3n_3 \qquad 15.$$

since $\mathbf{i} \cdot \mathbf{i} = \mathbf{j} \cdot \mathbf{j} = \mathbf{k} \cdot \mathbf{k} = 1$ and $\mathbf{i} \cdot \mathbf{j} = \mathbf{j} \cdot \mathbf{k} = \mathbf{i} \cdot \mathbf{k} = 0$, providing that the unit vectors, $\mathbf{i}$, $\mathbf{j}$, and $\mathbf{k}$ are orthogonal so that $\cos 0° = 1$ and $\cos 90° = 0$. We also see that the result of this multiplication is a scalar—that is, it has a magnitude but no unit vector assoicated with it. Hence the dot product is often called the scalar product of two vectors.

In terms of our generalized vectors, this means that the analog of a scalar product (the product of two ordered quantities) should be a quantity having only one element—that is, no order. Since in the classical case $\mathbf{b} \cdot \mathbf{b} = b_1^2 + b_2^2 + b_3^2$, we will define the product of a row vector and a column vector as

$$(b_1\ b_2\ b_3)\begin{pmatrix} b_1 \\ b_2 \\ b_3 \end{pmatrix} = b_1^2 + b_2^2 + b_3^2 \qquad\qquad 16.$$

Similarly, for two different vectors we have

$$(m_1\ m_2\ m_3)\begin{pmatrix} n_1 \\ n_2 \\ n_3 \end{pmatrix} = m_1n_1 + m_2n_2 + m_3n_3 \qquad\qquad 17.$$

Thus when we multiply a row vector times a column vector, we multiply the first element in the row by the first element in the column, add the result to the product of the second element in the row and the second element in the column, and add this result to the product of the last element in the row and the last element in the column. This rule will always work, providing that the number of elements in a row is the same as that in a column. Also, we can easily extend this definition to additional elements in the vectors and write

$$(m_1\, m_2\, m_3\, m_4\, m_5) \begin{pmatrix} n_1 \\ n_2 \\ n_3 \\ n_4 \\ n_5 \end{pmatrix} = m_1 n_1 + m_2 n_2 + m_3 n_3 + m_4 n_4 + m_5 n_5 \qquad 18.$$

When we reexamine Eq. 5, we find that we can write $r_1 = a_{11}x + a_{12}y + a_{13}z$ as this type of vector product

$$(a_{11}\, a_{12}\, a_{13}) \begin{pmatrix} x \\ y \\ z \end{pmatrix} = a_{11}x + a_{12}y + a_{13}z = r_1 \qquad 19.$$

Similarly,

$$r_2 = (a_{21}\, a_{22}\, a_{23}) \begin{pmatrix} x \\ y \\ z \end{pmatrix} \qquad r_3 = (a_{31}\, a_{32}\, a_{33}) \begin{pmatrix} x \\ y \\ z \end{pmatrix} \qquad 20.$$

Moreover, when we write the quantities r_1, r_2, and r_3 as the elements of a column vector on the left-hand side of Eq. 6, we note that each of these components can be written as a row vector multiplied by a *common* column vector. Therefore we will group all the row vectors together in a common array, called a *matrix*, and write

$$\begin{pmatrix} r_1 \\ r_2 \\ r_3 \end{pmatrix} = \begin{pmatrix} a_{11} & a_{12} & a_{13} \\ a_{21} & a_{22} & a_{23} \\ a_{31} & a_{32} & a_{33} \end{pmatrix} \begin{pmatrix} x \\ y \\ z \end{pmatrix} = \begin{pmatrix} b_1 \\ b_2 \\ b_3 \end{pmatrix} \qquad 21.$$

Letting

$$\mathbf{r} = \begin{pmatrix} r_1 \\ r_2 \\ r_3 \end{pmatrix}, \quad \mathbf{x} = \begin{pmatrix} x \\ y \\ z \end{pmatrix}, \quad \mathbf{b} = \begin{pmatrix} b_1 \\ b_2 \\ b_3 \end{pmatrix} \qquad 22.$$

and the matrix (square array) be

$$\mathbf{A} = \begin{pmatrix} a_{11} & a_{12} & a_{13} \\ a_{21} & a_{22} & a_{23} \\ a_{31} & a_{32} & a_{33} \end{pmatrix} \qquad 23.$$

we can write Eq. 6 in the form

$$\mathbf{r} = \mathbf{Ax} = \mathbf{b} \qquad 24.$$

It is clear that we could write exactly the same type of equation for any number of components merely by increasing the number of components in each of the vector or matrix quantities.

However, it is a simple matter to show that this procedure is not unique. Thus we could also write that

$$r_1 = (x\,y\,z)\begin{pmatrix} a_{11} \\ a_{12} \\ a_{13} \end{pmatrix} = a_{11}x + a_{12}y + a_{13}z \qquad\qquad 25.$$

using our previous definition of multiplication of a row vector by a column vector. Similarly,

$$r_2 = (x\,y\,z)\begin{pmatrix} a_{21} \\ a_{22} \\ a_{23} \end{pmatrix} \qquad r_3 = (x\,y\,z)\begin{pmatrix} a_{31} \\ a_{32} \\ a_{33} \end{pmatrix} \qquad\qquad 26.$$

Since the common vector in this case is a row vector, we could arrange the other elements in an ordered array of columns and write

$$(r_1\,r_2\,r_3) = (x\,y\,z)\begin{pmatrix} a_{11} & a_{21} & a_{31} \\ a_{12} & a_{22} & a_{32} \\ a_{13} & a_{23} & a_{33} \end{pmatrix} = (b_1\,b_2\,b_3) \qquad\qquad 27.$$

In order to prevent confusion between row and column vectors, we then define a transpose operation. We denote the transpose by a superscript T, and whenever we see this symbol we write a column vector as a row vector and vice versa. Thus

$$\mathbf{r} = \begin{pmatrix} r_1 \\ r_2 \\ r_3 \end{pmatrix} \qquad \mathbf{r}^T = (r_1\,r_2\,r_3) \qquad\qquad 28.$$

$$\mathbf{A} = \begin{pmatrix} a_{11} & a_{12} & a_{13} \\ a_{21} & a_{22} & a_{23} \\ a_{31} & a_{32} & a_{33} \end{pmatrix} \qquad \mathbf{A}^T = \begin{pmatrix} a_{11} & a_{21} & a_{31} \\ a_{12} & a_{22} & a_{32} \\ a_{13} & a_{23} & a_{33} \end{pmatrix}$$

where the first column in $\mathbf{A}$ has been written as the first row in $\mathbf{A}^T$, the second column as the second row, and so on. Thus Eq. 24 gives

$$\mathbf{r} = \mathbf{Ax} = \mathbf{b} \qquad\qquad 29a.$$

but Eq. 27 becomes

$$\mathbf{r}^T = \mathbf{x}^T\mathbf{A}^T = \mathbf{b}^T \qquad\qquad 29b.$$

This result reveals that the transpose of the product of two vector quantities is equal to the product of the transposed quantities taken in reverse order; that is,

$$(\mathbf{Ax})^T = \mathbf{x}^T\mathbf{A}^T \qquad\qquad 30.$$

Moreover, we find that the order of multiplication of vector quantities is extremely important and that the cumulative law of multiplication is not valid:

$$\mathbf{Ax} \neq \mathbf{xA} \qquad\qquad 31.$$

In fact, the quantity $\mathbf{xA}$, which for vectors having three components may be written as

$$\begin{pmatrix} x_1 \\ x_2 \\ x_3 \end{pmatrix} \begin{pmatrix} a_{11} & a_{12} & a_{13} \\ a_{21} & a_{22} & a_{23} \\ a_{31} & a_{23} & a_{33} \end{pmatrix} \qquad\qquad 32.$$

does not make sense because we cannot apply our rule of multiplication. Therefore we can formulate the general rule for multiplication of two vector quantities: there must be the same number of columns in the first quantity as there are rows in the second. For example, in Eq. 21 we see that $\mathbf{A}$ has 3 rows and 3 columns, while $\mathbf{x}$ has 3 rows and 1 column. The product $\mathbf{Ax}$ can be determined and is a $(3 \times 3)(3 \times 1) = (3 \times 1)$ column vector. Conversely, in Eq. 32 we observe that $\mathbf{x}$ has 3 rows and 1 column, while $\mathbf{A}$ has 3 rows and 3 columns, so that we cannot determine the product of $\mathbf{xA}$, $(3 \times 1)(3 \times 3) =$ not defined.

Matrix Operations

The preceding can be extended to situations where we consider two matrix quantities. Thus in order for two matrices to be added, they must have the same number of rows and columns. If this is the case, we merely add the elements together to form the sum

$$\mathbf{A} + \mathbf{B} = \mathbf{B} + \mathbf{A} = \begin{pmatrix} a_{11} + b_{11} & a_{12} + b_{12} & \cdots & a_{1m} + b_{1m} \\ a_{21} + b_{21} & a_{22} + b_{22} & \cdots & a_{2m} + b_{2m} \\ \cdot & \cdot & & \cdot \\ \cdot & \cdot & & \cdot \\ \cdot & \cdot & & \cdot \\ a_{n1} + b_{n1} & & \cdots & a_{nm} + b_{nm} \end{pmatrix} \qquad 33.$$

Similarly, the product of two matrices $\mathbf{A}$ and $\mathbf{B}$ may be formed if the number of columns in $\mathbf{A}$ is equal to the number of rows in $\mathbf{B}$. In this case, we say that the matrices are conformable. Also, if we say that $\mathbf{A}$ has i rows and n columns, while $\mathbf{B}$ has n rows and j columns, we find that the product $\mathbf{AB}$ will have $(i \times n)(n \times j) = (i \times j)$, i rows and j columns. The element in the ith row and jth column of the product matrix $\mathbf{C}$ is given by

$$c_{ij} = \sum_n a_{in} b_{nj} \qquad\qquad 34.$$

As a third illustration of this extension, we can show that

$$(\mathbf{AB})^T = B^T A^T \qquad\qquad 35.$$

It would be possible to enumerate many other results of this type, and the reader is encouraged to study one of the references listed in footnotes 4 to 6 on p. 86. Our purpose here is to illustrate the usefulness of matrix operations as an engineering tool rather than to derive all the important matrix operations.

Matrix Inverse

With this reminder, we return to our problem of solving a set of simultaneous equations, such as those given by Eq. 9. First we note that if we could find the inverse of the $\mathbf{A}$ matrix, which we will denote by the symbol $\mathbf{A}^{-1}$ and define as

$$\mathbf{A}^{-1}\mathbf{A} = \mathbf{I} = \begin{pmatrix} 1 & 0 & 0 & \cdots & 0 \\ 0 & 1 & 0 & \cdots & 0 \\ \cdot & & & & \cdot \\ \cdot & & & & \cdot \\ \cdot & & & & \cdot \\ 0 & & \cdots & & 1 \end{pmatrix} \qquad\qquad 36.$$

(where $\mathbf{I}$ is called the identity matrix and has the same number of rows and columns as the square matrix $\mathbf{A}$), we could premultiply Eq. 9 by this quantity to obtain

$$\mathbf{A}^{-1}\mathbf{A}\mathbf{x} = \mathbf{A}^{-1}\mathbf{b} \qquad\qquad 37.$$

or, since,

$$\mathbf{A}^{-1}\mathbf{A} = \mathbf{I}$$

and   38.

$$\mathbf{I}\mathbf{x} = \mathbf{x}$$

we find that

$$\mathbf{x} = \mathbf{A}^{-1}\mathbf{b} \qquad\qquad 39.$$

In other words, if we premultiply the column vector $\mathbf{b}$ by the inverse of the $\mathbf{A}$ matrix, we obtain the desired solution of the set of simultaneous equations. Our only question now is how do we determine $\mathbf{A}^{-1}$.

Determinant

Before describing the procedure for computing this inverse, however, it is necessary to review the definition of a determinant briefly. When we evaluate the determinant of a square matrix, which we denote as det A, or $|A|$, or

$$|A| = \begin{vmatrix} a_{11} & a_{12} & \cdots & a_{1n} \\ a_{21} & a_{22} & \cdots & a_{2n} \\ \cdot & \cdot & & \cdot \\ \cdot & \cdot & & \cdot \\ \cdot & \cdot & & \cdot \\ a_{n1} & a_{n2} & \cdots & a_{nn} \end{vmatrix} \qquad\qquad 40.$$

we obtain a single number. This can be written as

$$|A| = \sum_{i=1}^{n} a_{ij} A_{ij} = \sum_{j=1}^{n} a_{ij} A_{ij} \qquad\qquad 41.$$

where a_{ij} represents an element in the ith row and the jth column of the original matrix and A_{ij} is the cofactor of a_{ij}. The cofactor is defined as $(-1)^{i+j}$ multiplied by the determinant of the original matrix after the ith row and jth column have been deleted.

$$A_{ij} = (-1)^{i+j} \begin{vmatrix} a_{11} & a_{12} \cdots a_{1j} \cdots a_{1n} \\ a_{21} & a_{22} \cdots a_{2j} \cdots a_{2n} \\ \vdots & \vdots \qquad \vdots \\ a_{i1} & a_{i2} \quad a_{ij} \quad a_{in} \\ \vdots & \vdots \qquad \vdots \\ a_{n1} & a_{n2} \cdots a_{nj} \cdots a_{nn} \end{vmatrix} \qquad 42.$$

Thus, in order to determine each cofactor, we must evaluate an $(n-1)$st order determinant. Also, from Eq. 41 we see that the determinant of an nth-order matrix is defined in terms of n determinants, each of which is $(n-1)$st order. By continuing this procedure, eventually we obtain a second- or third-order determinant which can be evaluated by the diagonal method.

A great number of theorems can often be used to simplify the calculation of a determinant. Space precludes a discussion of these rules, so we suggest that the reader refer to one of the many excellent texts available. However, the foregoing definitions both appear in the definition of the inverse of a matrix.

Adjoint Matrix

When we select any element a_{ij} in a square matrix, we can use Eq. 42 to determine the cofactor of that element. Moreover, if we repeat this procedure for every element and then replace each element by its cofactor, we obtain a cofactor matrix. The transpose (switching the rows and columns) of this matrix is called the adjoint and is indicated as adj $\mathbf{A}$

$$\text{adj } \mathbf{A} = (A_{ij})^T = (A_{ji}) \qquad\qquad 43.$$

Inverse

Once we have evaluated the adjoint, we can find the inverse of the matrix by dividing the adjoint by the determinant of the original matrix **A**.

$$\mathbf{A}^{-1} = \frac{\text{adj } A}{|A|} = \frac{(A_{ij})^T}{|A|} = \frac{(A_{ji})}{|A|} \qquad 44.$$

Of course, in order for the inverse to exist, the determinant $|A|$ cannot be equal to zero. Thus only nonsingular matrices have an inverse.

Example A-1 Calculation of the inverse

Consider the matrix

$$\mathbf{A} = \begin{pmatrix} 1 & 2 & 0 \\ -1 & 3 & 2 \\ 4 & -2 & 0 \end{pmatrix} \qquad 45.$$

A simple way of evaluating the determinant is to apply Eq. 41 to the elements in the last column.

$$|A| = (a_{13}A_{13} + a_{23}A_{23} + a_{33}A_{33})$$
$$= 0(-1)^4 \begin{vmatrix} -1 & 3 \\ 4 & -2 \end{vmatrix} + 2(-1)^5 \begin{vmatrix} 1 & 2 \\ 4 & -2 \end{vmatrix} + 0(-1)^6 \begin{vmatrix} 1 & 2 \\ -1 & 3 \end{vmatrix} \qquad 46.$$
$$= -2(-2 - 8) = 20$$

The cofactors of the elements in the matirix are

$$A_{11} = (-1)^2 4 = 4, \quad A_{12} = (-1)^3(-8) = 8, \quad A_{13} = (-1)^4(2 - 12) = -10$$
$$A_{21} = (-1)^3 0 = 0, \quad A_{22} = (-1)^4 0 = 0, \quad A_{23} = (-1)^5(-2 - 8) = 10$$
$$A_{31} = (-1)^4 4 = 4, \quad A_{32} = (-1)^5 2 = -2, \quad A_{33} = (-1)^6(3 + 2) = 5$$

47.

so that the cofactor matrix is

$$(A_{ij}) = \begin{pmatrix} 4 & 8 & -10 \\ 0 & 0 & 10 \\ 4 & -2 & 5 \end{pmatrix} \qquad 48.$$

and the adjoint is

$$\text{adj } \mathbf{A} = (A_{ij})^T = \begin{pmatrix} 4 & 0 & 4 \\ 8 & 0 & -2 \\ -10 & 10 & 5 \end{pmatrix} \qquad 49.$$

Finally, the inverse is the adjoint divided by the determinant

$$\mathbf{A}^{-1} = \frac{1}{20} \begin{pmatrix} 4 & 0 & 4 \\ 8 & 0 & -2 \\ -10 & 10 & 5 \end{pmatrix} \qquad 50.$$

To show that this result is valid, we evaluate $\mathbf{A}^{-1}\mathbf{A}$.

$$\mathbf{A}^{-1}\mathbf{A} = \frac{1}{20}\begin{pmatrix} 4 & 0 & 4 \\ 8 & 0 & -2 \\ -10 & 10 & 5 \end{pmatrix}\begin{pmatrix} 1 & 2 & 0 \\ -1 & 3 & 2 \\ 4 & -2 & 0 \end{pmatrix}$$

$$= \frac{1}{20}\begin{pmatrix} (4+16) & (8-8) & 0 \\ (8-8) & (16+4) & 0 \\ (-10-10+20) & (-20+30-10) & 20 \end{pmatrix} = \begin{pmatrix} 1 & 0 & 0 \\ 0 & 1 & 0 \\ 0 & 0 & 1 \end{pmatrix}$$

51.

Matrix Differential Equations

If we consider the set of first-order, linear differential equations

$$\dot{x}_1 = a_{11}x_1 + a_{12}x_2 + \cdots + a_{1n}x_n$$
$$\dot{x}_2 = a_{21}x_1 + a_{22}x_2 + \cdots + a_{2n}x_n$$
$$.$$
$$.$$
$$.$$
$$\dot{x}_n = a_{n1}x_1 + a_{n2}x_2 + \cdots + a_{nn}x_n$$

52.

we can write these in matrix form as

$$\dot{\mathbf{x}} = \mathbf{A}\mathbf{x}$$

53.

One method we could use to solve the equations is to replace the derivative operation by the symbol D and write the equations in the form

$$(D - a_{11})x_1 - a_{12}x_2 - \cdots - a_{1n}x_n = 0$$
$$-a_{21}x_1 + (D - a_{22})x_2 - \cdots - a_{2n}x_n = 0$$
$$.$$
$$.$$
$$.$$
$$-a_{n1}x_1 - a_{n2}x_2 - \cdots + (D - a_{nn})x_n = 0$$

54.

Now we can eliminate all except one of the dependent variables to obtain

$$\begin{vmatrix} D - a_{11} & -a_{12} & \cdots & -a_{1n} \\ -a_{21} & D - a_{22} & \cdots & -a_{2n} \\ . & . & & . \\ . & . & & . \\ . & . & & . \\ -a_{n1} & -a_{n2} & \cdots & D - a_{nn} \end{vmatrix} x_1 = 0$$

55.

We could also write this expression in matrix notation as

$$|\mathbf{DI} - \mathbf{A}|x_1 = 0$$

56.

Moreover, we note that when we take the Laplace transform of the equations,

we obtain

$$(s\mathbf{I} - \mathbf{A})\tilde{\mathbf{x}} = \tilde{\mathbf{x}}_0 \qquad 57.$$

where $\tilde{\mathbf{x}}_0$ is the Laplace transform of the initial conditions. If we attempt to eliminate all but one of the state variables, or if we compute the inverse of $(s\mathbf{I} - \mathbf{A})$, we again encounter the determinant

$$|\lambda \mathbf{I} - \mathbf{A}| = 0 \qquad 58.$$

When we expand this determinant we obtain an nth-order polynomial in D, and the roots of this *characteristic equation*, providing that they are distinct, are the exponents, $\lambda_1, \lambda_2, \ldots, \lambda_n$, which appear in the solution

$$x_1 = c_{11}e^{\lambda_1 t} + c_{12}e^{\lambda_2 t} + \cdots + c_{1n}e^{\lambda_n t} \qquad 59.$$

where $c_{11}, c_{12}, \ldots, c_{1n}$ represent integration constants. The characteristic equation for each of the other variables is identical to that given above, and therefore the solutions for these variables will have the same form as Eq. 59. Thus we can write

$$\begin{aligned}
x_1 &= c_{11}e^{\lambda_1 t} + c_{12}e^{\lambda_2 t} + \cdots + c_{1n}e^{\lambda_n t} \\
x_2 &= c_{21}e^{\lambda_1 t} + c_{22}e^{\lambda_2 t} + \cdots + c_{2n}e^{\lambda_n t} \\
& \vdots \\
x_n &= c_{n1}e^{\lambda_1 t} + c_{n2}e^{\lambda_2 t} + \cdots + c_{nn}e^{\lambda_n t}
\end{aligned} \qquad 60.$$

There are only n initial conditions for the n equations, so that only n of the n^2 constants given above, $c_{11}, \ldots, c_{nn}$, can be independent. The standard procedure in ordinary differential equations for evaluating these integration constants is to substitute the solutions back into the set of differential equations and equate the coefficients of each of the $e^{\lambda_i t}$ terms in all but one of the equations. In this way we can develop $n(n-1)$ relationships between the constants. These expressions, together with the n initial conditions, provide enough information to determine all the unknown constants.

Rather than follow this procedure, we choose to write the solution, Eq. 60, in vector-matrix form

$$\begin{pmatrix} x_1 \\ x_2 \\ \vdots \\ x_n \end{pmatrix} = \begin{pmatrix} c_{11} & c_{12} & \cdots & c_{1n} \\ c_{21} & c_{22} & \cdots & c_{2n} \\ \vdots & \vdots & & \vdots \\ c_{n1} & c_{n2} & \cdots & c_{nn} \end{pmatrix} \begin{pmatrix} e^{\lambda_1 t} \\ e^{\lambda_2 t} \\ \vdots \\ e^{\lambda_n t} \end{pmatrix} \qquad 61.$$

or

$$\mathbf{x} = \mathbf{C}e^{\lambda t} \qquad 62.$$

Also, instead of choosing n independent constants to work with, we select

n linearly independent groups of constants. Although we could choose these in any manner we please, a convenient choice might be the columns of the $\mathbf{C}$ matrix. We retain these groups of constants in the form of column vectors and call them characteristic vectors, or eigenvectors, for each characteristic vector corresponds to one of the characteristic values, λ_j. Hence the jth eigenvector is

$$\mathbf{c}_j = \begin{pmatrix} c_{1j} \\ c_{2j} \\ \cdot \\ \cdot \\ \cdot \\ c_{nj} \end{pmatrix} \qquad\qquad 63.$$

and with this notation, we can write the solutions as

$$\mathbf{x} = \mathbf{c}_1 e^{\lambda_1 t} + \mathbf{c}_2 e^{\lambda_2 t} + \cdots + \mathbf{c}_n e^{\lambda_n t} \qquad\qquad 64.$$

This is just a linear combination of the n linearly independent eigenvectors, where the weighing factors are scalar exponential terms. The individual terms in this sum are called the natural modes of the system; that is, the jth term

$$\mathbf{c}_j e^{\lambda_j t} \qquad\qquad 65.$$

is the jth natural mode, or the mode associated with the jth eigenvalue.

Although we have been treating the eigenvectors as n linearly independent groups of constants, the n elements in each eigenvector are unknown, so that we still have n^2 unknown constants. Instead of developing $n(n-1)$ simultaneous relationships between these values, we will establish $(n-1)$ relationships n times by treating each eigenvector individually. This step can be accomplished because each of the exponential factors $e^{\lambda_j t}$ must satisfy the system equations separately, and according to Eq. 65, each of these factors is associated with only one eigenvector. Substituting the expression for the jth natural mode into the system equations, Eq. 53, we obtain

$$\lambda_j \mathbf{c}_j e^{\lambda_j t} = \mathbf{A}\mathbf{c}_j e^{\lambda_j t} \qquad\qquad 66.$$

or

$$\lambda_j \mathbf{c}_j = \mathbf{A}\mathbf{c}_j \qquad\qquad 67.$$

This expression provides n equations in the n unknowns $c_{1j}, c_{2j}, \ldots, c_{nj}$. However, the equations are homogeneous (i.e., they are not all linearly independent), so that actually they only give us expressions for $(n-1)$ of the constants in terms of one of the others. By repeating this procedure for all n of the modes, we obtain $n(n-1)$ relationships in all between the constants and we have n constants left to determine, one in each eigenvector. These constants are found by introducing the n initial conditions.

In many cases it is convenient to normalize the eigenvectors to unit length. We define the length of the vector as the square root of the sum of the squares of its elements. Calling the length of the jth eigenvector b_j, we have

$$\sqrt{c_{1j}^2 + c_{2j}^2 + \cdots + c_{nj}^2} = b_j \qquad 68.$$

and dividing each of the elements of $\mathbf{c}_j$ by this quantity, we obtain the normalized eigenvector $\mathbf{b}_j$

$$\mathbf{b}_j = \begin{pmatrix} c_{ij}/b_j \\ c_{2j}/b_j \\ \cdot \\ \cdot \\ \cdot \\ c_{nj}/b_j \end{pmatrix} \qquad 69.$$

This restriction of unit length, plus the $(n-1)$ relationships given by Eq. 67, will also completely define all the elements in the jth normalized eigenvector. Thus we must write the solution, Eq. 64, as

$$\mathbf{x} = c_1 \mathbf{b}_1 e^{\lambda_1 t} + c_2 \mathbf{b}_2 e^{\lambda_2 t} + \cdots + c_n \mathbf{b}_n e^{\lambda_n t} \qquad 70.$$

where $c_1, \ldots, c_n$ are n arbitrary constants, which can be determined from the initial conditions.

Example A-2 Eigenvectors of a second-order system

The linearized equations describing a CSTR, or any other second-order system, can be written as

$$\dot{x}_1 = a_{11} x_1 + a_{12} x_2 \qquad x_1(0) = x_{10} \qquad 71.$$
$$\dot{x}_2 = a_{21} x_1 + a_{22} x_2 \qquad x_2(0) = x_{20} \qquad 72.$$

or

$$\dot{\mathbf{x}} = \mathbf{A}\mathbf{x} \qquad\qquad \mathbf{x}(0) = \mathbf{x}_0 \qquad 73.$$

The eigenvalues are obtained from the characteristic equation

$$|\lambda \mathbf{I} - \mathbf{A}| = \left| \begin{pmatrix} \lambda & 0 \\ 0 & \lambda \end{pmatrix} - \begin{pmatrix} a_{11} & a_{12} \\ a_{21} & a_{22} \end{pmatrix} \right| = \begin{vmatrix} \lambda - a_{11} & -a_{12} \\ -a_{21} & \lambda - a_{22} \end{vmatrix}$$
$$= \lambda^2 - (a_{11} + a_{22})\lambda + a_{11}a_{22} - a_{12}a_{21} = 0 \qquad 74.$$

Thus

$$\lambda = \tfrac{1}{2}[(a_{11} + a_{22}) \pm \sqrt{(a_{11} + a_{22})^2 - 4(a_{11}a_{22} - a_{12}a_{21})}] \qquad 75.$$

We let λ_1 be the eigenvalue with the positive root and λ_2 be the value with the negative root. The general solution can be written as

$$x_1 = c_{11} e^{\lambda_1 t} + c_{12} e^{\lambda_2 t} \qquad x_1(0) = x_{10} \qquad 76.$$
$$x_2 = c_{21} e^{\lambda_1 t} + c_{22} e^{\lambda_2 t} \qquad x_2(0) = x_{20} \qquad 77.$$

or in terms of the eigenvectors as

$$\mathbf{x} = \mathbf{c}_1 e^{\lambda_1 t} + \mathbf{c}_2 e^{\lambda_2 t} \qquad 78.$$

The elements in each eigenvector are not independent, but since each of the natural modes (the two terms on the right-hand side of Eq. 78) must individually satisfy Eq. 43, we find that

$$\lambda_1 \mathbf{c}_1 e^{\lambda_1 t} = \mathbf{A}\mathbf{c}_1 e^{\lambda_1 t} \qquad \lambda_2 \mathbf{c}_2 e^{\lambda_2 t} = \mathbf{A}\mathbf{c}_2 e^{\lambda_2 t}$$

or

$$\lambda_1 \mathbf{c}_1 = \mathbf{A}\mathbf{c}_1 \qquad \lambda_2 \mathbf{c}_2 = \mathbf{A}\mathbf{c}_2 \qquad\qquad 79.$$

Considering the eigenvector $\mathbf{c}_1$ first, we obtain

$$\lambda_1 c_{11} = a_{11} c_{11} + a_{12} c_{21}$$
$$\lambda_1 c_{21} = a_{21} c_{11} + a_{22} c_{21} \qquad\qquad 80.$$

We cannot use these two equations to find both c_{11} and c_{21}, because if we use one of the equations to eliminate one of the unknowns from the other equation, we find that the coefficient of the remaining unknown is the characteristic equation, Eq. 74, which is identically equal to zero. However, we can use either one of the equations to solve for one of the unknowns in terms of the other; for example,

$$c_{21} = \frac{\lambda_1 - a_{11}}{a_{12}} c_{11} \qquad\qquad 81.$$

and then write the eigenvector $\mathbf{c}_1$ as

$$\mathbf{c}_1 = \begin{pmatrix} c_{11} \\ \dfrac{\lambda_1 - a_{11}}{a_{12}} c_{11} \end{pmatrix} \qquad\qquad 82.$$

Similarly, from Eq. 79 we can show that

$$c_{22} = \frac{\lambda_2 - a_{11}}{a_{12}} c_{21} \qquad\qquad 83.$$

and that the eigenvector $\mathbf{c}_2$ is

$$\mathbf{c}_2 = \begin{pmatrix} c_{21} \\ \dfrac{\lambda_2 - a_{11}}{a_{12}} c_{21} \end{pmatrix} \qquad\qquad 84.$$

These results are identical to those we obtained in Example 3.3-4, if we let $c_{11} = c_{12} = 1$ and introduce two arbitrary constants into the solution of the equations (which are then established by the initial conditions). Alternately, we could normalize the eigenvectors $\mathbf{c}_1$ and $\mathbf{c}_2$ by dividing each of the elements in Eq. 82 by the length

$$b_1 = c_{11}\sqrt{1 + [(\lambda_1 - a_{11})/a_{12}]^2} \qquad\qquad 85.$$

and each of the elements in Eq. 84 by the length

$$b_2 = c_{21}\sqrt{1 + [(\lambda_2 - a_{11})/a_{12}]^2} \qquad\qquad 86.$$

Whenever we use the normalized eigenvectors in the solution, Eq. 78, we again must introduce an arbitrary constant in front of each natural mode term.

Linear Transformation

If we consider our original set of matrix differential equations

$$\dot{\mathbf{x}} = \mathbf{A}\mathbf{x} \qquad\qquad 53.$$

and introduce the linear transformation

$$\mathbf{x} = \mathbf{P}\mathbf{y} \qquad\qquad 87.$$

we see that the equations can be put into the form

$$\dot{\mathbf{y}} = \mathbf{P}^{-1}\mathbf{A}\mathbf{P}\mathbf{y} \qquad\qquad 88.$$

We would like to be able to choose the transformation matrix $\mathbf{P}$ such that $\mathbf{P}^{-1}\mathbf{A}\mathbf{P}$ was a diagonal matrix because all the y equations would be uncoupled for this case. Before attempting to develop a procedure for selecting $\mathbf{P}$ to obtain this simplification, we had better be certain that a linear transformation does not change the characteristic roots, or eigenvalues, associated with the system. In other words, we want to show that $\lambda = \mu$ in the expressions

$$|\lambda\mathbf{I} - \mathbf{A}| = 0 \quad \text{and} \quad |\mu\mathbf{I} - \mathbf{P}^{-1}\mathbf{A}\mathbf{P}| = 0 \qquad\qquad 89.$$

This can be accomplished by first substituting the identity

$$\mathbf{I} = \mathbf{P}^{-1}\mathbf{I}\mathbf{P} \qquad\qquad 90.$$

into the second expression

$$|\mu\mathbf{I} - \mathbf{P}^{-1}\mathbf{A}\mathbf{P}| = |\mu\mathbf{P}^{-1}\mathbf{I}\mathbf{P} - \mathbf{P}^{-1}\mathbf{A}\mathbf{P}| = |\mathbf{P}^{-1}(\mu\mathbf{I} - \mathbf{A})\mathbf{P}| = 0 \qquad 91.$$

Now, remembering that the determinant of the product of three matrices is merely the product of their determinants, we can write

$$|\mu\mathbf{I} - \mathbf{P}^{-1}\mathbf{A}\mathbf{P}| = |\mathbf{P}^{-1}||\mu\mathbf{I} - \mathbf{A}||\mathbf{P}| = 0 \qquad\qquad 92.$$

Providing that $\mathbf{P}$ was taken to be a unique or nonsingular matrix, its determinant—and, similarly, that of $\mathbf{P}^{-1}$—cannot be equal to zero. Therefore we must have that

$$|\mu\mathbf{I} - \mathbf{A}| = 0 \qquad\qquad 93.$$

so that the eigenvalues μ must be identical to the values of λ

$$\mu = \lambda \qquad\qquad 94.$$

Canonical Transformation

If we consider the expressions we developed to provide relationships between the elements, Eq. 67, and expand them for a few eigenvalues

$$\lambda_1\mathbf{c}_1 = \mathbf{A}\mathbf{c}_1, \quad \lambda_2\mathbf{c}_2 = \mathbf{A}\mathbf{c}_2, \ldots, \lambda_n\mathbf{c}_n = \mathbf{A}\mathbf{c}_n \qquad\qquad 95.$$

or

$$\lambda_1 \begin{pmatrix} c_{11} \\ c_{21} \\ \cdot \\ \cdot \\ \cdot \\ c_{n1} \end{pmatrix} = \begin{pmatrix} a_{11} & a_{12} & \cdots & a_{1n} \\ a_{21} & a_{22} & \cdots & a_{2n} \\ \cdot & \cdot & & \cdot \\ \cdot & \cdot & & \cdot \\ \cdot & \cdot & & \cdot \\ a_{n1} & a_{n2} & \cdots & a_{nn} \end{pmatrix} \begin{pmatrix} c_{11} \\ c_{21} \\ \cdot \\ \cdot \\ \cdot \\ c_{n1} \end{pmatrix}$$

$$\lambda_2 \begin{pmatrix} c_{12} \\ c_{22} \\ \cdot \\ \cdot \\ \cdot \\ c_{2n} \end{pmatrix} = \begin{pmatrix} a_{11} & a_{12} & \cdots & a_{1n} \\ a_{21} & a_{22} & \cdots & a_{2n} \\ \cdot & \cdot & & \cdot \\ \cdot & \cdot & & \cdot \\ \cdot & \cdot & & \cdot \\ a_{n1} & a_{n2} & \cdots & a_{nn} \end{pmatrix} \begin{pmatrix} c_{12} \\ c_{22} \\ \cdot \\ \cdot \\ \cdot \\ c_{n2} \end{pmatrix}, \ldots$$

$$\lambda_n \begin{pmatrix} c_{1n} \\ c_{2n} \\ \cdot \\ \cdot \\ \cdot \\ c_{nn} \end{pmatrix} = \begin{pmatrix} a_{11} & a_{12} & \cdots & a_{1n} \\ a_{21} & a_{22} & \cdots & a_{2n} \\ \cdot & \cdot & & \cdot \\ \cdot & \cdot & & \cdot \\ \cdot & \cdot & & \cdot \\ a_{nn} & a_{n1} & \cdots & a_{nn} \end{pmatrix} \begin{pmatrix} c_{1n} \\ c_{2n} \\ \cdot \\ \cdot \\ \cdot \\ c_{nn} \end{pmatrix}$$

96.

obviously we obtain a set of vector equations. Suppose that we collect these column vectors into a matrix and factor out the **A** matrix from the right-hand side of the equation.

$$\begin{pmatrix} \lambda_1 c_{11} & \lambda_2 c_{12} & \cdots & \lambda_n c_{1n} \\ \lambda_1 c_{21} & \lambda_2 c_{22} & \cdots & \lambda_n c_{2n} \\ \cdot & \cdot & & \cdot \\ \cdot & \cdot & & \cdot \\ \cdot & \cdot & & \cdot \\ \lambda_1 c_{n1} & \lambda_2 c_{n2} & \cdots & \lambda_n c_{nn} \end{pmatrix} = \begin{pmatrix} a_{11} & a_{12} & \cdots & a_{1n} \\ a_{21} & a_{22} & \cdots & a_{2n} \\ \cdot & \cdot & & \cdot \\ \cdot & \cdot & & \cdot \\ \cdot & \cdot & & \cdot \\ a_{n1} & a_{n2} & \cdots & a_{nn} \end{pmatrix} \begin{pmatrix} c_{11} & c_{12} & \cdots & c_{1n} \\ c_{21} & c_{22} & \cdots & c_{2n} \\ \cdot & \cdot & & \cdot \\ \cdot & \cdot & & \cdot \\ \cdot & \cdot & & \cdot \\ c_{n1} & c_{2n} & \cdots & c_{nn} \end{pmatrix}$$

97.

Each column of this matrix equation, considering $\mathbf{Ac}_j$ on the right-hand side as a single column, is identical to what we had before. We also note that the matrix on the left-hand side of the equation can be written as the product of two matrices, one of which is diagonal.

$$\begin{pmatrix} \lambda_1 c_{11} & \lambda_2 c_{12} & \cdots & \lambda_n c_{1n} \\ \lambda_1 c_{21} & \lambda_2 c_{22} & \cdots & \lambda_n c_{2n} \\ \cdot & \cdot & & \cdot \\ \cdot & \cdot & & \cdot \\ \cdot & \cdot & & \cdot \\ \lambda_1 c_{n1} & \lambda_2 c_{n2} & \cdots & \lambda_n c_{2n} \end{pmatrix} = \begin{pmatrix} c_{11} & c_{12} & \cdots & c_{1n} \\ c_{21} & c_{22} & \cdots & c_{2n} \\ \cdot & \cdot & & \cdot \\ \cdot & \cdot & & \cdot \\ \cdot & \cdot & & \cdot \\ c_{n1} & c_{n2} & \cdots & c_{nn} \end{pmatrix} \begin{pmatrix} \lambda_1 & 0 & \cdots & 0 \\ 0 & \lambda_2 & \cdots & 0 \\ \cdot & \cdot & & \cdot \\ \cdot & \cdot & & \cdot \\ \cdot & \cdot & & \cdot \\ 0 & 0 & \cdots & \lambda_n \end{pmatrix}$$

98.

Calling the diagonal matrix $\mathbf{\Lambda}$, to denote that the diagonal elements are the eigenvalues of the system, and combining Eqs. 97 and 98 gives the result

$$C\Lambda = AC \qquad\qquad 99.$$

or that

$$\Lambda = C^{-1}AC \qquad\qquad 100.$$

However, if we now return to our original linear transformation, Eq. 87, and the transformed dynamic equations, Eq. 88, it becomes apparent that if we choose the transformation matrix P as the matrix of eigenvectors C, then Eq. 88 will have the form

$$\dot{y} = \Lambda y \qquad\qquad 101.$$

Our choice of Λ as a diagonal matrix means that all the equations will be uncoupled

$$\begin{pmatrix} \dot{y}_1 \\ \dot{y}_2 \\ \cdot \\ \cdot \\ \cdot \\ \dot{y}_n \end{pmatrix} = \begin{pmatrix} \lambda_1 & 0 & \cdots & 0 \\ 0 & \lambda_2 & \cdots & 0 \\ \cdot & \cdot & & \cdot \\ \cdot & \cdot & & \cdot \\ \cdot & \cdot & & \cdot \\ 0 & 0 & \cdots & \lambda_n \end{pmatrix} \begin{pmatrix} y_1 \\ y_2 \\ \cdot \\ \cdot \\ \cdot \\ y_n \end{pmatrix} \qquad 102.$$

This is a particularly simple case to handle; therefore the transformation $P = C$ is called a canonical transformation and the coordinates in the y domain are called canonical coordinates.

It is possible to prove that the n eigenvectors c_j are linearly independent and constitute what is called a *basis* in n space. This fact guarantees that the solution, Eq. 64, is unique and that the matrix C is nonsingular, so that Eq. 99 can be rewritten as Eq. 100. Also, our original linear transformation with $P = C$ can be written in terms of the eigenvectors as

$$x = Py = Cy = c_1 y_1 + c_2 y_2 + \cdots + c_n y_n \qquad 103.$$

After comparing this expression with the solution, we see that the canonical coordinates are equivalent to the x coordinates in the basis $c_1, c_2, \ldots, c_n$.

Left-Hand Eigenvectors

The eigenvectors just discussed are normally called right-hand eigenvectors. A related set of eigenvectors, which we will need when we describe the technique of modal analysis in Chapter 8, are called left-hand eigenvectors. These arise when we apply the preceding procedure to the system equations

$$\dot{x} = A^T x \qquad x(0) = x_0 \qquad 104.$$

First we must show that this set of system equations has the same eigenvalues as the original equations, despite the fact that we expect the solutions to be somewhat different. The eigenvalues are again calculated from the characteristic equation, which for this problem becomes

$$|\mu I - A^T| = 0 \qquad 105.$$

If we write out the terms in the determinant, we obtain

$$
\begin{vmatrix}
\begin{pmatrix}
\mu & 0 & \cdots & 0 \\
0 & \mu & \cdots & 0 \\
\cdot & \cdot & & \cdot \\
\cdot & \cdot & & \cdot \\
\cdot & \cdot & & \cdot \\
0 & 0 & \cdots & \mu
\end{pmatrix}
-
\begin{pmatrix}
a_{11} & a_{21} & \cdots & a_{n1} \\
a_{12} & a_{22} & \cdots & a_{n2} \\
\cdot & \cdot & & \cdot \\
\cdot & \cdot & & \cdot \\
\cdot & \cdot & & \cdot \\
a_{1n} & a_{2n} & \cdots & a_{nn}
\end{pmatrix}
\end{vmatrix}
$$

$$
=
\begin{vmatrix}
(\mu - a_{11}) & a_{21} & \cdots & a_{n1} \\
a_{12} & (\mu - a_{22}) & \cdots & a_{n2} \\
\cdot & \cdot & & \cdot \\
\cdot & \cdot & & \cdot \\
\cdot & \cdot & & \cdot \\
a_{1n} & a_{2n} & \cdots & (\mu - a_{nn})
\end{vmatrix}
= 0 \qquad 106.
$$

The value of a determinant is not affected if we interchange its rows and columns, so that

$$|\mu I - A^T| = |\mu I - A| = 0 \qquad 107.$$

which is identical to Eq. 89. Hence the eigenvalues for the two cases must be equal

$$\mu = \lambda \qquad 108.$$

The transpose of the new system equation, Eq. 104, is

$$\dot{x}^T = x^T A \qquad 109.$$

and we know that the solution of this equation must have the form

$$
\begin{aligned}
x_1(t) &= m_{11}e^{\lambda_1 t} + m_{21}e^{\lambda_2 t} + \cdots + m_{n1}e^{\lambda_n t} \qquad x_1(0) = x_{10} \\
x_2(t) &= m_{12}e^{\lambda_1 t} + m_{22}e^{\lambda_2 t} + \cdots + m_{n2}e^{\lambda_n t} \qquad x_2(0) = x_{20} \\
&\;\;\cdot \\
&\;\;\cdot \qquad\qquad\qquad\qquad\qquad\qquad\qquad\qquad\qquad\qquad 110.\\
&\;\;\cdot \\
x_n(t) &= m_{1n}e^{\lambda_1 t} + m_{2n}e^{\lambda_2 t} + \cdots + m_{nn}e^{\lambda_n t} \qquad x_n(0) = x_{n0}
\end{aligned}
$$

where $m_{11}, \ldots, m_{nn}$ are unknown constants and only n of these n^2 constants are linearly independent. Now, if we collect all the values of $x(t)$ on the left-hand side of the equation into a row vector, we obtain

$$
(x_1\, x_2 \ldots x_n) = (e^{\lambda_1 t}\, e^{\lambda_2 t} \ldots e^{\lambda_n t})
\begin{pmatrix}
m_{11} & m_{12} & \cdots & m_{1n} \\
m_{21} & m_{22} & \cdots & m_{2n} \\
\cdot & \cdot & & \cdot \\
\cdot & \cdot & & \cdot \\
\cdot & \cdot & & \cdot \\
m_{n1} & m_{n2} & \cdots & m_{nn}
\end{pmatrix}
\qquad 111.
$$

or

$$x^T = (e^{\lambda t})^T M \qquad 112.$$

Again, we pick out n linearly independent groups of constants to work with, and in this case we choose the rows of the matrix $\mathbf{M}$. These row vectors are called left-hand eigenvectors

$$\mathbf{m}_1^T = (m_{11}\, m_{12} \ldots m_{1n}) \qquad \mathbf{m}_2^T = (m_{21}\, m_{22} \ldots m_{2n}), \ldots \qquad 113.$$

Now, we can write Eq. 111 in terms of the natural modes of the system as

$$\mathbf{x}^T = \mathbf{m}_1^T e^{\lambda_1 t} + \mathbf{m}_2^T e^{\lambda_2 t} + \cdots + \mathbf{m}_n^T e^{\lambda_n t} \qquad 114.$$

Each of the natural modes must satisfy the original system equations, Eq. 109, so that

$$\lambda_1 \mathbf{m}_1^T e^{\lambda_1 t} = \mathbf{m}_1^T \mathbf{A} e^{\lambda_1 t} \qquad \lambda_2 \mathbf{m}_2^T e^{\lambda_2 t} = \mathbf{m}_2^T \mathbf{A} e^{\lambda_2 t}, \ldots,$$

$$\lambda_n \mathbf{m}_n^T e^{\lambda_n t} = \mathbf{m}_n^T \mathbf{A} e^{\lambda_n t}$$

or after dividing each expression by the exponential factors,

$$\lambda_1 \mathbf{m}_1^T = \mathbf{m}_1^T \mathbf{A} \qquad \lambda_2 \mathbf{m}_2^T = \mathbf{m}_2^T \mathbf{A}, \ldots, \lambda_n \mathbf{m}_n^T = \mathbf{m}_n^T \mathbf{A} \qquad 115.$$

Any one of the preceding equations provides n relationships between the n unknown constants in an eigenvector, but only $(n-1)$ of these expressions are linearly independent. Thus Eqs. 115 gives us $n(n-1)$ relationships in all, and the remaining n constants are determined from the initial conditions. Alternately, we could normalize each of the eigenvectors to unit length, so that all the n^2 elements of the matrix $\mathbf{M}$ would be specified, and then introduce n integration constants into the solution, Eq. 114.

If we arrange the row eigenvectors in Eq. 115 into a matrix, we obtain

$$\begin{pmatrix} \lambda_1 \mathbf{m}_1^T \\ \lambda_2 \mathbf{m}_2^T \\ \cdot \\ \cdot \\ \cdot \\ \lambda_n \mathbf{m}_n^T \end{pmatrix} = \begin{pmatrix} \lambda_1 m_{11} & \lambda_1 m_{12} & \cdots & \lambda_1 m_{1n} \\ \lambda_2 m_{21} & \lambda_2 m_{22} & \cdots & \lambda_2 m_{2n} \\ \cdot & \cdot & & \cdot \\ \cdot & \cdot & & \cdot \\ \cdot & \cdot & & \cdot \\ \lambda_n m_{n1} & \lambda_n m_{n2} & \cdots & \lambda_n m_{nn} \end{pmatrix} = \begin{pmatrix} \mathbf{m}_1^T \mathbf{A} \\ \mathbf{m}_2^T \mathbf{A} \\ \cdot \\ \cdot \\ \cdot \\ \mathbf{m}_n^T \mathbf{A} \end{pmatrix}$$

$$= \begin{pmatrix} m_{11} & m_{12} & \cdots & m_{1n} \\ m_{21} & m_{22} & \cdots & m_{2n} \\ \cdot & \cdot & & \cdot \\ \cdot & \cdot & & \cdot \\ \cdot & \cdot & & \cdot \\ m_{n1} & m_{n2} & \cdots & m_{nn} \end{pmatrix} \begin{pmatrix} a_{11} & a_{12} & \cdots & a_{1n} \\ a_{21} & a_{22} & \cdots & a_{2n} \\ \cdot & \cdot & & \cdot \\ \cdot & \cdot & & \cdot \\ \cdot & \cdot & & \cdot \\ a_{n1} & a_{n2} & \cdots & a_{nn} \end{pmatrix} \qquad 116.$$

Noting that

$$\begin{pmatrix} \lambda_1 m_{11} & \lambda_1 m_{12} & \cdots & \lambda_1 m_{1n} \\ \lambda_2 m_{21} & \lambda_2 m_{22} & \cdots & \lambda_2 m_{2n} \\ \cdot & \cdot & & \cdot \\ \cdot & \cdot & & \cdot \\ \cdot & \cdot & & \cdot \\ \lambda_n m_{n1} & \lambda_n m_{n2} & \cdots & \lambda_n m_{nn} \end{pmatrix} = \begin{pmatrix} \lambda_1 & 0 & \cdots & 0 \\ 0 & \lambda_2 & \cdots & 0 \\ \cdot & \cdot & & \cdot \\ \cdot & \cdot & & \cdot \\ \cdot & \cdot & & \cdot \\ 0 & 0 & \cdots & \lambda_n \end{pmatrix} \begin{pmatrix} m_{11} & m_{12} & \cdots & m_{1n} \\ m_{21} & m_{22} & \cdots & m_{2n} \\ \cdot & \cdot & & \cdot \\ \cdot & \cdot & & \cdot \\ \cdot & \cdot & & \cdot \\ m_{n1} & m_{n2} & \cdots & m_{nn} \end{pmatrix}$$

$$117.$$

and calling the diagonal matrix $\mathbf{\Lambda}$, we find that Eq. 116 can be written

$$\mathbf{\Lambda M} = \mathbf{M A} \qquad \text{118.}$$

or

$$\mathbf{\Lambda} = \mathbf{M A M}^{-1} \qquad \text{119.}$$

Suppose, now, that we return to our original system equation

$$\dot{\mathbf{x}}^T = \mathbf{x}^T \mathbf{A} \qquad \text{120.}$$

and introduce the linear transformation

$$\mathbf{x}^T = \mathbf{y}^T \mathbf{Q} \qquad \text{121.}$$

Then we obtain

$$\dot{\mathbf{y}}^T = \mathbf{y}^T \mathbf{Q A Q}^{-1} \qquad \text{122.}$$

By choosing the transformation matrix $\mathbf{Q}$ to be the matrix of left-hand eigenvectors $\mathbf{M}$, the term $\mathbf{Q A Q}^{-1}$ will become the diagonal matrix $\mathbf{\Lambda}$ and, again, the equations will be uncoupled

$$\dot{\mathbf{y}}^T = \mathbf{y}^T \mathbf{\Lambda} \qquad \text{123.}$$

or

$$\begin{aligned}
\dot{y}_1 &= \lambda_1 y_1 \\
\dot{y}_2 &= \lambda_2 y_2 \\
&\;\cdot \qquad \cdot \\
&\;\cdot \qquad \cdot \\
&\;\cdot \qquad \cdot \\
\dot{y}_n &= \lambda_n y_n
\end{aligned} \qquad \text{124.}$$

This set of equations has exactly the same form as we obtained using right-hand eigenvectors, and the eigenvalues are identical for both systems. However, the y variables are defined differently because the eigenvectors for the two problems are not the same unless the matrix $\mathbf{A}$ is symmetric, or $\mathbf{A} = \mathbf{A}^T$. In fact, it is a simple matter to show that the two sets of eigenvectors are orthogonal. From Eqs. 67 and 115 we had

$$\mathbf{A c}_j = \lambda_j \mathbf{c}_j \qquad \mathbf{m}_i^T \mathbf{A} = \lambda_i \mathbf{m}_i^T \qquad \text{125.}$$

Premultiplying the first expression by $\mathbf{m}_i^T$, postmultiplying the second by $\mathbf{c}_j$, and subtracting, we find that

$$(\lambda_j - \lambda_i) \mathbf{m}_i^T \mathbf{c}_j = 0 \qquad \text{126.}$$

At the beginning of our analysis we assumed that the eigenvalues were distinct, so that $\lambda_i \neq \lambda_j$ unless $i = j$. Thus, except for the case where $i = j$, we have

$$\mathbf{m}_i^T \mathbf{c}_j = 0 \qquad \text{127.}$$

or

$$m_{i1}c_{1j} + m_{i2}c_{2j} + \cdots + m_{in}c_{nj} = 0 \qquad \text{128.}$$

Since this is merely the scalar or dot product of two vectors and since this scalar product is equal to zero, the two vectors must be orthogonal. For the case where $i = j$, the product $\mathbf{m}_i^T \mathbf{c}_j$ will be equal to some constant, but if we normalize the vectors the scalar product will be equal to unity. Thus for the normalized case we can write

$$\mathbf{MC} = \mathbf{I} \qquad \text{129.}$$

where I is the identity matrix. We will use this result in our discussion of modal analysis.

Concluding Comments

Although we have described some matrix operations in considerable detail, we have barely scratched the surface of the information of interest to control engineers. For example, systems having repeated eigenvalues, the derivation of Sylvester's theorem, quadratic forms, etc., have not been mentioned. An interested reader is encouraged to read some of the numerous texts on these subjects.

Appendix B
Table of
Laplace Transforms

TABLE OF LAPLACE TRANSFORMS†

	$f(s)$	$F(t)$
1	$\dfrac{1}{s}$	1
2	$\dfrac{1}{s^2}$	t
3	$\dfrac{1}{s^n}$ $(n = 1, 2, \ldots)$	$\dfrac{t^{n-1}}{(n-1)!}$
4	$\dfrac{1}{\sqrt{s}}$	$\dfrac{1}{\sqrt{\pi t}}$
5	$s^{-\frac{3}{2}}$	$2\sqrt{\dfrac{t}{\pi}}$
6	$s^{-(n+\frac{1}{2})}$ $(n = 1, 2, \ldots)$	$\dfrac{2^n t^{n-\frac{1}{2}}}{1 \times 3 \times 5 \cdots (2n-1)\,\sqrt{\pi}}$
7	$\dfrac{\Gamma(k)}{s^k}$ $(k > 0)$	t^{k-1}
8	$\dfrac{1}{s-a}$	e^{at}
9	$\dfrac{1}{(s-a)^2}$	te^{at}
10	$\dfrac{1}{(s-a)^n}$ $(n = 1, 2, \ldots)$	$\dfrac{1}{(n-1)!}\,t^{n-1}e^{at}$
11	$\dfrac{\Gamma(k)}{(s-a)^k}$ $(k > 0)$	$t^{k-1}e^{at}$
12*	$\dfrac{1}{(s-a)(s-b)}$	$\dfrac{1}{a-b}(e^{at} - e^{bt})$
13*	$\dfrac{s}{(s-a)(s-b)}$	$\dfrac{1}{a-b}(ae^{at} - be^{bt})$
14*	$\dfrac{1}{(s-a)(s-b)(s-c)}$	$-\dfrac{(b-c)e^{at} + (c-a)e^{bt} + (a-b)e^{ct}}{(a-b)(b-c)(c-a)}$
15	$\dfrac{1}{s^2+a^2}$	$\dfrac{1}{a}\sin at$
16	$\dfrac{s}{s^2+a^2}$	$\cos at$

* Here a, b, and (in 14) c represent distinct constants.

† Reproduced from R.V., Churchill, *Operational Mathematics*, 2nd ed. McGraw-Hill, N.Y., 1958, by permission of the McGraw-Hill Book Company.

TABLE OF LAPLACE TRANSFORMS. (*Continued.*)

	$f(s)$	$F(t)$
17	$\dfrac{1}{s^2 - a^2}$	$\dfrac{1}{a}\sinh at$
18	$\dfrac{s}{s^2 - a^2}$	$\cosh at$
19	$\dfrac{1}{s(s^2 + a^2)}$	$\dfrac{1}{a^2}(1 - \cos at)$
20	$\dfrac{1}{s^2(s^2 + a^2)}$	$\dfrac{1}{a^3}(at - \sin at)$
21	$\dfrac{1}{(s^2 + a^2)^2}$	$\dfrac{1}{2a^3}(\sin at - at \cos at)$
22	$\dfrac{s}{(s^2 + a^2)^2}$	$\dfrac{t}{2a}\sin at$
23	$\dfrac{s^2}{(s^2 + a^2)^2}$	$\dfrac{1}{2a}(\sin at + at \cos at)$
24	$\dfrac{s^2 - a^2}{(s^2 + a^2)^2}$	$t \cos at$
25	$\dfrac{s}{(s^2 + a^2)(s^2 + b^2)}\ (a^2 \neq b^2)$	$\dfrac{\cos at - \cos bt}{b^2 - a^2}$
26	$\dfrac{1}{(s - a)^2 + b^2}$	$\dfrac{1}{b} e^{at}\sin bt$
27	$\dfrac{s - a}{(s - a)^2 + b^2}$	$e^{at}\cos bt$
28	$\dfrac{3a^2}{s^3 + a^3}$	$e^{-at} - e^{at/2}\left(\cos \dfrac{at\sqrt{3}}{2} - \sqrt{3}\sin \dfrac{at\sqrt{3}}{2}\right)$
29	$\dfrac{4a^3}{s^4 + 4a^4}$	$\sin at \cosh at - \cos at \sinh at$
30	$\dfrac{s}{s^4 + 4a^4}$	$\dfrac{1}{2a^2}\sin at \sinh at$
31	$\dfrac{1}{s^4 - a^4}$	$\dfrac{1}{2a^3}(\sinh at - \sin at)$
32	$\dfrac{s}{s^4 - a^4}$	$\dfrac{1}{2a^2}(\cosh at - \cos at)$
33	$\dfrac{8a^3s^2}{(s^2 + a^2)^3}$	$(1 + a^2t^2)\sin at - at \cos at$
34*	$\dfrac{1}{s}\left(\dfrac{s - 1}{s}\right)^n$	$L_n(t) = \dfrac{e^t}{n!}\dfrac{d^n}{dt^n}(t^n e^{-t})$
35	$\dfrac{s}{(s - a)^{\frac{3}{2}}}$	$\dfrac{1}{\sqrt{\pi t}} e^{at}(1 + 2at)$
36	$\sqrt{s - a} - \sqrt{s - b}$	$\dfrac{1}{2\sqrt{\pi t^3}}(e^{bt} - e^{at})$

* $L_n(t)$ is the Laguerre polynomial of degree n.

TABLE OF LAPLACE TRANSFORMS.   (*Continued.*)

	$f(s)$	$F(t)$
37	$\dfrac{1}{\sqrt{s}+a}$	$\dfrac{1}{\sqrt{\pi t}}-ae^{a^2t}\operatorname{erfc}(a\sqrt{t})$
38	$\dfrac{\sqrt{s}}{s-a^2}$	$\dfrac{1}{\sqrt{\pi t}}+ae^{a^2t}\operatorname{erf}(a\sqrt{t})$
39	$\dfrac{\sqrt{s}}{s+a^2}$	$\dfrac{1}{\sqrt{\pi t}}-\dfrac{2a}{\sqrt{\pi}}e^{-a^2t}\displaystyle\int_0^{a\sqrt{t}}e^{\lambda^2}\,d\lambda$
40	$\dfrac{1}{\sqrt{s}\,(s-a^2)}$	$\dfrac{1}{a}e^{a^2t}\operatorname{erf}(a\sqrt{t})$
41	$\dfrac{1}{\sqrt{s}\,(s+a^2)}$	$\dfrac{2}{a\sqrt{\pi}}e^{-a^2t}\displaystyle\int_0^{a\sqrt{t}}e^{\lambda^2}\,d\lambda$
42	$\dfrac{b^2-a^2}{(s-a^2)(b+\sqrt{s})}$	$e^{a^2t}[b-a\operatorname{erf}(a\sqrt{t})]$ $-be^{b^2t}\operatorname{erfc}(b\sqrt{t})$
43	$\dfrac{1}{\sqrt{s}\,(\sqrt{s}+a)}$	$e^{a^2t}\operatorname{erfc}(a\sqrt{t})$
44	$\dfrac{1}{(s+a)\sqrt{s+b}}$	$\dfrac{1}{\sqrt{b-a}}e^{-at}\operatorname{erf}(\sqrt{b-a}\,\sqrt{t})$
45	$\dfrac{b^2-a^2}{\sqrt{s}\,(s-a^2)(\sqrt{s}+b)}$	$e^{a^2t}\left[\dfrac{b}{a}\operatorname{erf}(a\sqrt{t})-1\right]$ $+e^{b^2t}\operatorname{erfc}(b\sqrt{t})$
46*	$\dfrac{(1-s)^n}{s^{n+\frac{1}{2}}}$	$\dfrac{n!}{(2n)!\,\sqrt{\pi t}}H_{2n}(\sqrt{t})$
47	$\dfrac{(1-s)^n}{s^{n+\frac{3}{2}}}$	$-\dfrac{n!}{\sqrt{\pi}\,(2n+1)!}H_{2n+1}(\sqrt{t})$
48†	$\dfrac{\sqrt{s+2a}}{\sqrt{s}}-1$	$ae^{-at}[I_1(at)+I_0(at)]$
49	$\dfrac{1}{\sqrt{s+a}\sqrt{s+b}}$	$e^{-\frac{1}{2}(a+b)t}I_0\left(\dfrac{a-b}{2}t\right)$
50	$\dfrac{\Gamma(k)}{(s+a)^k(s+b)^k}\ (k>0)$	$\sqrt{\pi}\left(\dfrac{t}{a-b}\right)^{k-\frac{1}{2}}e^{-\frac{1}{2}(a+b)t}$ $\times I_{k-\frac{1}{2}}\left(\dfrac{a-b}{2}t\right)$
51	$\dfrac{1}{(s+a)^{\frac{1}{2}}(s+b)^{\frac{3}{2}}}$	$te^{-\frac{1}{2}(a+b)t}\left[I_0\left(\dfrac{a-b}{2}t\right)\right.$ $\left.+I_1\left(\dfrac{a-b}{2}t\right)\right]$
52	$\dfrac{\sqrt{s+2a}-\sqrt{s}}{\sqrt{s+2a}+\sqrt{s}}$	$\dfrac{1}{t}e^{-at}I_1(at)$

* $H_n(x)$ is the Hermite polynomial, $H_n(x)=e^{x^2}\dfrac{d^n}{dx^n}(e^{-x^2})$.

† $I_n(x)=i^{-n}J_n(ix)$, where J_n is Bessel's function of the first kind.

TABLE OF LAPLACE TRANSFORMS. (*Continued.*)

	$f(s)$	$F(t)$
53	$\dfrac{(a-b)^k}{(\sqrt{s+a}+\sqrt{s+b})^{2k}}$ $(k>0)$	$\dfrac{k}{t}e^{-\frac{1}{2}(a+b)t}I_k\left(\dfrac{a-b}{2}t\right)$
54	$\dfrac{(\sqrt{s+a}+\sqrt{s})^{-2\nu}}{\sqrt{s}\sqrt{s+a}}$ $(\nu>-1)$	$\dfrac{1}{a^\nu}e^{-\frac{1}{2}at}I_\nu\left(\dfrac{1}{2}at\right)$
55	$\dfrac{1}{\sqrt{s^2+a^2}}$	$J_0(at)$
56	$\dfrac{(\sqrt{s^2+a^2}-s)^\nu}{\sqrt{s^2+a^2}}$ $(\nu>-1)$	$a^\nu J_\nu(at)$
57	$\dfrac{1}{(s^2+a^2)^k}$ $(k>0)$	$\dfrac{\sqrt{\pi}}{\Gamma(k)}\left(\dfrac{t}{2a}\right)^{k-\frac{1}{2}}J_{k-\frac{1}{2}}(at)$
58	$(\sqrt{s^2+a^2}-s)^k$ $(k>0)$	$\dfrac{ka^k}{t}J_k(at)$
59	$\dfrac{(s-\sqrt{s^2-a^2})^\nu}{\sqrt{s^2-a^2}}$ $(\nu>-1)$	$a^\nu I_\nu(at)$
60	$\dfrac{1}{(s^2-a^2)^k}$ $(k>0)$	$\dfrac{\sqrt{\pi}}{\Gamma(k)}\left(\dfrac{t}{2a}\right)^{k-\frac{1}{2}}I_{k-\frac{1}{2}}(at)$
61	$\dfrac{e^{-ks}}{s}$	$S_k(t)=\begin{cases}0\text{ when }0<t<k\\1\text{ when }t>k\end{cases}$
62	$\dfrac{e^{-ks}}{s^2}$	$\begin{cases}0\quad\text{when }0<t<k\\t-k\text{ when }t>k\end{cases}$
63	$\dfrac{e^{-ks}}{s^\mu}$ $(\mu>0)$	$\begin{cases}0\quad\text{when }0<t<k\\\dfrac{(t-k)^{\mu-1}}{\Gamma(\mu)}\text{ when }t>k\end{cases}$
64	$\dfrac{1-e^{-ks}}{s}$	$\begin{cases}1\text{ when }0<t<k\\0\text{ when }t>k\end{cases}$
65	$\dfrac{1}{s(1-e^{-ks})}=\dfrac{1+\coth\frac{1}{2}ks}{2s}$	$1+[t/k]=n$ when $(n-1)k<t<nk$ $(n=1,2,\ldots)$ (Fig. 5)
66	$\dfrac{1}{s(e^{ks}-a)}$	$\begin{cases}0\quad\text{when }0<t<k\\1+a+a^2+\cdots+a^{n-1}\\\quad\text{when }nk<t<(n+1)k\\\quad\quad(n=1,2,\ldots)\end{cases}$
67	$\dfrac{1}{s}\tanh ks$	$M(2k,t)=(-1)^{n-1}$ when $2k(n-1)<t<2kn$ $(n=1,2,\ldots)$ (Fig. 9)
68	$\dfrac{1}{s(1+e^{-ks})}$	$\dfrac{1}{2}M(k,t)+\dfrac{1}{2}=\dfrac{1-(-1)^n}{2}$ when $(n-1)k<t<nk$
69	$\dfrac{1}{s^2}\tanh ks$	$H(2k,t)$ (Fig. 10)

TABLE OF LAPLACE TRANSFORMS. *(Continued.)*

	$f(s)$	$F(t)$
70	$\dfrac{1}{s \sinh ks}$	$F(t) = 2(n - 1)$ when $(2n - 3)k < t < (2n - 1)k$ $(t > 0)$
71	$\dfrac{1}{s \cosh ks}$	$M(2k, t + 3k) + 1 = 1 + (-1)^n$ when $(2n - 3)k < t < (2n - 1)k$ $(t > 0)$
72	$\dfrac{1}{s} \coth ks$	$F(t) = 2n - 1$ when $2k(n - 1) < t < 2kn$
73	$\dfrac{k}{s^2 + k^2} \coth \dfrac{\pi s}{2k}$	$\lvert \sin kt \rvert$
74	$\dfrac{1}{(s^2 + 1)(1 - e^{-\pi s})}$	$\begin{cases} \sin t \text{ when} \\ \qquad (2n - 2)\pi < t < (2n - 1)\pi \\ 0 \quad \text{when} \\ \qquad (2n - 1)\pi < t < 2n\pi \end{cases}$
75	$\dfrac{1}{s} e^{-(k/s)}$	$J_0(2 \sqrt{kt})$
76	$\dfrac{1}{\sqrt{s}} e^{-(k/s)}$	$\dfrac{1}{\sqrt{\pi t}} \cos 2 \sqrt{kt}$
77	$\dfrac{1}{\sqrt{s}} e^{k/s}$	$\dfrac{1}{\sqrt{\pi t}} \cosh 2 \sqrt{kt}$
78	$\dfrac{1}{s^{\frac{3}{2}}} e^{-(k/s)}$	$\dfrac{1}{\sqrt{\pi k}} \sin 2 \sqrt{kt}$
79	$\dfrac{1}{s^{\frac{3}{2}}} e^{k/s}$	$\dfrac{1}{\sqrt{\pi k}} \sinh 2 \sqrt{kt}$
80	$\dfrac{1}{s^\mu} e^{-(k/s)} \ (\mu > 0)$	$\left(\dfrac{t}{k}\right)^{(\mu-1)/2} J_{\mu-1}(2 \sqrt{kt})$
81	$\dfrac{1}{s^\mu} e^{k/s} \ (\mu > 0)$	$\left(\dfrac{t}{k}\right)^{(\mu-1)/2} I_{\mu-1}(2 \sqrt{kt})$
82	$e^{-k\sqrt{s}} \ (k > 0)$	$\dfrac{k}{2 \sqrt{\pi t^3}} \exp\left(-\dfrac{k^2}{4t}\right)$
83	$\dfrac{1}{s} e^{-k\sqrt{s}} \ (k \geqq 0)$	$\text{erfc}\left(\dfrac{k}{2 \sqrt{t}}\right)$
84	$\dfrac{1}{\sqrt{s}} e^{-k\sqrt{s}} \ (k \geqq 0)$	$\dfrac{1}{\sqrt{\pi t}} \exp\left(-\dfrac{k^2}{4t}\right)$
85	$s^{-\frac{3}{2}} e^{-k\sqrt{s}} \ (k \geqq 0)$	$2 \sqrt{\dfrac{t}{\pi}} \exp\left(-\dfrac{k^2}{4t}\right)$ $\qquad - k \, \text{erfc}\left(\dfrac{k}{2 \sqrt{t}}\right)$
86	$\dfrac{ae^{-k\sqrt{s}}}{s(a + \sqrt{s})} \ (k \geqq 0)$	$-e^{ak}e^{a^2 t} \, \text{erfc}\left(a \sqrt{t} + \dfrac{k}{2 \sqrt{t}}\right)$ $\qquad + \text{erfc}\left(\dfrac{k}{2 \sqrt{t}}\right)$

TABLE OF LAPLACE TRANSFORMS.   (*Continued.*)

	$f(s)$	$F(t)$
87	$\dfrac{e^{-k\sqrt{s}}}{\sqrt{s}\,(a + \sqrt{s})}$ $(k \geqq 0)$	$e^{ak}e^{a^2t}\,\text{erfc}\left(a\,\sqrt{t} + \dfrac{k}{2\,\sqrt{t}}\right)$
88	$\dfrac{e^{-k\sqrt{s(s+a)}}}{\sqrt{s(s + a)}}$	$\begin{cases} 0 & \text{when } 0 < t < k \\ e^{-\frac{1}{2}at}I_0(\frac{1}{2}a\,\sqrt{t^2 - k^2}) \\ & \text{when } t > k \end{cases}$
89	$\dfrac{e^{-k\sqrt{s^2+a^2}}}{\sqrt{s^2 + a^2}}$	$\begin{cases} 0 & \text{when } 0 < t < k \\ J_0(a\,\sqrt{t^2 - k^2}) & \text{when } t > k \end{cases}$
90	$\dfrac{e^{-k\sqrt{s^2-a^2}}}{\sqrt{s^2 - a^2}}$	$\begin{cases} 0 & \text{when } 0 < t < k \\ I_0(a\,\sqrt{t^2 - k^2}) & \text{when } t > k \end{cases}$
91	$\dfrac{e^{-k(\sqrt{s^2+a^2}-s)}}{\sqrt{s^2 + a^2}}$ $(k \geqq 0)$	$J_0(a\,\sqrt{t^2 + 2kt})$
92	$e^{-ks} - e^{-k\sqrt{s^2+a^2}}$	$\begin{cases} 0 & \text{when } 0 < t < k \\ \dfrac{ak}{\sqrt{t^2 - k^2}}\,J_1(a\,\sqrt{t^2 - k^2}) \\ & \text{when } t > k \end{cases}$
93	$e^{-k\sqrt{s^2-a^2}} - e^{-ks}$	$\begin{cases} 0 & \text{when } 0 < t < k \\ \dfrac{ak}{\sqrt{t^2 - k^2}}\,I_1(a\,\sqrt{t^2 - k^2}) \\ & \text{when } t > k \end{cases}$
94	$\dfrac{a^\nu e^{-k\sqrt{s^2+a^2}}}{\sqrt{s^2 + a^2}\,(\sqrt{s^2 + a^2} + s)^\nu}$ $(\nu > -1)$	$\begin{cases} 0 & \text{when } 0 < t < k \\ \left(\dfrac{t - k}{t + k}\right)^{\frac{1}{2}\nu} J_\nu(a\,\sqrt{t^2 - k^2}) \\ & \text{when } t > k \end{cases}$
95	$\dfrac{1}{s}\log s$	$\Gamma'(1) - \log t \quad [\Gamma'(1) = -0.5772]$
96	$\dfrac{1}{s^k}\log s \;(k > 0)$	$t^{k-1}\left\{\dfrac{\Gamma'(k)}{[\Gamma(k)]^2} - \dfrac{\log t}{\Gamma(k)}\right\}$
97*	$\dfrac{\log s}{s - a} \;(a > 0)$	$e^{at}[\log a - \text{Ei}\,(-at)]$
98†	$\dfrac{\log s}{s^2 + 1}$	$\cos t\,\text{Si}\,t - \sin t\,\text{Ci}\,t$
99	$\dfrac{s \log s}{s^2 + 1}$	$-\sin t\,\text{Si}\,t - \cos t\,\text{Ci}\,t$
100	$\dfrac{1}{s}\log(1 + ks) \;(k > 0)$	$-\,\text{Ei}\left(-\dfrac{t}{k}\right)$

* Ei(t) is the exponential-integral function, Si(t) is the sine-integral, and Ci(t) is the cosine integral.

† The cosine-integral function is defined in Sec. 33. Si t is defined in Sec. 18.

TABLE OF LAPLACE TRANSFORMS. *(Continued.)*

	$f(s)$	$F(t)$
101	$\log \dfrac{s-a}{s-b}$	$\dfrac{1}{t}\,(e^{bt} - e^{at})$
102	$\dfrac{1}{s} \log\,(1 + k^2 s^2)$	$-2\,\mathrm{Ci}\left(\dfrac{t}{k}\right)$
103	$\dfrac{1}{s} \log\,(s^2 + a^2)\ (a > 0)$	$2 \log a - 2\,\mathrm{Ci}\,(at)$
104	$\dfrac{1}{s^2} \log\,(s^2 + a^2)\ (a > 0)$	$\dfrac{2}{a}\,[at \log a + \sin at - at\,\mathrm{Ci}\,(at)]$
105	$\log \dfrac{s^2 + a^2}{s^2}$	$\dfrac{2}{t}\,(1 - \cos at)$
106	$\log \dfrac{s^2 - a^2}{s^2}$	$\dfrac{2}{t}\,(1 - \cosh at)$
107	$\arctan \dfrac{k}{s}$	$\dfrac{1}{t} \sin kt$
108	$\dfrac{1}{s} \arctan \dfrac{k}{s}$	$\mathrm{Si}\,(kt)$
109	$e^{k^2 s^2}\,\mathrm{erfc}\,(ks)\ (k > 0)$	$\dfrac{1}{k\sqrt{\pi}} \exp\left(-\dfrac{t^2}{4k^2}\right)$
110	$\dfrac{1}{s}\,e^{k^2 s^2}\,\mathrm{erfc}\,(ks)\ (k > 0)$	$\mathrm{erf}\left(\dfrac{t}{2k}\right)$
111	$e^{ks}\,\mathrm{erfc}\,\sqrt{ks}\ (k > 0)$	$\dfrac{\sqrt{k}}{\pi\sqrt{t}\,(t + k)}$
112	$\dfrac{1}{\sqrt{s}}\,\mathrm{erfc}\,(\sqrt{ks})$	$\begin{cases} 0 & \text{when } 0 < t < k \\ (\pi t)^{-\frac{1}{2}} & \text{when } t > k \end{cases}$
113	$\dfrac{1}{\sqrt{s}}\,e^{ks}\,\mathrm{erfc}\,(\sqrt{ks})\ (k > 0)$	$\dfrac{1}{\sqrt{\pi(t + k)}}$
114	$\mathrm{erf}\left(\dfrac{k}{\sqrt{s}}\right)$	$\dfrac{1}{\pi t} \sin\,(2k\sqrt{t})$
115	$\dfrac{1}{\sqrt{s}}\,e^{k^2/s}\,\mathrm{erfc}\left(\dfrac{k}{\sqrt{s}}\right)$	$\dfrac{1}{\sqrt{\pi t}}\,e^{-2k\sqrt{t}}$
116*	$K_0(ks)$	$\begin{cases} 0 & \text{when } 0 < t < k \\ (t^2 - k^2)^{-\frac{1}{2}} & \text{when } t > k \end{cases}$
117	$K_0(k\sqrt{s})$	$\dfrac{1}{2t} \exp\left(-\dfrac{k^2}{4t}\right)$
118	$\dfrac{1}{s}\,e^{ks}K_1(ks)$	$\dfrac{1}{k}\sqrt{t(t + 2k)}$
119	$\dfrac{1}{\sqrt{s}}\,K_1(k\sqrt{s})$	$\dfrac{1}{k} \exp\left(-\dfrac{k^2}{4t}\right)$
120	$\dfrac{1}{\sqrt{s}}\,e^{k/s}K_0\left(\dfrac{k}{s}\right)$	$\dfrac{2}{\sqrt{\pi t}}\,K_0(2\sqrt{2kt})$

* $K_n(x)$ is Bessel's function of the second kind for the imaginary argument.

TABLE OF LAPLACE TRANSFORMS. *(Continued.)*

	$f(s)$	$F(t)$
121	$\pi e^{-ks} I_0(ks)$	$\begin{cases} [t(2k - t)]^{-\frac{1}{2}} & \text{when } 0 < t < 2k \\ 0 & \text{when } t > 2k \end{cases}$
122	$e^{-ks} I_1(ks)$	$\begin{cases} \dfrac{k - t}{\pi k \sqrt{t(2k - t)}} & \text{when } 0 < t < 2k \\ 0 & \text{when } t > 2k \end{cases}$
123	$-e^{as}\,\text{Ei}\,(-as)$	$\dfrac{1}{t + a}\ (a > 0)$
124	$\dfrac{1}{a} + s e^{as}\,\text{Ei}\,(-as)$	$\dfrac{1}{(t + a)^2}\ (a > 0)$
125	$\left(\dfrac{\pi}{2} - \text{Si}\,s\right)\cos s + \text{Ci}\,s\,\sin s$	$\dfrac{1}{t^2 + 1}$

Appendix C
Contents
for Volume 2

Preface

Part III Process Control

Author Index

Subject Index

RETURN
TO ➡ CHEMISTRY LIBRARY
100 Hildebrand Hall   642-3753

LOAN PERIOD 1	2	3
7 DAYS	1 MONTH	
4	5	6

ALL BOOKS MAY BE RECALLED AFTER 7 DAYS
Renewable by telephone

DUE AS STAMPED BELOW

UNIVERSITY OF CALIFORNIA, BERKELEY
FORM NO. DD5, 3m, 12/80   BERKELEY, CA 94720

(P)s